NONLINEAR DYNAMICS IN CIRCUITS

NONLINEAR DYNAMICS IN CIRCUITS

editors

T. Carroll
L. Pecora
Naval Research Laboratory, USA

World Scientific
Singapore • New Jersey • London • Hong Kong

Published by

World Scientific Publishing Co. Pte. Ltd.

P O Box 128, Farrer Road, Singapore 9128

USA office: Suite 1B, 1060 Main Street, River Edge, NJ 07661

UK office: 57 Shelton Street, Covent Garden, London WC2H 9HE

NONLINEAR DYNAMICS IN CIRCUITS

ISBN 981-02-2438-9

Printed in Singapore.

Contents

Contents

Preface

One of the side effects of nonlinear dynamics research over the last 20 years is that simple electronic circuits have become important tools in the study of the fundamentals of physics. This is in contrast to other fields of physics, where experiments become more expensive as time goes by and require larger and larger collaborations.

In many ways, nonlinear dynamics is not really a new science but the study of basic problems in Newtonian mechanics that have been too difficult to solve earlier. The reason that circuits have been so useful in the study of nonlinear dynamics is that nonlinear dynamics is a universal science: it is the study of basic principles of the time evolution of any dynamical system, from planetary motion to fluid dynamics. It is not so much a matter of "Why use circuits?" as "Why not?". Pretty much any system will do to confirm that theories and numerical simulations from nonlinear dynamics do indeed apply to the real world, and electronic circuits are simply a convenient physical system to work with.

One could think of all of the experiments described in this book as analog computers, and as such, decades of experience with analog computation are available to help circuit designers. There are occasional objections to the use of circuits in understanding dynamics which usually go something like: "but your circuit is just an analog computer". There is nothing wrong with this- any real physical system can be used as an analog computer if one can modify the rules of evolution of the system toward some new desired set of rules. The important point is that in an analog computer, real physical quantities are changing over time, while a digital computer manipulates finite precision numbers that represent physical quantities. There are many advantages to digital computation, but there can also be subtle effects that distort the results. Building an electronic circuit model guarantees that the results apply to real physical systems.

Besides being a cheap way to study real physical systems, electronic circuits have become useful in themselves as speculation increases on possible applications for chaos and nonlinear dynamics. The field of communications already depends on electronic circuits, so the study of nonlinear electronic circuits is naturally related to this field. Control systems also depend on electronics, so there is a close connection between that field and new work in nonlinear dynamics.

This book is divided into two sections. Part I contains circuits. These chapters are mostly about basic nonlinear science, with circuits used as a tool to study the physics. Some of the work in Part I describes new ways to build a particular chaotic circuit. Part II focuses on applications of nonlinear dynamics to science and technology. In this part, the emphasis is not on the circuits themselves, but on some end goal (such as understanding a class of physical systems or transmitting information) for which the nonlinear circuits are useful. There is, of course, much mixing between Parts I and II.

Part I: Circuits and Theory

Nonlinear Dynamics in Driven and Autonomous Electronic Circuits: From Period Doubling to Systems of Multiple Oscillators

Paul S. Linsay
Plasma Fusion Center
Massachusetts Institute of Technology
Cambridge, MA 02139, USA

ABSTRACT

Electronic circuits provide an ideal means for investigating the theory and phenomena of nonlinear dynamics. I review the application of electronic circuits to a range of issues that have been important in the development of nonlinear dynamics over the past dozen years: period doubling, two frequency quasiperiodicity, three frequency quasiperiodicity, and systems of coupled oscillators.

1. Introduction

Electronic circuits have been fundamental in understanding theories of chaotic dynamics and for studying properties of dynamical systems. A variety of electronic circuits have been used for these studies. Many of the electronic circuits used in experimental dynamics have been created using simple nonlinear components such as diodes and transistors. This type of circuit has been used in experiments meant to study general properties of nonlinear dynamics that do not depend on the specifics of the equations of motion. In fact, that is one of the great fascinations of this field, so much behavior is "universal" and does not depend on details of the differential equations. Of course, the physics of pn-junctions is extremely well understood and it is possible to write down and solve the equations of motion for these circuits. Typically, while the equations are necessary to full understanding, they are only of secondary importance to these experiments. Other types of nonlinear circuits have been created where the nonlinearity is not in the circuit elements but in the "boundary conditions." An op-amp relaxation oscillator with positive feedback that alternately charges and discharges a capacitor is an example of this. The charge and discharge is linear, and the nonlinearity is the switching action that changes the state from charge to discharge.

Electronic systems have a natural built in advantage over many other ways of exploring nonlinear phenomena. They can be made to be very fast so that a great deal of data is generated in a very short period of time. Combine this with modern data collection technology and computer control with real-time analysis and it becomes possible to explore very large regions of parameter space in a very short time. Depending on the experimental parameter resolution, one can also explore small areas of parameter space in some detail. What is more, electronic circuits are quite noise free and precise so that the data are clean and reproducible. They are at a disadvantage with respect to computer modeling of specific equations if one is interested in fine detail or high precision, nor are they as flexible as a computer program. But, given their advantages, electronic systems are a popular alternative to solving nonlinear equations on a digital computer.

The development of many circuits was motivated by the desire to study various transitions to chaos. In this article I will discuss how several different kinds of circuits were used to investigate the period-doubling transition to chaos, two frequency quasi-periodicity, three frequency quasi-periodicity, and systems of coupled oscillators.

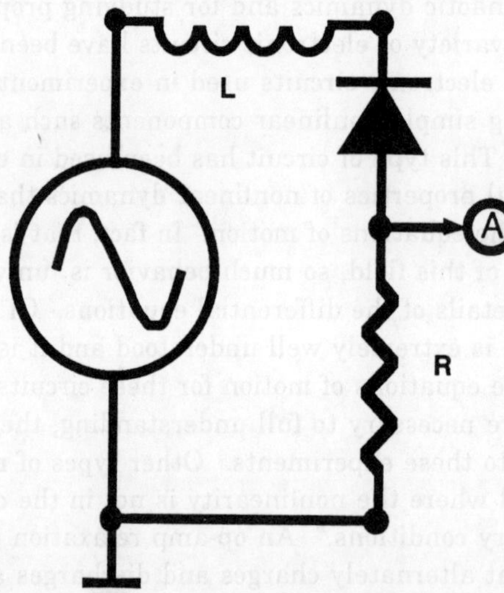

Figure 1. Diode resonator circuit, a simple RL-diode circuit driven by a sine wave generator. Output is at A, which measures the current through the circuit.

2. The Diode Resonator

One of the earliest circuits[1] used to study nonlinear dynamics is the diode resonator shown in Figure 1. The nonlinearity is provided by the pn-junction while the inductor in conjunction with the internal capacitance of the diode makes this a second order system that is capable of producing chaotic behavior when driven by a time dependent signal. The signal generator is simply a low impedance sine wave source with variable frequency and amplitude. These circuits have been made with a variety of components that operated in several different frequency regimes. Some typical circuit parameters are given in Table 1.

Table 1. Some representative RL-diode circuit parameters. Resistances marked [i] are the reported d.c. internal resistance of the inductor. The characteristic drive frequency is given in the column labeled f.

Diode	L	R	f	reference
1N5470A	100 μH	180 Ω	1.78 MHz	1
1N953	10 mH	28 Ω [i]	93 kHz	2
1N1221	14 mH	14 Ω [i]	100 kHz	3
1N2858	100 mH	25 Ω [i]	53 kHz	4
BB109G	150 μH	50 Ω	1.5 MHz	5

2.1 *The Period Doubling Route to Chaos*

Diode resonators were first used to study the period-doubling route to chaos. In this phenomena the period of the wave form doubles successively as a control parameter is varied until the time series eventually becomes chaotic. Suppose we look at the current flowing through the circuit as a function of time and with the frequency of the driving sine wave fixed. When the amplitude of the sine wave is below some value, A_0, the time series has the same period as the drive. If the amplitude is now increased to be in the range $[A_0, A_1]$ the current will only repeat itself every second cycle of the drive. A further increase in amplitude into the range $[A_1, A_2]$ will now produce a time series that only repeats itself every fourth drive cycle, and so on. This is illustrated in Figure 2 with data from an actual circuit showing the period 2, 4, and 8 period doublings. The differences become quite subtle as the period grows geometrically. The power spectrum of the time series shows new components that appear at subharmonics of the fundamental frequency. The frequencies of the subharmonics are multiples of $f_0/2^P$ where f_0 is the frequency of the fundamental and 2^P is the period of the response.

This transition to chaos is quite common in low dimensional systems, especially those with one or two dimensions, where dimension means the effective number of variables that are active in producing the system dynamics. Feigenbaum[6-9] did a great deal of work showing that

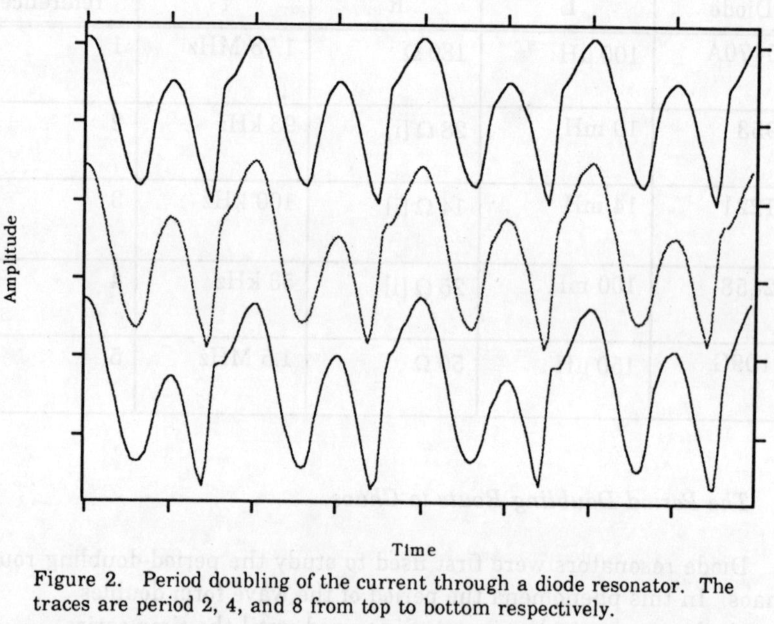

Figure 2. Period doubling of the current through a diode resonator. The traces are period 2, 4, and 8 from top to bottom respectively.

this transition had universal scaling properties, independent of the particular system in which it occurred. He showed that changes in the control parameter, A, needed to create successive period doublings decrease at a geometrical rate $\delta = 4.669...$ In practice, this means that if the measured change in A needed to go from period 2 to period 4 is ΔA, then chaos will set in if A is increased by about $(4.669^{-1} + 4.669^{-2} + \ldots)\,\Delta A = 0.27\,\Delta A$ more. The geometrically decreasing spacing between period doublings makes observations of high order doublings, typically period 16, 32, and higher, difficult to observe experimentally. A second universal constant, $\alpha = 2.5029\ldots$, governs scaling of time series amplitude changes as does the trajectory scaling function σ, which is discontinuous at all the rationals.

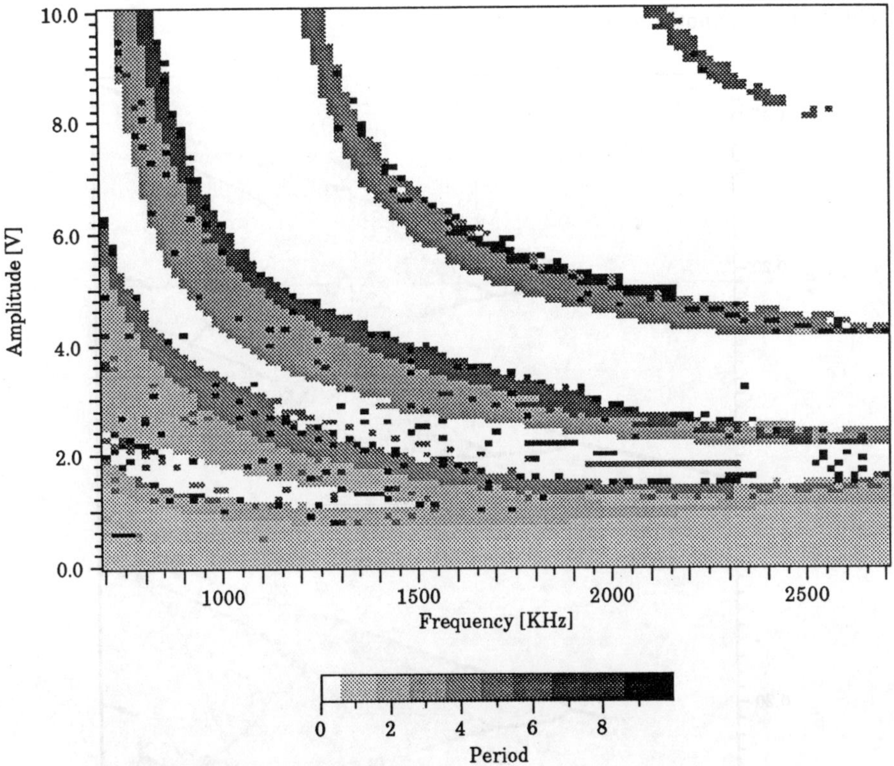

Figure 3. Parameter scan of a diode resonator. The circuit response is chaotic in the white areas and a periodic subharmonic response in the gray bands.

2.2 *Basic Behavior of a Diode Resonator*

Despite the apparent simplicity of the basic circuit of Figure 1 the output is quite rich and one observes a variety of phenomena that depend on the choice of amplitude and frequency of the sinusoidal drive. The plot in Figure 3 was generated by stepping the amplitude and frequency of the drive and recording the behavior of the circuit. White areas are chaos and gray bands are periodic response at a subharmonic of the drive. Low order period doubling is obvious at this resolution for the period 2→4→chaos, 3→6→chaos, and 4→8→chaos transitions. The plot shows the general features that can be observed with any diode resonator circuit, but the details will vary with the circuit components. Typically, larger resistances move the periodic bands to higher amplitudes and reduce the width of the chaotic zones.

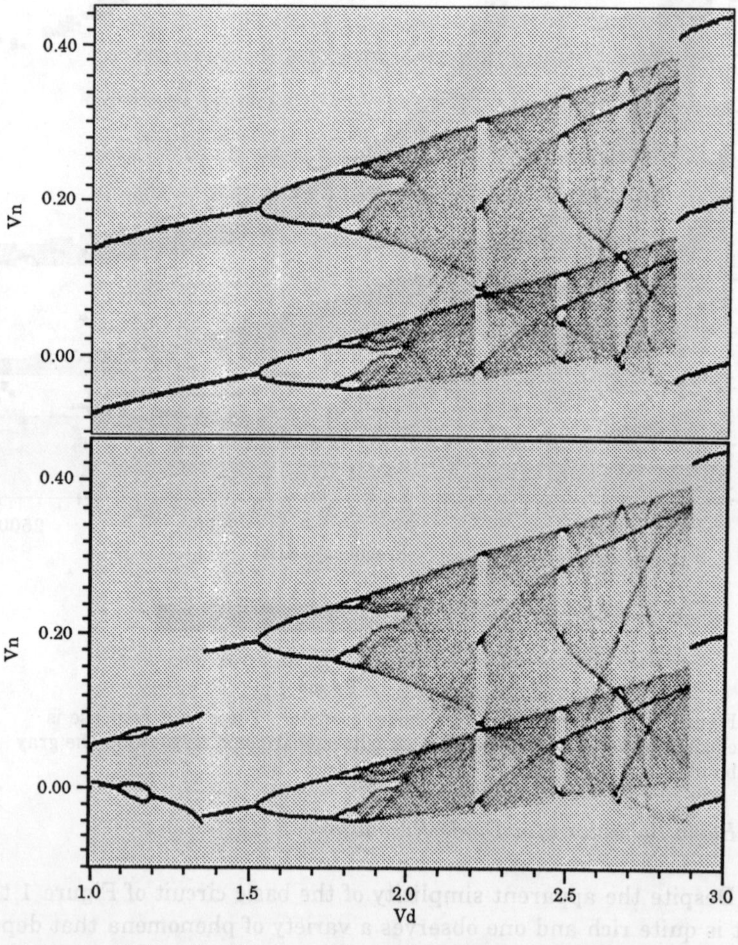

Figure 4. Bifurcation diagram of the diode resonator at a fixed drive frequency of 1.75 MHz. In the lower panel the drive voltage was stepped up from 1 V while in upper panel it was stepped down from 3 V.

More detailed information appears in the bifurcation diagram of Figure 4. This was created by stepping the drive voltage, V_d, and taking 1000 voltage samples of the circuit waveform at each step. The samples, V_n, were taken once per drive period with the phase chosen to coincide with the maxima of the output. All 1000 samples are plotted for each step of V_d. In the lower panel, Vd was stepped from 1 to 3 volts in 4 mV steps while in the upper panel it was reduced from 3 to 1 volt, also in 4 mV steps. Several interesting features are apparent here. There is a period doubling sequence of 2→4→8 in the voltage range [1.5, 1.8] that becomes chaotic at about 1.85 V. The chaotic band from 1.85 V to about 2.9 V is broken up by periodic windows: period 5 at 2.2 V; period 7 at 2.45 V; period 8 at 2.65 V; and period 10 at 2.75 V. There is clear evidence of two types of "crisis" in the plot10, 11. The sudden expansion from two chaotic bands to one at 2.05 V is an "interior crisis" and the abrupt transition to period 3 at 2.9 V is a "boundary crisis." Both cases can be explained as the result of the collision of an unstable fixed point with a chaotic attractor. Despite the fact that the signal is chaotic, some voltages are more probable than others as revealed by the dark lines that weave through the figure, evidence for the existence of unstable periodic orbits.

The most dramatic difference between the two scans is the small amplitude period 2 below 1.35 V that is obvious in the lower figure but has disappeared in the upper one. This is an example of two co-existing attractors, in this case, two different period 2 orbits. For the same set of system parameters either attractor is stable, but only one is observable at a time depending on the initial conditions for the dynamical variables. Here they are the current through the circuit and voltage across the diode. Past history determines that the small amplitude period 2 is visited when V_d increases from small values but is not seen at all when V_d decreases from a large enough initial value. A more subtle form of the same phenomena and accompanying hysteresis can be seen at the right edge of the diagram where the period 3 orbit and chaos coexist over a narrow band.

2.3 *Period Doubling and Scaling*

As discussed above, the period doubling transition to chaos exhibits universal scaling behavior. This can be tested out using the diode circuit data from figures 2 and 3. The simplest quantity to measure is δ, the control parameter scaling. Dividing the width of the period 4 zone in Fig. 3

by the width of the period 8 zone gives $\delta = 3.4 \pm 0.3$, which is to be compared to the theoretical value of 4.669... We also compute the scaling function, σ, and compare it to its theoretical value in Fig. 5. The scaling function was computed by subtracting the second half of the period 8 waveform from the first half and dividing it into the front-back difference of the period 4 time series. The scaling function computed from the period 512 and 1024 cycles of the logistic map has also been plotted for comparison. [Actually the shifted mirror image of the theoretical function

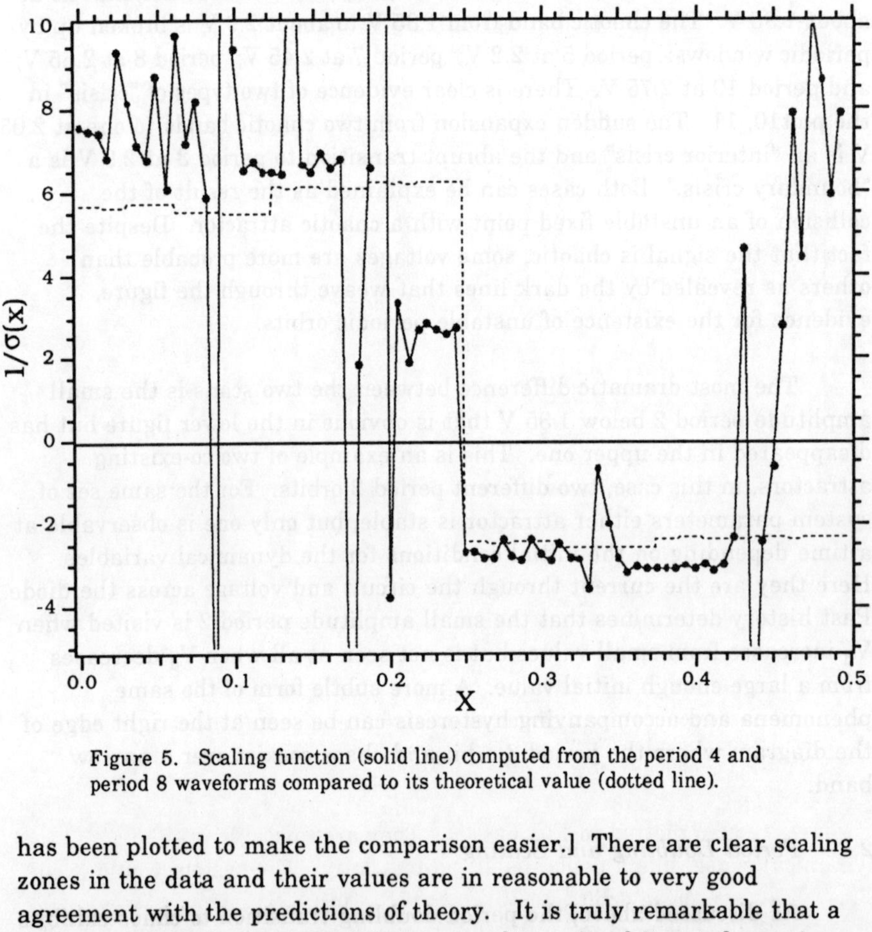

Figure 5. Scaling function (solid line) computed from the period 4 and period 8 waveforms compared to its theoretical value (dotted line).

has been plotted to make the comparison easier.] There are clear scaling zones in the data and their values are in reasonable to very good agreement with the predictions of theory. It is truly remarkable that a theoretical function computed from a one dimensional iterated map is so well able to describe the properties of a continuous physical system that must be described by at least a driven second order nonlinear differential equation.

2.4 *The Strange Attractor and an Interior Crisis*

The interior crisis that broadens the bifurcation diagram near 2.05 V in Fig. 4 can be investigated in more detail. By using a time delay embedding of the time series we can construct the attractor of the circuit[12, 13]. A two dimensional embedding of the circuit attractor appears in Fig. 6 in which the x-coordinate is the voltage sample , V_n, and the y-coordinate is the sample V_{n+2}. Clearly the attractor must lie in a higher dimension than two since it intersects itself, which is not allowed for a deterministic system. In fact, three dimensions are a high enough embedding to prevent the attractor from intersecting itself (for examples of this see Brorson, *et. al.*[14]). Several low order unstable periodic orbits

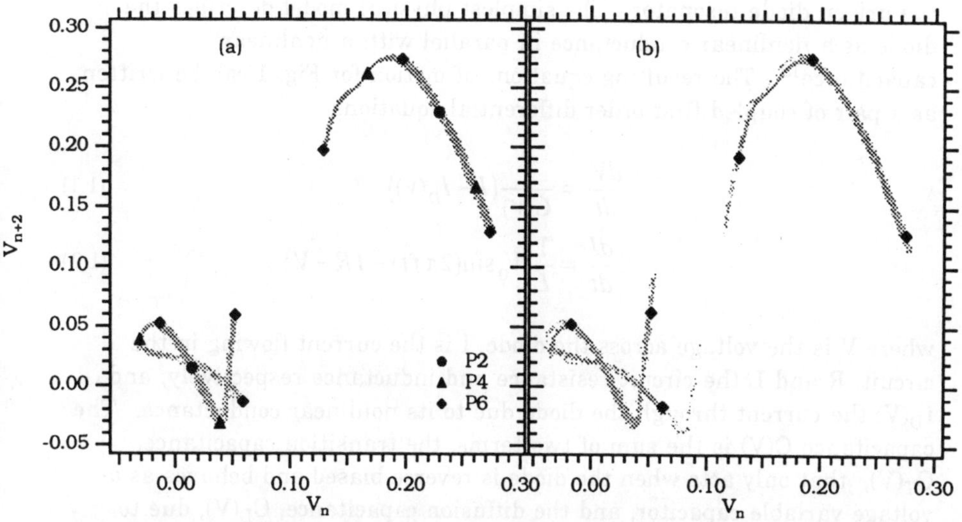

Figure 6. Strange attractor of the diode resonator. In (a) V_d = 2.05 V, just at the point where the bifurcation diagram expands. Period 2, 4, and 6 UPOs are plotted. (b) V_d = 2.06 V. The attractor has expanded and the P6 UPO is no longer touching the tips of the attractor.

(UPO's) are also plotted in Fig. 6a[15-17]. They are unstable in the usual dynamical sense, like a pencil balanced on its tip. The system would cycle through any of these orbits indefinitely if the initial conditions were precisely right but any noise or other perturbation will eventually cause the system motion to drift away to other parts of the attractor. Fig. 6a is taken at a drive voltage of V_d = 2.05V, just where the bifurcation diagram of Fig. 4 begins to fill out internally. The system spends about 2% of its time the vicinity of the P2 and P4 UPO's and less than 0.1% near P6. In fact, P6 has just appeared at this drive voltage. But P6 can be seen to be colliding with the tips of the attractor and will cause the attractor to expand in size as can be seen in Fig. 6b taken at 2.06 V. The attractor has spread away from the P6 cycle points at the tips and very fine new branches have appeared where the internal cycle points touched the attractor. P6 has also become quite dominant in the dynamics, the system spending about 5% of it time in its vicinity versus a little more than 1% near the P2 and P4 UPO's (not shown).

2.5 *Physics and Dynamical Equations of Motion*

Several models have been developed to understand the dynamics of the driven diode resonator. The simplest physical model describes the diode as a nonlinear conductance in parallel with a nonlinear capacitance[18]. The resulting equations of motion for Fig. 1 can be written as a pair of coupled first order differential equations

$$\frac{dV}{dt} = \frac{1}{C(V)}\left(I - I_D(V)\right) \tag{1.1}$$

$$\frac{dI}{dt} = \frac{1}{L}\left(V_0 \sin(2\pi f t) - IR - V\right) \tag{1.2}$$

where V is the voltage across the diode, I is the current flowing in the circuit, R and L the circuit resistance and inductance respectively, and $I_D(V)$ the current through the diode due to its nonlinear conductance. The capacitance C(V) is the sum of two terms, the transition capacitance, $C_T(V)$, that only acts when the diode is reverse biased and behaves as a voltage variable capacitor, and the diffusion capacitance, $C_D(V)$, due to minority carrier charge injection under forward bias of the junction. The voltage dependence of these quantities is given in eq. 2.

$$I_D(V) = I_s\left(\exp(V/V_t) - 1\right) \tag{2.1}$$

$$C_T(V) = C_0\left(1 - \frac{V}{\phi}\right)^{-\gamma}, \quad V < \phi \tag{2.2}$$

$$C_D(V) = \frac{\tau}{2V_t} I_s \exp(V/V_t) \tag{2.3}$$

$$C(V) = C_T(V) + C_D(V) \tag{2.4}$$

The measured coefficients for a typical circuit are in Table 2. Note that the diffusion capacitance depends on the minority carrier lifetime, τ. It has been established experimentally that unless the period of the driving signal, $2\pi/\omega$, is of the same order as τ no bifurcations take place[19, 20]. Models based on Eq. 1 and 2 give good agreement with the experimental data including the bifurcation diagram, hysteresis at the transition between periodic windows and chaotic bands, and the shape of the attractor.

Table 2. Typical coefficients for a diode resonator with a 1N5470A diode

I_S	3.6 pA
V_t	34 mV
C_0	80 pF
γ	0.44
τ	1.36 ms
R	25 W
L	50 μH
ϕ	0.6 V
f	2.5 MHz

The capacitances C_T and C_D are really differential capacitances, dQ/dV, where Q is the charge on the diode. The first order equations of motion can be rewritten in terms of the charge and combined to form a second order differential equation for the circuit[14],

$$L\frac{d^2Q}{dt^2}+\left(L\frac{dI_D}{dQ}+R\right)\frac{dQ}{dt}+V(Q)+RI_D(Q)=V_0\sin(2\pi ft). \tag{3}$$

This equation can be approximated by the linear ODE,

$$L\frac{d^2Q}{dt^2}+\left(\frac{2L}{\tau}+R\right)\frac{dQ}{dt}+\phi=V_0\sin(2\pi ft),\quad Q\geq 0 \tag{4}$$

where the nonlinearity is in the boundary condition, $Q \geq 0$, due to the diode shut off when it is reverse biased. The effect of the minority carrier lifetime is quite clear in this formulation. A small lifetime creates a large damping term that effectively makes the equation first order, thereby eliminating the possibility of chaotic motion. This equation can be put in dimensionless form,

$$\frac{d^2\zeta}{d\varphi^2}+\mu\frac{d\zeta}{d\varphi}+1=\rho\sin(2\pi\chi\varphi),\quad \zeta\geq 0 \tag{5}$$

where the dimensionless quantities are defined by the relations,

$$\zeta\equiv\frac{Lf^2}{\phi}Q \tag{6.1}$$

$$\varphi\equiv\frac{2t}{\tau} \tag{6.2}$$

$$\mu\equiv\frac{\tau}{2}\left(\frac{R}{L}+\frac{2}{\tau}\right) \tag{6.3}$$

$$\rho\equiv\frac{V_0}{\phi}. \tag{6.4}$$

$$\chi\equiv\frac{f\tau}{2} \tag{6.5}$$

From this we see that the transition capacitance voltage scale, ϕ, determines the amplitude of the driving function. Equation 5 is a straightforward linear differential equation with the solution given by,

$$\zeta(\varphi)=-\frac{\rho}{2\pi\sigma\chi}\sin 2\pi(\chi\varphi+\theta)+\alpha\exp(-\mu\varphi)-\frac{\varphi}{\mu}+\beta \tag{7.1}$$

$$\alpha=-\frac{1}{\mu}\left[\zeta_0'+\frac{1}{\mu}+\frac{\rho}{\sigma}\cos 2\pi(\chi\varphi_0+\theta)\right]\exp(\mu\varphi_0) \tag{7.2}$$

$$\beta=\zeta_0+\frac{\varphi_0}{\mu}+\frac{\rho}{2\pi\sigma\chi}\sin 2\pi(\chi\varphi_0+\theta)-\alpha\exp(-\mu\varphi_0) \tag{7.3}$$

$$\sigma=\sqrt{\mu^2+4\pi^2\chi^2} \tag{7.4}$$

$$\theta=\frac{1}{2\pi}\arctan\frac{\mu}{2\pi\chi} \tag{7.5}$$

with the initial conditions and phases defined by,

$$\zeta_0 = \zeta(\varphi_0) = 0 \tag{8.1}$$

$$\zeta_0' = \frac{d\zeta}{d\varphi}(\varphi_0). \tag{8.2}$$

The system is solved by finding successive zeroes of Eq. 7.1. The iteration is started by choosing a positive value for ζ'_0 at $\varphi_0 = 0$. Thereafter, the replacement $\zeta'_0 \rightarrow |\zeta'_0|$ is made at the start of the search for the next root of Eq. 7.1 since ζ'_0 must be less than or equal to zero at each root but positive at the start of each iteration. Exact periodic solutions can be found by satisfying the conditions $\zeta(\varphi_k) = \zeta(\varphi_k+k/\chi)$ and $\zeta'(\varphi_k) = -\zeta'(\varphi_k+k/\chi)$ where the period, k, is an integer. The model exhibits bifurcation diagrams, hysteresis, and periodic orbits very similar to the data. More details can be found in Linsay [21]. Rollins and Hunt[3, 22] developed an iterated model for the diode resonator that is quite similar to Eq. 7. A simpler one dimensional model that explains some of the observed features has been derived by other authors[23, 24].

3. Relaxation Oscillators and Quasiperiodicity

Simple relaxation oscillators have played a major role in studying quasiperiodic routes to chaos. A quasiperiodic system is one that has a power spectrum made of distinct spectral lines that are not related to each other by rational numbers. If all the frequencies in the power spectrum can be written as rational sums and differences of two base frequencies the system is two frequency quasiperiodic, QP2. Similarly, if only three base frequencies are needed to describe all the spectral lines the system is three frequency quasiperiodic, QP3, and so on. The number of independent spectral lines determines the dimension of the system and the dynamics can be thought of as lying on a torus of appropriate dimension. For QP2, the motion lies on the surface of a torus suspended in three dimensions. Two dynamical variables are needed to describe the motion, corresponding to the two angles in Fig. 7. The motion traces a curve on the surface of the torus. If the two frequencies are irrationally related, the curve will eventually cover the entire surface of the torus and never close on itself, but if phase locking occurs and the frequencies become rationally related the curve will close on itself after a finite time. In this case the variable Θ will go through 2π, N times and Φ will rotate through 2π, M times before the curve reconnects with itself. The system is said to have a winding number, $\rho = M/N$. The motion can also be thought of as being confined to a unit square for which the right and left edges are identified as the same, and the top and bottom edges are identified as the same, because joining these edges of the square will form a torus. Similarly for QP3, the motion is described by 3 angles on the surface of a 3 torus suspended in 4 dimensions, which of course cannot be drawn. However, a cube with opposite faces identified as identical has the same dynamics as the 3 torus by the same reasoning that equates the unit square to the 2 torus.

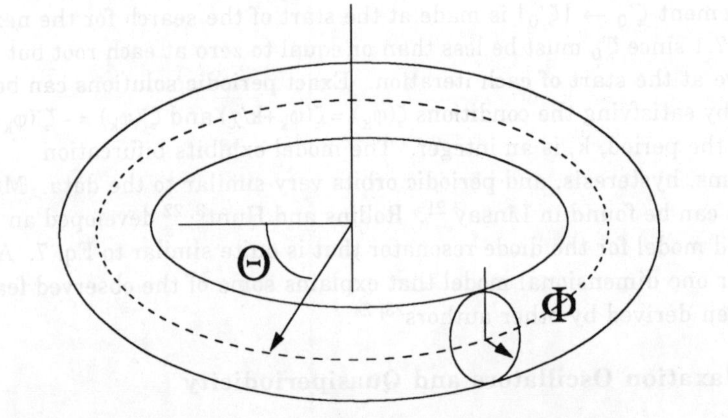

Figure 7. Schematic of a torus showing the toroidal angle, Θ, and the poloidal angle, Φ.

3.1 *Two Frequency Quasiperiodicity*

These concepts are illustrated by a driven van der Pol oscillator with a cubic nonlinearity as given by Eq. 9.

$$\frac{d^2x}{dt^2} + \gamma\left(x^2 - 1\right)\frac{dx}{dt} + \omega_0^2 x + g x^3 = A\sin(2\pi f t) \tag{9}$$

The circuit was synthesized using standard analog techniques with op-amp integrators and commercial multipliers to generate the nonlinear terms in the equation[25]. Fig. 8 is a parameter space plot of the behavior of the circuit when the amplitude, A, and frequency, f, are scanned. When the parameters are in the black tongues, the Arnold tongues[26], the circuit and drive phase-lock and x is periodic with time. The winding number labels on the plot mean, in the 3/5 tongue for example, the van der Pol oscillator variable x executes 3 oscillations in the same time the drive executes 5 oscillations. This is true throughout the 3/5 tongue, even though the frequency and amplitude vary considerably. Only a few of the

Arnold tongues appear at the resolution of the figure but many more can be found. In fact, if two tongues have winding numbers p/q and p'/q' then there is a tongue between them that has winding number (p+p')/(q+q'). In general, the width of the tongues decreases as the denominator increases, making high period Arnold tongues hard to find. The circuit response is quasiperiodic in the large light gray zone along the bottom of the figure. In this region, the power spectrum of x is composed of sums and differences of only two frequencies. Chaotic behavior occurs in the dark gray zones. Notice that there is a sharp dividing line, the critical line, between the quasiperiodic and chaotic zones. No chaotic motion is possible below this line, and no quasiperiodic motion appears above this line. [The apparent mixing of QP2 into the chaotic zone, and vice versa in the figure is due to inefficiencies in the realtime computer algorithm used to detect the different types of behavior.]

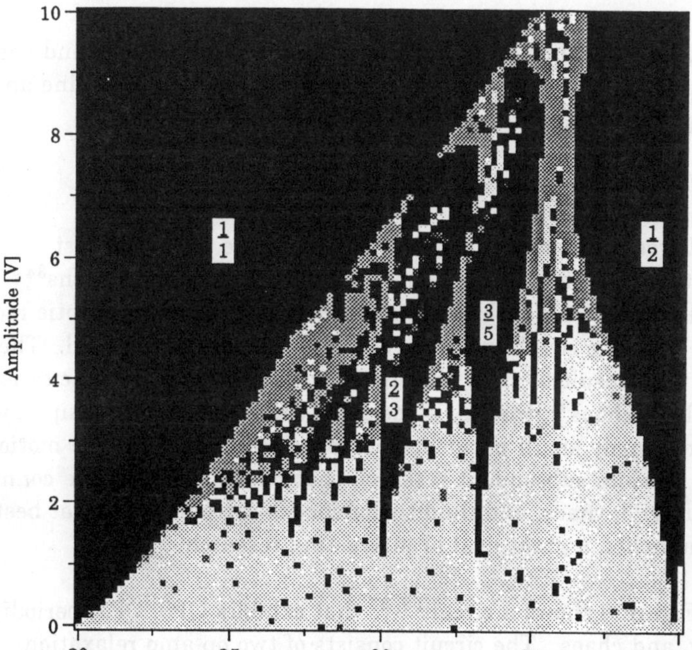

Figure 8. Parameter space plot of a driven van der Pol oscillator. Black zones are phase locked periodic response, light gray zones are QP2, and dark gray zones are chaos. The fractions indicate the winding number of four of the periodic bands.

Two frequency quasiperiodicity has an extensive literature about universality but the nature of the universality is much more subtle than in the case of period doubling and more difficult to observe. As the amplitude, A, is increased the Arnold tongues increase in width and eventually squeeze out all the quasiperiodic motion at the critical line. Theory[27, 28] predicts that the quasiperiodic set remains behind as a fractal set with dimension of approximately 0.87. At the critical line and in the vicinity of a winding number with the value of the golden mean, (√5 - 1)/2, the power spectrum has certain scaling properties and the widths of the approximating Arnold tongues, those with winding numbers 2/3, 3/5, 5/8..., converge at a specific rate[29-32]. All these theoretical predictions are derived from properties of the sine-circle map,

$$\Theta_{n+1} = \Theta_n + \Omega - \frac{K}{2\pi}\sin(2\pi\Theta_n). \tag{10}$$

Other properties, in particular, the generalized dimensions and the $f(\alpha)$ spectrum, are also universal[33] but again just at the critical line and for a winding number equal to the golden mean.

3.2 *Three Frequency Quasiperiodicity*

Three frequency quasiperiodicity was one of the earliest inspirations for the study of chaos. In 1971, Ruelle and Takens[34] suggested that fluid turbulence might have its origin in chaotic motion arising from the coupling of four oscillatory modes of the fluid. They later refined this result to show that three frequency quasiperiodicity is unstable to chaos[35], that is, if one finds three frequency quasiperiodicity, then arbitrarily small parameter changes will transform the motion into chaos. Despite subsequent research that has shown that the connection between QP3, chaos, and fluid turbulence is extremely weak at best, these ideas launched a great deal of interesting research.

Figure 9 shows a circuit[36, 37] that exhibits QP3, QP2, periodic mode locking, and chaos. The circuit consists of two op-amp relaxation oscillators coupled through a resistor (R_y) and driven by a sine wave generator. The upper oscillator has a fixed frequency and sets the time scale of the circuit. The lower oscillator has a variable frequency that can be changed via the programmable feedback resistor R_2 and the sine wave generator can be varied in both frequency and amplitude. The coupling between oscillators is zero when R_y is zero and increases linearly with its value. The nonlinearity in this oscillator is a boundary condition nonlinearity and is due to the output voltages, U_i, switching between two different values. This is a different type of nonlinearity from the smooth functional nonlinearities of the previous circuits.

The operation of an individual oscillator is quite simple. For the purposes of explanation hold the point V_y at ground. Now, suppose that U_1 is high, then 1/2 U_1 appears at the positive input terminal of the op-amp. The parallel capacitors C_{11} and C_{12} charge up through the feedback resistor R_1 until V_1 equals 1/2 U_1 when U_1 flips to its low value, -U_1. Now, -1/2 U_1 appears at the positive input terminal of the op-amp. The parallel capacitors C_{11} and C_{12} discharge through the feedback resistor R_1 until V_1

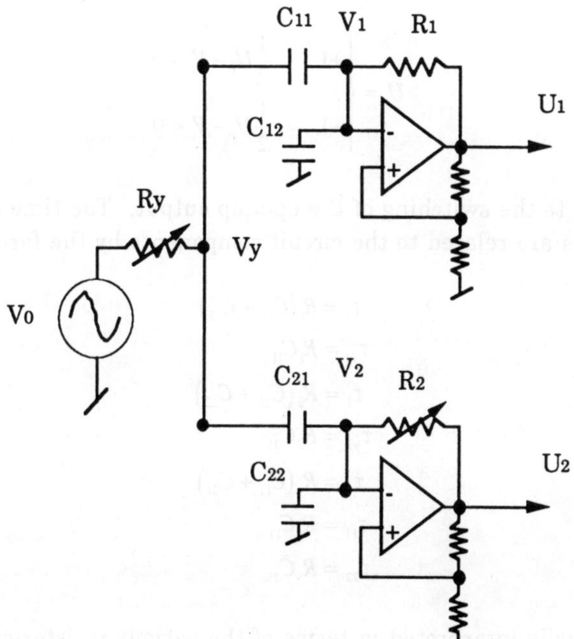

Figure 9. Relaxation oscillator circuit for generating three frequency quasiperiodicity. The upper oscillator sets the frequency scale, the frequency of the lower oscillator is variable as are the frequency and amplitude of the driving sine wave oscillator. The outputs may be inspected individually or summed.

equals -1/2 U_1 when U_1 flips to its high value, U_1. The cycle is now complete and repeats itself. The voltage V_1 rises and falls exponentially with a characteristic time constant of $R_1 \times (C_{11}+C_{12})$ and the output, U_1, is a square wave.

The equations of motion can be found by straightforward application of Kirchoff's laws and are the three coupled linear ordinary differential equations,

$$V_1(t) + \tau_1 \dot{V}_1(t) = U_1 + \tau_{y1} \dot{V}_y(t) \tag{11.1}$$

$$V_2(t) + \tau_2 \dot{V}_2(t) = U_2 + \tau_{y2} \dot{V}_y(t) \tag{11.2}$$

$$V_y(t) + \tau_y \dot{V}_y(t) = V_0 \sin(\omega t) + \tau_{11} \dot{V}_1(t) + \tau_{22} \dot{V}_2(t) \tag{11.3}$$

with the nonlinear boundary condition,

$$U_i = \begin{cases} +1, & \frac{1}{2}U_i - V_i > 0 \\ -1, & \frac{1}{2}U_i - V_i < 0 \end{cases} \tag{12}$$

which is due to the switching of the op-amp output. The time constants in the equations are related to the circuit components by the formulas,

$$\tau_1 = R_1(C_{11} + C_{12}) \tag{13.1}$$

$$\tau_{y1} = R_1 C_{11} \tag{13.2}$$

$$\tau_2 = R_2(C_{21} + C_{22}) \tag{13.3}$$

$$\tau_{y2} = R_2 C_{21} \tag{13.4}$$

$$\tau_y = R_y(C_{11} + C_{21}) \tag{13.5}$$

$$\tau_{11} = R_y C_{11} \tag{13.6}$$

$$\tau_{22} = R_y C_{21} \tag{13.7}$$

which are easily interpreted in terms of the various resistor-capacitor combinations.

There is a far larger parameter space to scan for this circuit since there are two variable frequencies and an amplitude available as control parameters. By changing the values of the control parameters it is possible to get a wide variety of behavior. The attractor in figure 10a is typical when the parameters have been set for QP2. The complexity can vary from a simple loop to many turns, depending on the settings. Figure 10b is a typical attractor when the circuit is exhibiting QP3 or chaos. There is little distinction between chaos and QP3 in this system since chaos is typically extremely weak. We distinguished between chaos and QP3 by inspecting a highly undersampled autocorrelation function for the time series. Rather than sampling once per drive period we sampled only once every 64 drive periods and then computed the autocorrelation. A time series was considered chaotic if the envelope of the autocorrelation function had an amplitude of less than 0.5 for a time delay of greater than 8192 drive periods.

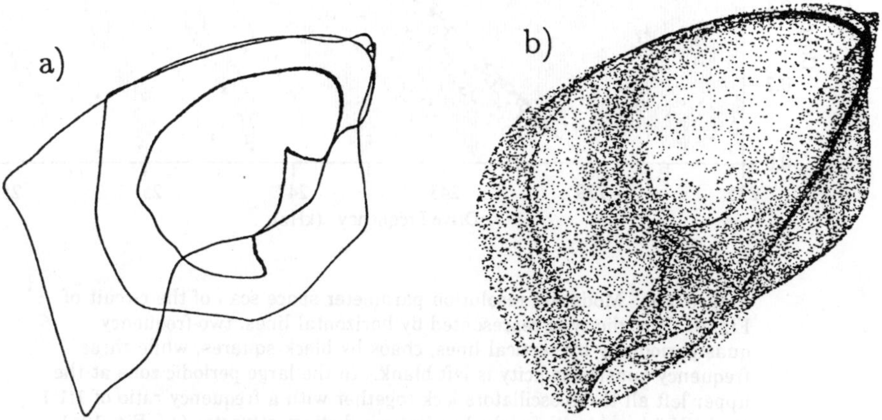

Figure 10. Poincaré section, or attractor, of the circuit of Fig. 9 made by sampling the sum of the outputs U_1 and U_2 at the period of the drive, $V_y(t)$. Successive samples are plotted as x-y pairs to generate the attractors. (a) Two frequency quasiperiodicity, (b) three frequency quasiperiodicity.

The purpose of this circuit was to test the Ruelle-Takens scenario, that is, whether three frequency quasiperiodicity is unstable and always makes a transition to chaos. Figure 11 is a moderate resolution scan of the circuit parameter space in amplitude and frequency of the driving sine wave oscillator. The fixed frequency oscillator was set to run at 240 KHz and the variable frequency oscillator at 288 KHz so that it did not automatically mode lock to the first oscillator through the coupling. All four possible responses appear in the plot. Periodic response, without regard to the actual period, is marked by horizontal lines, chaos by black squares, QP2 by vertical lines, and white areas are QP3. The existence of large zones of QP3 already shows that the Ruelle-Takens scenario cannot be strongly true.

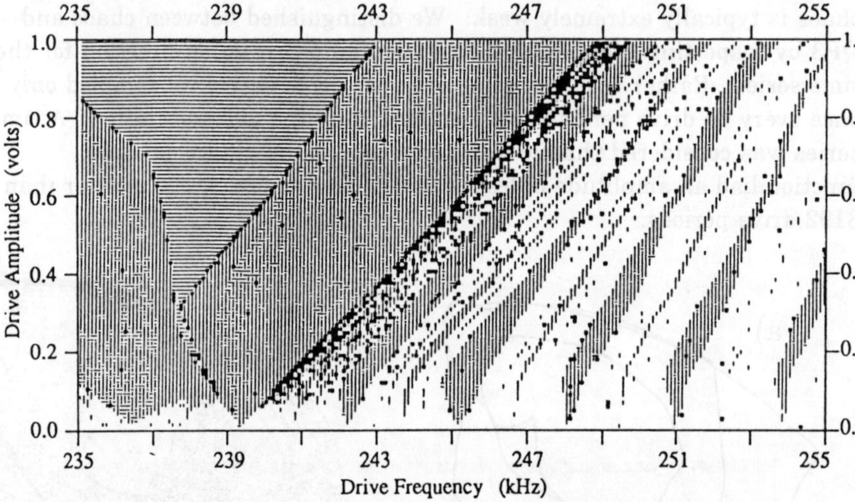

Figure 11. Moderate resolution parameter space scan of the circuit of Fig. 9. Periodicity is represented by horizontal lines, two-frequency quasiperiodicity by vertical lines, chaos by black squares, while three frequency quasiperiodicity is left blank. In the large periodic zone at the upper left all three oscillators lock together with a frequency ratio of 1:1:1. Each QP2 band on the right has its own distinct attractor (c.f. Fig. 10a).

By increasing the resolution, we see in Fig. 12 a region in which the scenario is true. The chaotic bands on the left alternate with QP2 bands, displacing the QP3 zones that would be there if QP3 were stable. In general, the prominent chaotic zones are strongly correlated with regions of strong periodic motion and occur at the junction between QP2 motions of different types (see below for an explanation of this). For example, on the right side of the plot QP3 appears to displace chaos and is not unstable. Most likely, the difference here between chaos and QP3 is so small as to make them indistinguishable. Despite the fact that at best the Ruelle-Takens scenario is only weakly true, it is one of those interesting ideas that provoked a great deal of interest and created a new direction in the study of turbulence and chaotic dynamics.

The circuit parameter space is better understood by scanning in frequency-frequency space and holding the amplitude of the drive oscillator fixed, essentially by making a scan perpendicular to the plane of Fig. 11 or 12. This shows the intricate interlocking structure of the periodic, two frequency quasiperiodic and three frequency quasiperiodic zones. Figure 13 was produced by setting the drive amplitude to a large value where the coupling among the oscillators is strong and systematically varying the frequency of the variable oscillator and the drive oscillator. The white zones in the figure are QP3, the light gray bands represent QP2, the dark gray regions are periodic, and the black regions are chaotic. The most striking feature of this figure is the large number of QP2 bands (light gray) of differing thicknesses crisscrossing the parameter space at many different angles. Close inspection reveals that where ever two QP2 bands cross there is a dark gray zone, that is, the intersection is a periodic zone.

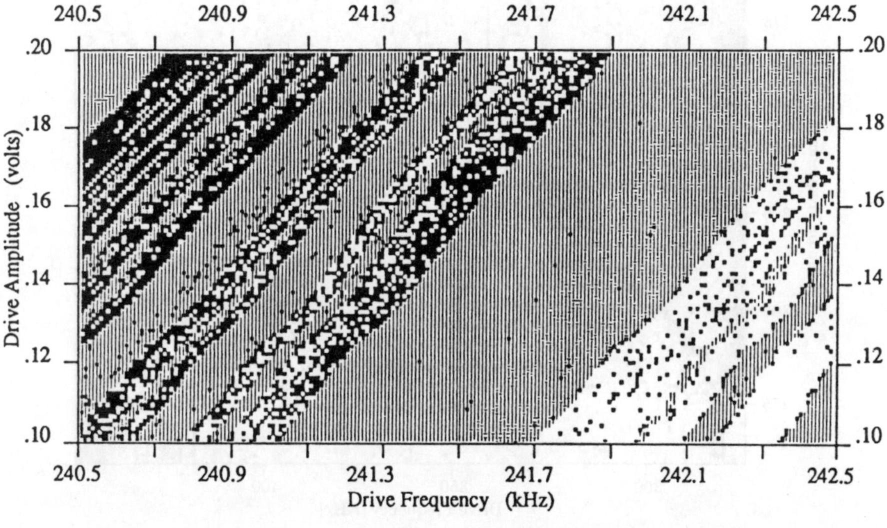

Figure 12. High resolution scan of a portion of Fig. 11.

In a periodic zone, all three oscillators have phase-locked and there is a definite relation among the frequencies of the oscillators. This is best understood by considering the winding *numbers* for a three frequency system. For a two frequency quasiperiodic system there was one winding number, the ratio of the two oscillator frequencies. Clearly, we need two winding numbers here, which can be chosen as the oscillator to the drive ratios, $\rho_i \equiv f_i / f_{drive}$. If the behavior is QP3 there is no rational relationship between the winding numbers. If it is QP2 then they are related by a linear equation of the form,

$$\rho_2 = p/q + (r/s)\rho_1 \tag{14}$$

where p and q are relatively prime integers, as are r and s. The winding numbers have this relationship inside a gray band with p,q,r, and s constant. It is now obvious why there is a periodic phase locking when bands intersect. The winding numbers must simultaneously satisfy two different relations of the form given in eq. 14, hence they must both be rational numbers since the coefficients in eq. 14 are rational. This allows

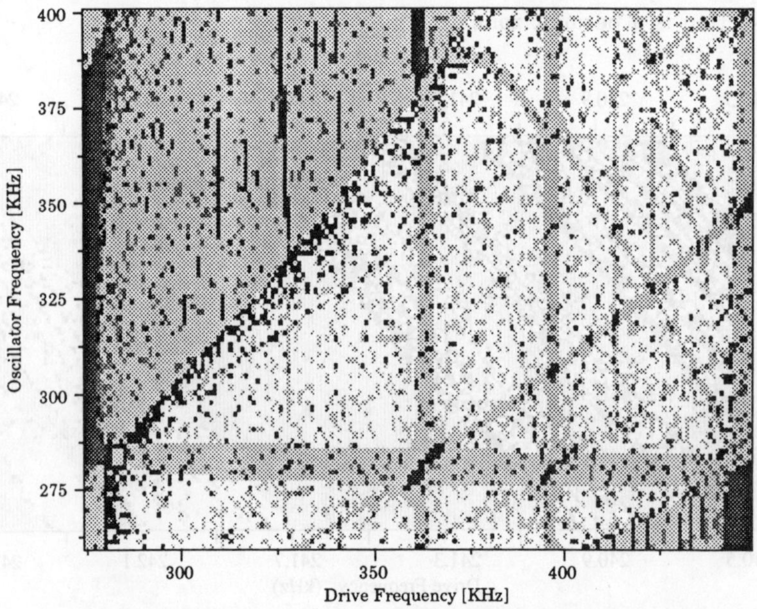

Figure 13. Low resolution scan of the frequency-frequency parameter plane for the three frequency circuit. White zones are QP3, light gray bands are QP2, dark gray zones are periodic, and black squares are chaos. The dark gray periodic band at the left is the (1,1,1) mode locking and the one in the lower right corner is (1,1,2). The dark diagonal band at drive = 365 KHz and Oscillator = 285 Khz is the (2,2,3) mode locking.

a labeling of the periodic zones by a set of integers. The winding numbers must both be of the form $\rho_i = P_i/Q$ where Q is the period of the resonance and P_i is an integer. Consequently every periodic zone can be labeled by a vector of integers (P_1, P_2, Q). If we plot the measured winding numbers of the identified periodic resonances against each other, the result is Fig. 14 (N.B. the plot in winding number space exchanges left for right in physical space, Fig. 13). The mode locking zones are the dots labeled with resonance triplets and the lines correspond to the gray bands in Fig. 13 where QP2 occurs. This data was taken at low resolution but this structure maintains itself as the resolution is increased with ever higher periodic resonances and finer QP2 bands appearing. The structure is a "devil's cobweb" that repeats itself at all scales.

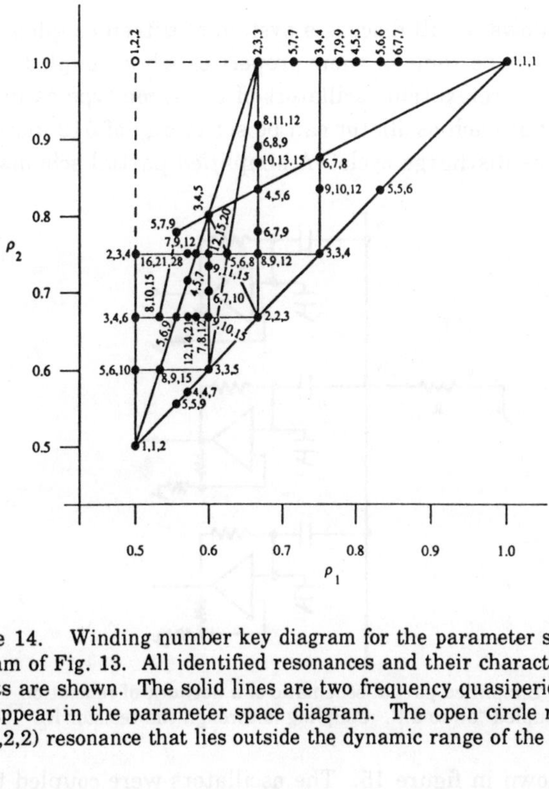

Figure 14. Winding number key diagram for the parameter space diagram of Fig. 13. All identified resonances and their characteristic triplets are shown. The solid lines are two frequency quasiperiodic bands that appear in the parameter space diagram. The open circle represents the (1,2,2) resonance that lies outside the dynamic range of the experiment.

3.3 *Multiple Oscillators*

The circuits described up to now have dealt with a maximum of only three oscillators, yet there are a wide variety of important systems that consist of a large number of oscillators. Josephson junction arrays are of technological importance for their possible use as wide tuning range millimeter wave sources, broadband mixers, parametric amplifiers, and voltage standards. Hadley, Beasley, and Wiesenfeld have discussed the nonlinear dynamics of systems of coupled Josephson junctions[38, 39] with interesting results. It is highly desirable that the electric power grid act in a well behaved manner. Yet, studies of simple swing equations that govern the motion of two coupled power generators show that chaotic motion is possible[40]. Biological systems have been modeled by coupled relaxation oscillators, so called integrate-and-fire models, for many years[41]. These are only a few examples of many.

In what follows, I will discuss a system of fifteen coupled relaxation oscillators and describe some of their properties when coupled "all-to-all". The oscillators were relaxation oscillators of the same type as in Fig. 9. but modified so that each oscillator can be set to any of 512 distinct initial states in its charge-discharge cycle. A simplified partial schematic of the

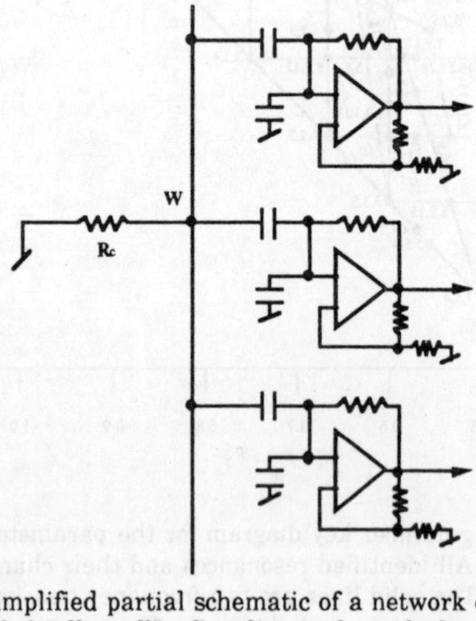

Figure 15. Simplified partial schematic of a network of relaxation oscillators coupled "all-to-all". Coupling is through the resistor R_c.

experiment is shown in figure 15. The oscillators were coupled together in an all-to-all coupling configuration with the strength of the coupling determined by the resistor Rc: there was no coupling when Rc was 0 and maximum coupling when it was large. In the experiments we have done up to now, an isolated oscillator has a frequency 5 KHz when coupled to the network. The spread of free running oscillator frequencies was usually about 10% to ensure that in the locked state all frequency ratios would be 1:1. For simplicity, only one bit of information was collected from each oscillator: 1 when it was charging or 0 when discharging. This allowed data from the oscillators to be collected by a Macintosh II computer at speeds up to 500 KHz via a digital interface board with typical sampling speeds of 100 KHz.

The fundamental measurement was the computation of locking time that established when the oscillators were in phase and allowed direct computation of the period and relative oscillator phases. Oscillator zero was designated as the reference oscillator and all measurements were made with respect to it. The locking time was measured by tracking the time difference between peaks of equal phase of the reference oscillator and the measured oscillator. When the time difference as a function of peak phase became constant the oscillators were locked. The oscillator periods and phases were measured with data from the locked portion of a time series. For simplicity, the relative oscillator phases were normalized to be between zero and one with the reference oscillator always assigned a phase of zero. We identified different final states by examining the phase vector whose coordinates were the N oscillator phases. Two final states were considered the same if the absolute value of the coordinate differences was always less than some preassigned error. That is, given two final states ϕ and Φ, they were the same if

$$\max\{ |\phi_1 - \Phi_1|, |\phi_2 - \Phi_2|, \ldots, |\phi_N - \Phi_N| \} < \varepsilon. \tag{15}$$

A value of 0.010 was normally used for ε. This is to be compared with the measured phase resolution of $\pm$ 0.001. Reducing ε to 0.005 did not change the number of final states observed.

Before discussing the results in detail, we note that there are two basic operating regimes observed for this system. If the coupling was small, then the oscillators were unlocked and the power spectrum indicated that the system was quasiperiodic. Once the coupling was large enough, the oscillators always phase-lock to each other to create a periodic state. Just below the critical coupling, the minimum required to achieve phase-locking, there was a region where the system could be quasiperiodic or periodic depending on initial conditions. But there was no coupling that produced chaotic behavior.

The behavior of the transients carries interesting information about the final states and the basin structure. We studied this by measuring the locking time distribution as a function of both the coupling strength and the number of coupled oscillators. The distribution was measured by initializing the oscillators with 1000 random initial conditions and accumulating the locking times. The shape of this distribution varied with the number of oscillators. With small numbers of oscillators, two to four, the distributions tended to be sharply peaked but gradually developed an exponential background as the number of coupled oscillators increased. Once the number of oscillators was as large as eight, the distribution was purely exponential with no trace of the peaks that appeared with smaller numbers of oscillators. The slope of the exponential was a function of the coupling among the oscillators. Just above the critical coupling the slope varied inversely with the strength of the coupling resistance resulting in very long locking times close to the critical coupling and much shorter times as the coupling strength increased.

The number and character of the final states is also interesting. Using eq. 15 as a discriminant, we counted the number of final states for a given number of oscillators and coupling strength. For N oscillators, the number of states increased from a few near the critical coupling to (N-1)! once the coupling was strong enough. Since this number increases so rapidly, we were only able to verify this directly for seven or fewer oscillators. However, this is the number of permutations of N objects arranged in a circle, and what one expects since no oscillator occupies a preferred position physically in the all-to-all coupling scheme. For three, four, and five oscillators it was also clear that each final state had its own period and locking time distribution. As the numbers of oscillators increase, these distinctions wash out since the different periods become very close to each other and the distributions overlap.

The dynamics of the network can be understood in a straightforward manner. The individual oscillators have a waveform at the negative port of the op-amp that rises and falls exponentially. The nonlinearity is the boundary condition that switches the direction of the exponential from rising (falling) to falling (rising) when the upper (lower) threshold is hit. The coupling network differentiates this signal and turns it into an approximate square wave that is high when the exponential is increasing and low when it is decreasing. It can be shown that the net effect of the mutual coupling between two oscillators drives them 180° out of phase. This creates the interesting dynamics when several oscillators are connected together since it is impossible for more than two oscillators to all simultaneously be 180° out of phase with respect to each other. In fact, for small numbers of oscillators, the oscillators lock in 180° pairs and the phase among pairs is determined by residual coupling terms. Once there are more than six oscillators, the residual couplings become large enough that the phases in the final locked state tend to be distributed evenly over a full 360°. The details of the final state are determined by the free running frequencies of the individual oscillators and the initial phases.

The basic dynamics and other features of the system were clarified by a numerical model developed to understand how the oscillators interacted. Since this was a numerical model, we could explore in much finer detail than in the actual experiment. The basins of attraction of the final states, the set of initial conditions that lead to a particular phase-locking, were interesting features to study since it has been observed in other systems that they can be fractal. Figure 16 is a highly magnified section of the initial state phase space for a system of four oscillators.

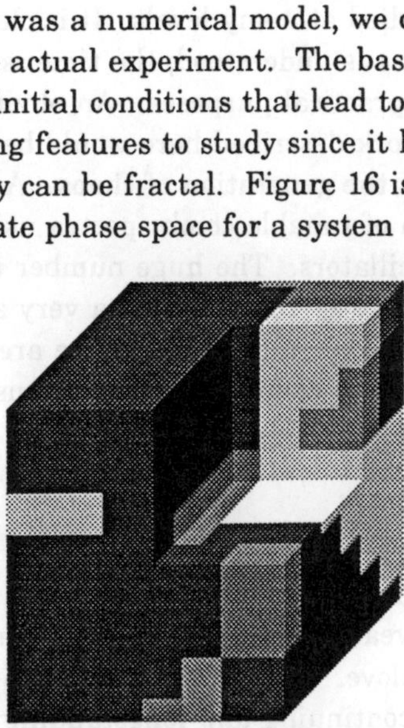

Figure 16. Magnification of the basins of attraction for a system of four coupled oscillators. The three axes represent different initial phases for each of three oscillators, the reference oscillator is always initialized with a phase of zero. Each shade of gray corresponds to a different set of relative phases between the oscillators.

Only three axes are needed since there are only three relative phases for a set of four oscillators. Each initial condition (each voxel) has been labeled by a color that represents which final state it eventually becomes. In this section of initial conditions most evolve to the periodic state represented by the black voxels, and some to the other two states indicated by the white and gray voxels. What is important is that small changes in initial conditions lead to very different final periodic states. Even though this system exhibits no chaos, it does show a form of sensitive dependence on initial conditions[42, 43]. More details about this experiment can be found elsewhere[44].

4. Conclusion

Simple electronic circuits have proven to be an excellent laboratory for testing and understanding theories about nonlinear dynamics. Even the simplest circuit, a driven diode resonator, has a very rich variety of behavior and can be studied at many levels. It is clear, now that so much about nonlinear dynamics is understood, the time is ripe for exploiting nonlinear dynamics for practical purposes. An obvious avenue of development from the work described here would be noise sources with colored spectra based on the generation of chaos. Another application might be the generation of tunable comb spectra using quasiperiodic systems of nonlinear oscillators. The huge number of accessible and stable resonances that appear in Fig. 13 make this a very attractive way of generating these signals. One of the most active areas of development is the use of chaotic carriers in communications discussed elsewhere in this volume.

5. Acknowledgements

I would like to thank my collaborators who have worked with me on these projects over the years, Andrew Cumming, Dan Dewey, Stuart Brorson, and Adam Brailove. I would especially like to thank the Office of Naval Research for its continuing and longstanding support for this research.

6. References

1. P. S. Linsay, *Physical Review Letters* **47** (1981) 1349

2. J. Testa, J. Perez and C. Jeffries, *Physical Review Letters* **48** (1982) 714

3. E. R. Hunt and R. W. Rollins, *Physical Review A* **29** (1984) 1000

4. E. R. Hunt, *Physical Review Letters* **67** (1991) 1953

5. T. Klinker, W. Meyer-Ilse and W. Lauterborn, *Physics Letters A* **101** (1984) 371

6. M. J. Feigenbaum, *Los Alamos Science* **1** (1980) 4

7. M. J. Feigenbaum, *Journal of Statistical Physics* **19** (1978) 25

8. M. J. Feigenbaum, *Journal of Statistical Physics* **21** (1979) 669

9. M. J. Feigenbaum, *Commun. Math. Phys.* **77** (1980) 65

10. C. Grebogi, E. Ott and J. A. Yorke, *Physical Review Letters* **48** (1982) 1507

11. C. Grebogi, E. Ott and J. A. Yorke, *Physica D* **7** (1983) 181

12. N. H. Packard, J. P. Crutchfield, J. D. Farmer and R. S. Shaw, *Physical Review Letters* **45** (1980) 712

13. F. Takens, in *Dynamical Systems and Turbulence, Warwick 1980*, eds. D. A. Rand and L.-S. Young (Springer-Verlag, Berlin, 1980) p. 366

14. S. D. Brorson, D. Dewey and P. S. Linsay, *Physical Review A* **28** (1983) 1201

15. D. Auerbach, P. Cvitanovic, J.-P. Eckmann, G. Gunaratne and I. Procaccia, *Physical Review Letters* **58** (1987) 2387

16. G. H. Gunaratne, P. S. Linsay and M. J. Vinson, *Physical Review Letters* **63** (1989) 1

17. G. B. Mindlin, X.-J. Hou, H. G. Solari, R. Gilmore and N. B. Tufillaro, *Physical Review Letters* **64** (1990) 2350

18. *Electronic Designer's Handbook*, 2nd ed., ed. L. J. Giacoletto (McGraw-Hill, New York, 1977)

19. J. Testa, J. Perez and C. Jeffries, *Physical Review Letters* **49** (1982)

1055

20. E. R. Hunt, *Physical Review Letters* **49** (1982) 1054

21. P. S. Linsay, in *Perspectives in Nonlinear Dynamics*, eds. M. F. Shlesinger, R. Cawley, A. W. Saenz and W. Zachery (World Scientific, Singapore, 1986) p. 138

22. R. W. Rollins and E. R. Hunt, *Physical Review Letters* **49** (1982) 1295

23. T.-H. Yoon, J.-W. Song, S. Shin and J.-W. Ra, *Physical Review A* **30** (1984) 3347

24. J. M. Perez, *Physical Review A* **32** (1985) 2513

25. G. Qin, D. Gong, R. Li and X. Wen, *Physics Letters A* **141** (1989) 412

26. V. I. Arnold, *Geometrical Methods in the Theory of Ordinary Differential Equations* (Springer-Verlag, New York, Heidelberg, Berlin, 1983) p. 108

27. M. H. Jensen, P. Bak and T. Bohr, *Physical Review A* **30** (1984) 1970

28. M. H. Jensen, P. Bak and T. Bohr, *Physical Review Letters* **50** (1983) 1637

29. S. J. Shenker, *Physica D* **5** (1982) 405

30. M. J. Feigenbaum, L. P. Kadanoff and S. J. Shenker, *Physica D* **5** (1982) 370

31. D. Rand, S. Ostlund, J. Sethna and E. D. Siggia, *Physical Review Letters* **49** (1982) 132

32. S. Ostlund, D. Rand, J. Sethna and E. D. Siggia, *Physica D* **8** (1983) 303

33. T. C. Halsey, M. H. Jensen, L. P. Kadanoff, I. Procaccia and B. I. Shraiman, *Physical Review A* **33** (1986) 1141

34. D. Ruelle and F. Takens, *Commun. Math. Phys.* **20** (1971) 167

35. S. Newhouse, D. Ruelle and F. Takens, *Commun. Math. Phys.* **64** (1978) 35

36. A. Cumming and P. S. Linsay, *Physical Review Letters* **26** (1988) 2719

37. P. S. Linsay and A. W. Cumming, *Physica D* **40** (1989) 196

38. P. Hadley, M. R. Beasley and K. Wiesenfeld, *Physical Review B* **38** (1988) 8712

39. P. Hadley, M. R. Beasley and K. Wiesenfeld, *Appl. Phys. Lett.* **52** (1988) 1619

40. N. Kopell and R. B. W. Jr, *IEEE Transactions on Circuits and Systems* **CAS-29** (1982) 738

41. L. Glass and M. C. Mackey, *Journal of Mathematical Biology* **7** (1979) 339

42. C. Grebogi, S. W. McDonald, E. Ott and J. A. Yorke, *Physics Letters A* **99** (1983) 415

43. S. W. McDonald, C. Grebogi, E. Ott and J. A. Yorke, *Physica D* **17** (1985) 125

44. P. S. Linsay and A. Brailove, in *Proceedings of the 1st Experimental Chaos Conference*, eds. S. Vohra, M. Spano, M. Shlesinger, L. Pecora and W. Ditto (World Scientific, Singapore, 1992) p. 315

SYMMETRIC DYNAMICS AND ELECTRONIC CIRCUITS

PETER ASHWIN
Mathematics Institute, University of Warwick, Coventry CV4 7AL, UK.

Present address: INLN, 1361 route des Lucioles, 06560 Valbonne, France.

ABSTRACT

We give a brief introduction to concepts from symmetric dynamics and discuss applications to electronic circuits using two examples. (a) Symmetric networks of identical limit cycle oscillators; we discuss bifurcation behaviour, the detection of symmetries and observations of bifurcations of symmetric chaotic attractors. (b) Synchronisation of two chaotic circuits; focussing on the stability of synchronised chaotic states, we discuss riddled basins of attraction and stability, with examples of where this has been observed experimentally.

1. Introduction

Electronic engineers often work with nonlinear circuits that can be thought of as consisting of a set of identical cells coupled together in some way. Often, the coupling is invariant under some set of symmetries. If the cells can display oscillatory dynamics then the system is called a *coupled oscillator network*.

Seen from the mathematician's or theoretical physicist's viewpoint, electronic networks provide clean examples of possible types of dynamic behaviour in systems that are (to a high degree) finite dimensional with relatively fast dynamics. For example, it is easy to run circuits in the kilohertz to megahertz range of frequencies, whereas laboratory fluid dynamics experiments typically have frequencies of a few Hertz at fastest and so waiting for transients to die away can take a very long time.

There are a few methods from the bifurcation theory of symmetric systems that we are going to discuss in this article. Until recently most of the effort in symmetric dynamics has been put into understanding bifurcation of steady or oscillatory solutions in a systematic way. In the last few years, general properties of attractors in symmetric systems have been a subject of research, although the theory is still in its infancy compared to local bifurcation theory.

The *theory of groups* (and their representations) was initially developed as a theory of symmetries; one might say that the theory of groups provides a language with which to discuss symmetries. Moreover, since bifurcation of steady and oscillatory solutions can be thought of as local problems, so-called *singularity theory*, can be used to discuss the behaviour near a bifurcation point in terms of (hopefully only a few low order terms in) a Taylor expansion near a so-called *bifurcation point.*

There are now many techniques to rigorously analyse such local bifurcations. These give a detailed classification and description with considerable predictive power. Important tools for this theory are normal forms, first developed by Poincaré and Birkhoff, and center manifold theory developed in the 1960s. The theory of unfolding of degenerate normal forms gained a rigorous grounding during the 1970s, and this gave rise to the development of catastrophe theory. For a comprehensive survey of the mathematics of bifurcation theory and catastrophe theory, we refer the reader to Arnol'd. [4]

If the system is symmetric (in a sense that will be made precise later) we find that a theory for local bifurcations is substantially modified; behaviour that is generic for non-symmetric systems can become impossible for symmetric systems, whereas behaviour that is generic for symmetric systems can be non-generic for non-symmetric systems. This means that the theory of bifurcations of symmetric systems has had to be developed using tools that do not have much application to non-symmetric systems. In particular, a combination of group theory and singularity theory are necessary to classify what might be called the *bifurcation scenarios* for a particular symmetry; moreover, there seems to be little hope of a theory that covers all possible symmetry groups which can say anything of use. To a large extent the symmetry groups have to be considered in a case-by-case (or at least, family-by-family) way.

For the most comprehensive survey of symmetric local bifurcation theory yet, we refer the reader to the two volumes Golubitsky *et al.*,[19, 20] while for taste of recent research in this area, the collections [2,3,13] have a selection of research articles. For a more detailed mathematical introductory article than this, see Stewart.[37]

In Section 2 we introduce some ideas from symmetric dynamics and apply them in section 3 to an example system of three identical coupled oscillators in two limits; near Hopf bifurcation (onset of oscillation) and for weak coupling. In the former case there are solid predictions about which patterns are and are not permissible, whereas for the second case, we can discuss the stability of the in-phase solution in some detail.

During the last decade there has been some investigation into just how much can be said about bifurcation of chaotic attractors in systems with symmetry. Starting with some numerical experiments[14,16] on mappings of the plane with the symmetry of regular polygons, there is now a rudimentary understanding at least of which subsymmetries are realisable as symmetries of attractors. We shall briefly review this rapidly developing area with a view to discussing an application in section 4 to synchronisation of two chaotic oscillators (see also the article by Cuomo and Oppenheim in this volume). This leads us to consider the phenomenon of *basin riddling* of attractors in synchronised systems.

2. Symmetric Dynamics

We shall restrict our discussion to symmetries that are technically described as compact Lie groups whose elements are $n \times n$ orthogonal matrices (i.e. subgroups of $\mathbf{O}(n)$), where the phase space of the system is $\mathbf{R}^n$ (It is possible to show that this is a large class of groups that includes, for instance, all finite groups. An ordinary differential equation (ODE) is symmetric (more precisely, *equivariant*) with group Γ if it satisfies the following property:

Definition 2.1 *An ordinary differential equation* $\dot{x} = f(x)$ $(x \in \mathbf{R}^n)$ *is* Γ-equivariant *if*

$$f(\gamma x) = \gamma f(x)$$

for all $\gamma \in \Gamma$.

Note that f is a generally a nonlinear function of x. A typical example for a network of n coupled electronic oscillators is to observe that the ODE

$$\dot{x}_i = f(x_i) + g_i(x_1, \ldots, x_n)$$

on $x \in \mathbf{R}^{nm}$ ($\mathbf{R}^m$ is the phase space for one oscillator) is symmetric under some subgroup of $\mathbf{S}_n$, the group of all permutations on n objects represented by matrices in $\mathbf{R}^{nm \times nm}$. This would mean that the ODE is equivariant for matrices γ with an $n \times n$ block structure such that each row and each column has the $m \times m$ identity appearing precisely once; other entries are zero.

The next two definitions are motivated respectively by the questions: (a) Which points in $\mathbf{R}^n$ have symmetry at least Σ, where Σ is a subgroup of Γ? (b) How much symmetry does a point $v \in \mathbf{R}^n$ have?

Definition 2.2 *The* fixed point space *of a subgroup* Σ *of* Γ *is the set:*

$$Fix(\Sigma) = \{v \in \mathbf{R}^n : \sigma(v) = v, \ \forall\, \sigma \in \Sigma\}.$$

Definition 2.3 *The* isotropy subgroup *of a point* $v \in \mathbf{R}^n$ *is the subgroup:*

$$\Sigma_v = \{\sigma \in \Sigma : \sigma(v) = v, \ \forall\, v \in \mathbf{R}^n\}.$$

Considering all the isotropy subgroups of a group leads to an ordering of them given by containment. This is referred to this as the *isotropy diagram* of the action of a group. Note that not all subgroups of the symmetry group are isotropy subgroups, and moreover, that the "instantaneous symmetry" of a point on a trajectory must be one of these subgroups.

2.1. Local bifurcations

If an ODE has a parameter $\lambda \in \mathbf{R}$ such that $\dot{x} = f(x, \lambda)$ we can investigate the change of the solution set of the ODE with λ; sudden changes in the number of, say steady solutions are *bifurcations* of the ODE. When symmetries of the ODE also present, this means we need to study families of equivariant maps.

Local bifurcations correspond to changes in the number of steady state or periodic solutions locally in both the phase space and the parameter space. Consider a smooth path of fixed points $x_0(\lambda)$ such that $f(x_0(\lambda), \lambda) = 0$. It follows from the implicit function theorem that where a fixed point has a linearisation that is nontrivial in the sense that all the eigenvalues of the Jacobian $Df(x_0)$ have non-zero real parts, there can be no other fixed points (or small amplitude periodic solutions) nearby. When this condition is violated, i.e. *marginal stability*, when there is an eigenvalue of $Df(x_0)$ with zero real part, we typically get other solutions that are nearby. In order to find these, there is a standard procedure that we now outline.[21]

The first step usually involves application of the so-called *centre manifold theory*; this says that in a small enough neighbourhood of the origin, the asymptotic dynamics are effectively just the dynamics restricted to a manifold of dimension equal to the number of eigenvalues of $Df(x_0)$ with zero real part (the span of such eigenvectors is a linear subspace called the *centre eigenspace*.

There are two typical ways in which an eigenvalue can have zero real part; either it is zero, or there is a complex conjugate pair of eigenvalues that pass through zero. The first case typically creates other steady solutions, and is referred to as a *steady state* bifurcation; the second case typically creates nearby small amplitude periodic solutions and is named a *Hopf* bifurcation.

For systems without symmetry, this typically allows us to reduce to a first or second order ODE, as it exceptional for more than two eigenvalues to pass through zero simultaneously on varying one parameter. If the system does have a symmetry, it becomes generic for higher dimensional centre manifolds to appear; this is because the symmetry can cause eigenvalues to come in sets.

More precisely, it is possible to characterise the dimension of a centre eigenspace of a Γ-symmetric system in the following way:

Theorem 2.4 *Generically (for one parameter systems) the center eigenspace V_0 at a* steady-state *bifurcation point is such that*

$$\gamma L v = L \gamma v$$

for all $v \in N_0$ and $\gamma \in \Gamma$ (with L an $n \times n$ real *matrix) implies that $L = \lambda Id$ (with $\lambda \in \mathbf{R}$).*

This property corresponds to there being an *absolutely irreducible representation* of Γ on V_0 in the language of group representation theory. We do not intend to explain this at any length, but wish to say that for a given group, the possible dimensions are known. Knowing these dimensions gives us precisely the admissible dimensions of the centre eigenspace. We can then reduce the bifurcation problem to studying an ODE with this dimension.

Theorem 2.5 *Generically (for one parameter) the center eigenspace V_0 at a* Hopf *bifurcation point is such that*

$$\gamma L v = L \gamma v$$

for all $v \in N_0$ and $\gamma \in \Gamma$ (with L an $n \times n$ complex *matrix) implies that $L = \lambda Id$.*

The only difference here is that we consider complex matrices; this property corresponds to there being a *complex irreducible representation* of Γ on V_0 (the complex structure has its root in the phase-shift symmetry of periodic solutions).

One of the most important principles that comes from the study of symmetric bifurcations is that of *spontaneous symmetry breaking.* This is the observation that many bifurcations of symmetric solutions are forced to create branches of solutions with less symmetry than that of the original solution. We make the concept 'many' a bit more precise:

Suppose that we have an (irreducible) action of the symmetry group Γ on $\mathbf{R}^n$. It is possible to find the set of polynomial functions I which are invariant and those E which are equivariant under the group action, such that a general Taylor expansion of the singularity can be written

$$f(x,\lambda) = \sum_{v\in E} v f_v(I)$$

where the f_v are smooth functions of the invariants. As an example, consider steady state bifurcation with circular symmetry ($\Gamma = \mathbf{O}(2)$ symmetry in group theory terminology). Because absolutely irreducible representations of $\mathbf{O}(2)$ are at most two dimensional, we can write down a general $\mathbf{O}(2)$-equivariant germ of a vector field (with $\mathbf{O}(2)$ acting irreducibly) in the form:

$$\dot{z} = z f(z\bar{z})$$

for $z \in \mathbf{C}$ and some arbitrary function f. Expanding this as a Taylor series

$$\dot{z} = a_0 z + a_1 z^2 \bar{z} + a_2 z^3 \bar{z}^2 + \cdots$$

where the coefficients are real valued and can be assumed not to take 'exceptional' values, i.e we are allowed to exclude specific values that might cause degeneracies on the grounds that small perturbations will destroy them anyway. (N.B. This example is particularly simple, because z is the only equivariant and $z\bar{z}$ is the only invariant of the action of $\mathbf{O}(2)$ on $\mathbf{C}$. Usually we wish to do this for *bifurcation equations*, where we need to perform two or more variable Taylor expansions, and the exercise becomes non-trivial.)

A similar process is possible for Hopf bifurcation, except that we need to do a change of coordinates to put the bifurcation into so-called *normal form.*

2.2. Symmetry and attractors

For a large class of dynamical systems (in particular, those that dissipate energy in some sense) there are subsets in the phase space that are "attractors" for the dynamics. By this, we mean that for some initial conditions, the trajectory gets arbitrarily close to some points in phase space infinitely often. The set of points is called an *attractor* if there are many initial conditions that approach it in this way.

In a limited sense, it is possible to carry across some concepts from local bifurcations with symmetry to examine the symmetry and changes in the symmetry of chaotic attractors; not just steady states or periodic solutions. In particular, Chossat and Golubitsky[14] suggested that there may be *symmetry breaking or increasing* bifurcations for arbitrary (possibly chaotic) attractors, just as there are for the simpler solutions.

Suppose we have a map $f : \mathbf{R}^n \to \mathbf{R}^n$ equivariant under a symmetry group Γ which we shall assume is finite. We say that a set A is an *attractor* if there is a 'large' set of initial conditions that evolve towards A, i.e. whose ω-limit sets are contained within A. [1] By large, we mean that that there is a basin of attraction,

$$\mathcal{B}(A) = \{x \in \mathbf{R}^n \ : \ A \text{ contains the } \omega\text{-limit set of } x\},$$

and this basin of attraction has positive measure. There are many other possible definitions for the concept of attractor[28] but the definition we give here is natural in that we assume that we pick the initial conditions for the system with respect to a natural 'background' Lebesgue measure.

We shall assume that our attractors satisfy the following technical *isolation* condition; given $x \in A$, we assume that given any $0 < \epsilon < 1$ we can find a $\delta > 0$ such that $\epsilon\ell(B_\delta(x)) < \ell(B_\delta(x) \cap \mathcal{B}(A))$ where $\ell(.)$ is Lebesgue measure on $\mathbf{R}^n$. In other words, we assume that the basin of attraction of A takes up an asymptotically large volume of small balls around points in the attractor. This will be automatically satisfied if the basin of attraction includes an open neighbourhood of the attractor; however due to the phenomenon of riddled basins we cannot expect this to be generally the case, especially for symmetric systems (see below).

For a given attractor A there are two important subgroups determined by A; they are the *symmetry on average*,

$$\Sigma_A = \{\sigma \in \Gamma \mid \sigma A = A\}$$

and the *instantaneous symmetry*,

$$T_A = \{\sigma \in \Gamma \mid \sigma x = x \ \forall x \in A\}.$$

Note that T_A is a subgroup of Σ_A (in fact it is a normal subgroup); on average, an attractor must have at least as much symmetry as it does at any point in time. It is relatively simple to see that any point on a trajectory must have at least as much symmetry as its initial point due to the fact that there is a unique trajectory passing through any initial point.

For attractors as described above, we can characterise their symmetry on average via the following dichotomy[14] (which is a consequence of the isolation condition):

$$\text{either } \sigma A = A$$
$$\text{or } \sigma A \cap A \text{ is empty.}$$

This enables us to classify the subgroups of Γ that can be realised as symmetries of attractors, and provides an analogue to the isotropy subgroups used for steady-state solutions. For the case of maps with finite symmetry groups, this has been done by Ashwin and Melbourne[8] and extended to flows by Field, Melbourne and Nicol.[17]

[1] The ω-limit set of an initial condition x is the set of all limits of subsequences of $\{f^n(x)\}$; that is, all points that the trajectory gets arbitrarily close to infinitely often.

Once we have information about which groups can be realised by attractors, we are in a position to start discussing the following questions: (a) How does one detect the symmetry on average of a particular attractor? (b) What transitions in the symmetry group do we expect to see.

To be able to correctly or faithfully reconstruct the dynamics and/or the symmetries of an attractor it is necessary to take measurements from the system in a way that respects the symmetries. King and Stewart[25] show that there are certain necessary conditions on the symmetric observation that must be fulfilled to enable to dynamics to be reconstructed; a prerequisite to enable one to measure statistical properties of the attractor. This implicitly requires that we already know what the symmetry group of the problem is. For example, if we have n oscillators with full permutation (or $\mathbf{S}_n$) symmetry, we need to observe all oscillators simultaneously, which requires taking a time series of n dimensional vectors. Once provided with such a time series it is possible to construct so-called *detectives*[12,27] which are averages that indicate the presence or absence of specific symmetries.

With regard to changes in the symmetry of families of attractors, this is considerably complicated by the fact that attractors do not appear as continuous families on varying a parameter. Typically, they are destroyed by taking arbitrarily small perturbations; however, they do seem often to exist for *positive measure* subsets of the parameter space.[29] There has been some work[15] on general properties of transitions in Σ_A, and also some work on transitions in T_A;[9,10] where attractors in fixed-point subspaces lose transverse stability. This is closely related to *on-off intermittency*,[35] recently observed in an electronic oscillator system by Hammer *et al.*.[22] We shall discuss an example of this behaviour in more detail in Section 4.

3. Identical oscillators; examples

Suppose that we have, for example, three oscillators coupled such that the oscillators are indistinguishable (note that this does not imply that for a given initial condition they must behave in dynamically the same way). We write an ODE on $x_1 \in \mathbf{R}^2$ governing the behaviour of the first oscillator as

$$\dot{x}_1 = f(x_1) + g(x_1; x_2, x_3)$$

and obtain the equations for the other two oscillators by permuting the indices $(1, 2, 3)$. If we have $g(a; b, c) = g(a; c, b)$ this implies that the oscillators are symmetric under the transformations

$$\begin{array}{c}(x_1, x_2, x_3) \mapsto (x_2, x_3, x_1)\\(x_1, x_2, x_3) \mapsto (x_1, x_3, x_2)\end{array} \tag{1}$$

which in all possible combinations generate the group of permutations $\mathbf{S}_3$ (this is the same as the group of symmetries of reflections of the equilateral triangle, $\mathbf{D}_3$).

3.1. Example (a): Hopf bifurcation for three identical oscillators

Assume that the ODE described by (1) has a trivial steady solution, $(x_1, x_2, x_3) = (0,0,0)$ and that on varying a parameter λ (that preserves the symmetry) this trivial steady state goes from being linearly stable to being linearly unstable. Moreover, assume that the loss of linear stability is via a complex pair of eigenvalues of the linearisation passing through the imaginary axis. For simplicity, assume that at we have a pair of eigenvalues equal to $\lambda \pm i\omega$ and that all others have negative real part.

It is simple to see that as λ passes through 0 there is a loss of stability of the steady state. What is not so obvious is what (if anything) is an attractor for $\lambda > 0$. It is at this point that $\mathbf{D}_3$-symmetric Hopf bifurcation can be applied to tell us what is possible. By looking at generic Hopf bifurcations, Golubitsky and Stewart[18] predict the symmetry types of all branches of solutions coming from the trivial solution. If the Hopf bifurcation corresponds to a complex pair of eigenvalues with eigenvectors in the space of in-phase solutions, then there is only one branch of periodic solutions, and these are in-phase, $(x_1, x_2, x_3) = (X(t), X(t), X(t))$ with $X(t) = X(t+T)$. If not, then there will be three types of solution created at the bifurcation:

> **Rotating waves**; periodic solutions with a phase shift of one third of a period between the oscillators; i.e. solutions of the form $(x_1, x_2, x_3) = (X(t), X(t+T/3), X(t+2T/3))$ where $X(t) = X(T+t)$.
>
> **Standing waves (1)**: these are periodic solutions of the form solutions of the form $(x_1, x_2, x_3) = (X(t), X(t), Y(t))$ where $X(t+T) = X(t)$ and $Y(t+T) = Y(t)$.
>
> **Standing waves (2)**: these are periodic solutions of the form solutions of the form $(x_1, x_2, x_3) = (X(t), X(t+T/2), Y(t))$ where $X(t+T) = X(t)$ and $Y(t+T/2) = Y(t)$. Note that the frequency of the third oscillator is twice the frequency of the other two.

All this can be obtained using symmetry. However, it is necessary to perform a calculation (centre manifold and normal form reduction) in order to find out: (a) whether these periodic solutions exist while the trivial steady state is stable or only after it has lost stability and (b) which periodic solutions (if any) are stable near the bifurcation point. We refer the interested reader to the books by Golubitsky *et al.*[19,20] for a detailed discussion of theory, application to three oscillators and generalisations.

3.2. *Example (b): Weak coupling for three identical oscillators*

Away from the small amplitude limit considered for Hopf bifurcation, there is another asymptotic limit where we can say a lot about the behaviour of the system: the weak coupling limit. An appropriate setting for this is if the dynamics is governed by an ODE of the form

$$\dot{x}_i = f(x_i) + \epsilon g_i(x), \tag{2}$$

(where $\dot{x}_i = f(x_i)$ has a (linearly) attracting limit cycle and $i = 1, \cdots, n$ indexes the oscillators). In the limit of $\epsilon \to 0$ (the *weak coupling limit*) this can be reduced to a set of *phase equations* of the form

$$\dot{\theta}_i = \omega_i + \epsilon G_i(\theta), \quad (i = 1, \ldots, 3) \tag{3}$$

where θ_i are defined mod 2π. This is because the asymptotic dynamics for the uncoupled case $\epsilon = 0$ are parameterised by three angles (a three-torus, $\mathbf{T}^3$). The theory of normal hyperbolicity can be applied to show that this description in terms of the phase angles persists for small enough ϵ.[11,7]

Furthermore, these phase equations can be averaged to get an approximation to the dynamics. We define

$$\Theta = \frac{1}{3}(\theta_1 + \theta_2 + \theta_3)$$

and

$$\phi = \omega\theta_1 + \omega^2\theta_2 + \theta_3$$

where $\omega = e^{2i\pi/3}$ (so that Θ is invariant and ϕ is equivariant under permutations of the oscillators). The ODE on $\mathbf{T}^3$ can be written in these coordinates as:

$$\dot{\Theta} = 1 + \epsilon g(\Theta, \phi)\dot{\phi} = \epsilon h(\Theta, \phi)$$

where g (resp. h) is invariant (resp. equivariant) under the action of permutations on (Θ, ϕ). Averaging allows us to rewrite the equation for ϕ as

$$\dot{\phi} = \epsilon\overline{h}(\phi) + O(\epsilon^2)$$

where $\overline{h}(\phi) = \frac{1}{2\pi}\int_{\Theta=0}^{2\pi} h(\Theta, \phi)$ and ignoring the second order terms allows us to model the dynamics by an ODE on $\mathbf{T}^2$ (identified with $\mathbf{C}$ modulo translations by 2π in the directions 1, ω and ω^2) equivariant under $\mathbf{S}_3$ symmetry. Rather than detail the necessary calculations, we mention that there can be bifurcation directly from the attractor being the in-phase solution to the attractor being large amplitude two frequency quasiperiodic oscillations. This route to loss of stability of the in-phase solution has been called the $\mathbf{S}_3$ transcritical homoclinic bifurcation ($\mathbf{S}_3$THB).[6]

It is possible to observe such a bifurcation scenario in a system of three coupled van der Pol oscillators.[6] Each oscillator consists of a parallel inductor-capacitor-resistor (LCR) network. The nonlinear negative resistor is synthesised using a 741 operational amplifier with diodes providing the nonlinearity. The coupling is achieved by linking the oscillators through a low-pass network. Each oscillator is connected via a resistor R to an earthed variable capacitor C.

The output from the three oscillators is visualised on an oscilloscope by plotting the voltages $V_x = \frac{1}{2}(-V_1 - V_2 + 2V_3)$ against $V_y = \frac{\sqrt{3}}{2}(V_1 - V_2)$ (this symmetric observation enables one to observe the symmetry of the trajectory as a symmetry of the patterns directly). Note that $V_x + iV_y = \omega V_1 + \omega^2 V_2 + V_3$ and so a cyclic permutation of the three oscillators gives a rotation by $2\pi/3$ in the (V_x, V_y) plane whereas a permutation of two oscillators gives a reflection.

Observation of a bifurcation sequence for this system is shown in Fig. 1. For large C, the two rotating waves are the only stable solutions. As C is decreased each rotating

(a) (b) (c)

Figure 1: Plots of (V_x, V_y) showing the transition from a stable rotating wave to stable in-phase oscillation. (a) The rotating wave, a periodic solution with $\mathbf{Z}_3$ cyclic symmetry. (b) A transition state with two frequency quasiperiodic behaviour. (c) The in-phase oscillation.

wave undergoes a supercritical Hopf bifurcation to a stable torus. (In the ϕ plane, this is a normal Hopf bifurcation from a fixed point to a limit cycle, while in original system this is a Hopf bifurcation from a periodic orbit to a torus.) This torus grows until it comes very close to the in-phase solution. The in-phase solution is uniquely stable at low C.

The equation of motion for the system of oscillators is

$$\begin{aligned} C_0\frac{d^2V_i}{dt^2} + \tilde{\Gamma}(V_i)\frac{dV_i}{dt} + \frac{V_i}{L} &= \frac{1}{R}\left[\frac{dV_c}{dt} - \frac{dV_i}{dt}\right] \\ C\frac{dV_c}{dt} &= \frac{1}{R}\left(\sum_{i=1}^{3} V_i - 3V_c\right). \end{aligned} \tag{4}$$

where C_0 and L are the capacitances and inductances in each oscillator, and approximately $\tilde{\Gamma}(V) = a + bV + cV^2$, with $b > 0$.

3.3. *Example (c): Symmetric chaos of four forced oscillators*

By forcing four oscillators in a circuit similar to those in Example (b) with a sinusoidal signal at the coupling point of the oscillators, it is possible to find chaotic attractors with a variety of symmetries on average. Sampling the equivariant projection of the oscillator voltages (V_x, V_y) in synchronisation with the forcing signal, it is possible to get a Poincaré return map; from this map it is then possible to find the symmetries of attractors.

The ODE governing the system is (4) except that there is an extra oscillator and an additional forcing term $A\cos\omega t$ in the equation for dV_c/dt; two examples from an electronic circuit are shown in Fig. 2.

The mechanism by which chaos is reached in this system is a combination of many different types of bifurcation (some involving changes in symmetries) to give chaotic attractors of low dimension. In particular, on increasing the amplitude of forcing one sees

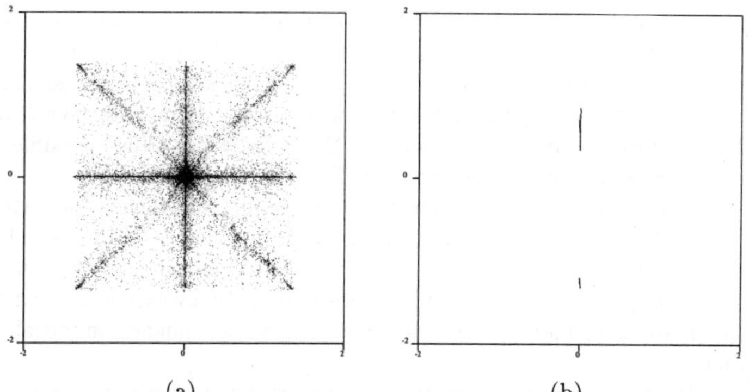

Figure 2: Two symmetric chaotic attractors for a system of three forced coupled oscillators. Projections of Poincaré sections at the forcing frequency are shown plotted in the $(V_1 - V_2, V_3 - V_4)$ plane. The coupling capacitance is $C = 30\text{nF}$ and the forcing frequency is 6.158kHz. The symmetries of attractors are as follows: (a) $A = 6.20\text{V}$ and $\Sigma_A = \mathbf{S}_4$, $T_A = \text{Id}$. (b) $A = 5.77\text{V}$ and $\Sigma_A = \mathbf{D}_3$, $T_A = \mathbf{D}_3$.

a breakdown of a two-torus via saddle-node bifurcations of limit cycles, Hopf and 2-torus doubling bifurcations.[26] It is possible to identify tori with up to three incommensurate frequencies, circle map behaviour, and chaotic attractors with dimensions between 2 and 3, but there are transitions which indicate that not all of the dynamics can be embedded on a 3-torus.[5] This necessarily implies that using the averaged equations will not predict all the transitions; in particular, we are out of the weak coupling regime of Example (b).

4. Synchronisation of Chaotic Circuits

A question of great importance for electronics is the synchronisation or otherwise of sets of electronic circuits. We wish to show how in many cases this can be interpreted as a question about the symmetry of attractors.[2]

For example, consider two identical chaotic circuits coupled by two-way coupling. Specifically, consider a circuit whose dynamics are governed by an ODE

$$\begin{aligned} \dot{x}_1 &= f(x_1) + D(x_2 - x_1) \\ \dot{x}_2 &= f(x_2) + D(x_1 - x_2). \end{aligned} \qquad (5)$$

[2] This is not always the case. For example, synchronisation of one system forced by another does not have to have any symmetry.

where D is a scalar constant and $x_i \in \mathbf{R}^n$ is the state variable for one unit. This system has a symmetry $\mathbf{Z}_2$ acting by interchange of x_1 and x_2. The *synchronised manifold* of this system consists of all points such that $x_1 = x_2$. This is an n dimensional subspace of the $2n$ dimensional phase space. Thus we can (trivially) characterise synchronised but possibly chaotic states as attractors that have $\Sigma_A = T_A = \mathbf{Z}_2$. Other possible attractors can be classified as *synchronised on average* if $\Sigma_A = \mathbf{Z}_2$, $T_A =$Id.

There has been a lot of recent interest in the stability of synchronised attractors in equations of the form (5). Instrumental in many studies has been the use of *Liapunov exponents* to characterise stability. These are quantities that indicate the average rate of exponential growth of nearby trajectories. A positive Liapunov exponent for a trajectory approaching an attractor indicates that there is sensitive dependence on initial conditions in the attractor.

For example, if the equation $\dot{x} = f(x)$ has an attractor whose largest Liapunov exponent is λ then it can be shown that (5) has a synchronised attractor if $D > \lambda/2$.[38] More generally, the stability of a synchronised attractor is determined by the Liapunov exponents of the attractor for directions transverse to the synchronised manifold. These transverse Liapunov exponents correspond to the subsystem Liapunov exponents investigated by Pecora and Carroll.[33]

However, something that is only now being fully appreciated is that typically for chaotic synchronised attractors there is not just one set of Liapunov exponents but many sets corresponding to all dynamics (possibly unstable) contained within the attractor. The stability of each invariant set (or more precisely, each ergodic invariant measure supported on the attractor) is determined by the largest transverse Liapunov exponent. In particular, it is possible to have two dense sets contained within an attractor, one of which is transversely stable while the other is transversely unstable!

This complicated intertwining of transversely stable and unstable sets gives rise to some interesting effects noticed by several authors.[1,32,9,34] Namely, the basin of attraction of synchronised attractors (or more generally, just attractors in invariant subspaces) can have great complexity with 'fat fractal' structure (fat fractals have integer Hausdorff dimension but have e.g. self-similar structure and may contain a dense set of holes). In contrast, many examples where a basin boundary is fractal, the basin itself is still an open set. In the case of riddled basins, it is possible to get basins whose complements are open dense sets in phase space, but whose measure is still positive.

Putting this mathematically, suppose we have an attractor A with basin of attraction $\mathcal{B}(A)$. The basin of A is *riddled*[1] with that of another attractor C if for all points $x \in \mathcal{B}(A)$ and all $\delta > 0$ then

$$\ell(B_\delta(x) \cap \mathcal{B}(A)).\ell(B_\delta(x) \cap \mathcal{B}(C)) > 0$$

where $\ell(.)$ denote Lebesgue measure. This means that arbitrarily small open sets around points in the basin of attraction also intersect the basin of another attractor in a set of positive measure.

A weaker form of basin riddling is possible even if $\mathcal{B}(A)$ is not fractal. We say the basin of A is *locally riddled*[10] if there exists an open set V containing A such that for all

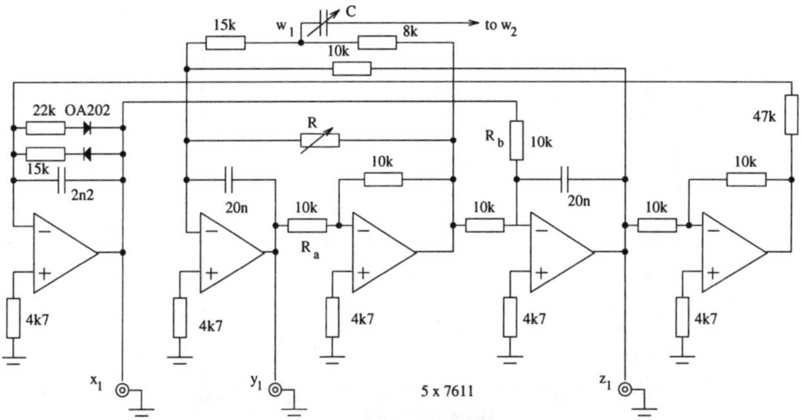

Figure 3: Circuit diagram of one chaotic oscillator, consisting of three integrators and two inverters with associated feedback components. By varying R it is possible to observe a breakdown to chaos via period doubling of a limit cycle, giving a Rössler-type attractor. Coupling is via the variable capacitor C.

$x \in A$ and $\delta > 0$ we have

$$\ell(B_\delta(x) \cap \bigcap_{k=0}^{\infty} f^{-k}(V)) > 0.$$

This means that, near any point on the attractor, there is a positive measure set of points that leave the neighbourhood V at some time in the future (even though almost all points may eventually come back to the attractor at some future iterate).

The property of basin riddling is a global property, whereas local riddling can be predicted from the Liapunov exponents; it is a local property of the attractor. The presence of riddling in the basin can be thought of as differences in the transverse Liapunov exponents of unstable dynamics contained in the chaotic attractor. For example, if there are unstable periodic orbits that are contained in the attractor, these go transversely unstable at different parameter values, and these parameter values are different from where the *natural measure* of the attractor goes unstable.

This effect can be illustrated in a system of two coupled electronic chaotic systems.[9,10] The circuit of one chaotic oscillator is shown in Fig. 3. This three degree of freedom system has nonlinearity provided by two diodes and is powered by $\pm 5V$. The circuit has an attracting limit cycle for small values of R which undergoes a period-doubling cascade to chaos, producing a Rössler type attractor for larger values of R. A typical attractor for $R = 39.5k\Omega$ projected in the (x, y) plane looks like a Roessler attractor. The underlying frequency of the periodic orbit before period doubling is of the order of $700Hz$.

Coupling is via a capacitance decade box and the coupling provides an extra degree of freedom, so the coupled system has seven degrees of freedom altogether and a typical

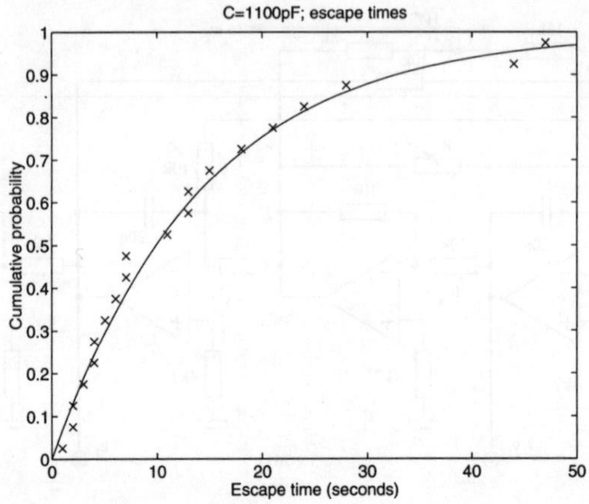

Figure 4: Distribution of times of escape from synchrony at $C = 1100pF$. The solid line shows a fit to an exponential distribution with mean escape time 14 seconds.

oscillation frequency of 700Hz. Observations in the region of interest show that the asymptotic flow is effectively on a four dimensional branched manifold, the product of two Roessler attractors. The synchronised manifold is of three dimensions in the seven dimensional phase space.

Both circuits are set to the same chaotic state with $R = 39.5k\Omega$ and the coupling capacitance was varied between $C = 0$ and $10nF$. For low values of the coupling, the lack of synchronisation of the oscillators shows in the (x_1, x_2) plane as seemingly random motion of the trace over a square region. Above $C = 30pF$ there is evidence of some degree of synchronisation. This gives way via a reverse cascade of period doublings to a stable periodic orbit at about $C = 1100pF$ that is not synchronous. This persists until C is $1139pF$, where this behaviour loses stability to a synchronous chaotic attractor. Decreasing the coupling capacitance by a small amount shows that this in-phase attractor coexists hysteretically with a non-in-phase periodic orbit. For C lower than about $1400pF$, small deviations away from synchronicity are noticeable at irregular intervals. For C at about $1100pF$, the in-phase chaotic attractor is observable only as a transient, although it can persist for over a minute; Fig. 4 shows a distribution of escape times for $C = 1100pF$; the mean escape time is 14 seconds and there is a good fit to an exponential distribution. Note that this means that on average there are many tens of thousands of oscillations before a typical escape.

For the case of an in-phase attractor exhibiting occasional excursions from the in-phase subspace, it is possible to reconstruct the dynamics on the symmetric subspace and

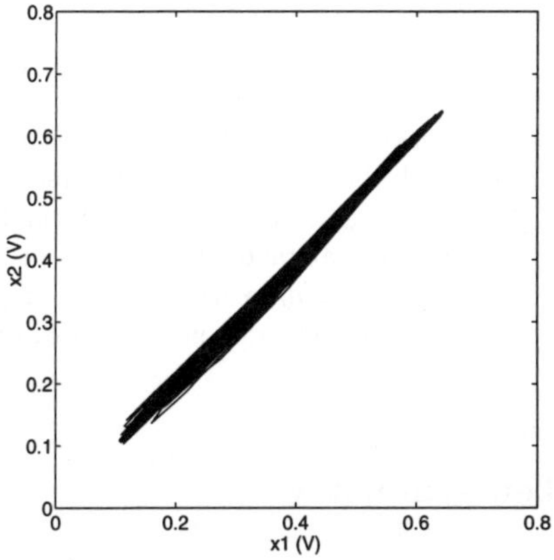

Figure 5: For $R = 39.5k$ and $C = 1110pF$, the system evolves towards a state which is close to synchronisation for most of the time, but makes sudden deviations away from it in a seemingly irregular way. Plotted in the (x_1, x_2) plane, the transients going away from the synchronised state are visible deviations of the trajectory from the line $x_1 = x_2$.

approximate the transverse dynamics on a Poincaré surface of section.[9] A return map for the synchronous state very close to a unimodal map of the unterval is obtained, as expected for a Roessler attractor. We can approximate the expansion and contraction of trajectories near the synchronous state by fitting to the data obtained.

It is then possible to approximately find the natural measure and various unstable periodic orbits contained within the attractor; we can then compute the transverse Liapunov exponents for these measures. The largest transverse Liapunov exponent $\sigma_{max} = 0.277$ for the sample in Fig. 5 is positive and corresponds to the unique (unstable) fixed point of the map f. The largest transverse Liapunov exponent for the natural measure, $\sigma_{nat} = -0.2442$, is negative as expected by the observed attractivity of the fixed point space. What is more, there are periodic points (of period less than 6) which have all transverse Liapunov exponents smaller than σ_{nat}.

Although for an idealised dynamical description we expect to see no fluctuations away from the synchronised state, the presence of low level noise unavoidably causes these to be present: a state named bubbling.[9] This phenomenon also observed by Platt *et al.*.[36] In this case, there will be differences in the scaling of the attractor with low level noise to what one might expect for stronger, asymptotically stable attractors. In a recent publication, Heagy *et al.*[23] have very neatly and *directly* shown riddled basin structure of a synchronised attractor in an electronic system, using an automated scan through the space of initial conditions.

The general message of this is that loss of stability of a chaotic synchronised state will typically happen 'gradually'; the first loss of stability will occur close to where the largest transverse Liapunov exponent for *any* dynamics contained within the attractor passes through zero. This will only be observable in the fractal structure of the basin or on addition of noise. When the largest Liapunov exponent for the natural measure on the attractor goes through zero, the attractor loses all its basin of attraction in what Ott and Sommerer have called a *blowout bifurcation.*[32] What happens here and whether there exists a nearby attractor or not, depends on the global dynamics away from the synchronised state.[31,10]

5. Effects of broken symmetry

One can perform an analysis of an electronic, or any other system, by assuming that any symmetries are perfect. However this is not the case for real systems; if we have a tolerance of less than one percent in the parameters of two supposedly identical systems then technologically we are doing well. Thus, it is important to carefully consider the robustness of any predictions to perturbations that break the symmetries.

The effect of small symmetry breaking perturbations becomes most important in the neighbourhood of bifurcations and the so-called *unfolding theory* for symmetric bifurcations[20] enables a systematic study of this. If the symmetry breaking perturbations are small compared to the distance of an attractor to bifurcation we can typically expect the predictions of symmetric bifurcation theory to be valid.

In the example discussed in Section 3, the theory predicts a direct transition from a two frequency quasiperiodic state to an in-phase oscillation via the $\mathbf{S_3}$THB. Because the $\mathbf{S_3}$ symmetry is not perfect, the $\mathbf{S_3}$THB is replaced by a sequence of bifurcations via a number of states that are forbidden if perfect symmetry were present. The closer we are to perfect symmetry, the smaller the range of parameter values over which the sequence of bifurcations happens.

For the example discussed in Section 4, the effect of arbitrarily small symmetry breaking perturbations on an attractor with riddled basins is not so clear. Typically, they seem to explode the attractor out to something much larger. This is because on perturbation, the subspace containing the dynamics is no longer invariant. However, for small amounts of symmetry breaking it seems that this effect is qualitatively similar to the effect of the addition of noise to a perfect system. There is as yet no systematic way to quantify this, to the author's knowledge.

References

[1] J. C. Alexander, I. Kan, J. A. York and Zhiping You. Riddled Basins. *Intl. Journal of Bifurcations and Chaos*, **2**:795–813 (1992).

[2] E. L. Allgower, K. Böhmer and M. Golubitsky (Eds). *Bifurcation and Symmetry*, ISNM 104, Birkhäuser, Basel (1992).

[3] E. L. Allgower, K. Georg, R. Miranda (Eds). *Exploiting Symmetry in Applied and Numerical Analysis*, Lectures in Applied Mathematics **29**, AMS, Providence (1993).

[4] V.I. Arnol'd (Ed.). Dynamical Systems V, *Encyclopaedia of Math. Sciences* **5** (1993)

[5] P. Ashwin. Symmetric chaos in systems of three and four coupled oscillators. *Nonlinearity*, **3**:604–618 (1990).

[6] P. Ashwin, G.P King, and J.W. Swift. Three identical oscillators with symmetric coupling. *Nonlinearity*, 3:585–603 (1990).

[7] P. Ashwin and J.W. Swift. The dynamics of n identical oscillators with symmetric coupling. *Journal of Nonlinear Science*, **2**:69–108 (1992).

[8] P. Ashwin and I. Melbourne. Symmetry groups of attractors. *Arch. Rat. Mech. Anal.* **126**:59–78 (1994).

[9] P. Ashwin, J. Buescu and I.N. Stewart. Bubbling of attractors and synchronisation of oscillators. *Phys. Lett. A* **193**: 126–139 (1994).

[10] P. Ashwin, J. Buescu and I. Stewart, Transverse stability of attractors in invariant subspaces Preprint, University of Warwick (1994).

[11] C. Baesens, J. Guckenheimer, S. Kim, and R.S. MacKay. Three coupled oscillators: mode-locking, global bifurcations and toroidal chaos. *Physica D*, **42**:387–475 (1991).

[12] E. Barany, M. Dellnitz and M. Golubitsky. Detecting the symmetry of attractors. *Physica D* **67**:66-87 (1993).

[13] P. Chossat (Ed.). *Dynamics, Bifurcation and Symmetry*, Kluwer Academic, Dordrecht (1994).

[14] P. Chossat and M. Golubitsky. Symmetry-increasing bifurcation of chaotic attractors, *Physica D* **32**:423–436 (1988).

[15] M. Dellnitz and C. Heinrich. *Admissible symmetry increasing bifurcations* Preprint, Univeristy of Houston (1994).

[16] M. Field and M. Golubitsky. *Symmetry and Chaos* Oxford University Press (1992).

[17] M. Field, I. Melbourne and M. Nicol. Symmetric attractors for diffeomorphosms and flows. *Proc. London. Math. Soc.* Submitted. (1994).

[18] M. Golubitsky and I.N. Stewart. Hopf bifurcation with dihedral group symmetry. In *Multiparameter bifurcation theory*, volume 56 of *Contemporary Maths*, pages 131–173. AMS, Providence, RI (1985).

[19] M. Golubitsky and D. Schaeffer. *Groups and Singularities in Bifurcation Theory. Volume 1*, Appl. Math. Sci. **51**, Springer-Verlag, New York (1986).

[20] M. Golubitsky, I.N Stewart, and D. Schaeffer. *Groups and Singularities in Bifurcation Theory. Volume 2.* Appl. Math. Sci. **69**, Springer-Verlag, New York (1988).

[21] J. Guckenheimer and P. Holmes. *Nonlinear oscillations, dynamical systems and bifurcations of vector fields.* App. Math. Sci. **42**. Springer, New York (1983).

[22] P.W. Hammer, N. Platt, S.M. Hammel, J.F. Heagy and B.D. Lee. Experimental observation of on-off intermittency. *Phys. Rev. Lett.* **73**:1095–1098 (1994).

[23] J.F. Heagy, T.L. Carroll and L.M. Pecora. Experimental and numerical evidence for riddled basins in coupled chaotic systems. *Phys. Rev. Lett.* **73**:3528–3531 (1994).

[24] G.P. King and I.N. Stewart. Symmetric chaos. In Ames W. and Rogers C., editors, *Nonlinear Equations in the Applied Sciences.* Academic press (1991).

[25] G.P. King and I. Stewart. Phase space reconstruction for symmetric dynamical systems. *Physica D* **58**: 216–228 (1992).

[26] R.S. MacKay and C. Tresser. Transition to topological chaos for circle maps. *Physica D*, 19:206–237 (1986).

[27] I. Melbourne, M. Dellnitz and M. Golubitsky. The Structure of Symmetric Attractors, *Arch. Rational. Mech. Anal.* **123**:75—98 (1993).

[28] J. Milnor. On the concept of attractor. *Commun. Math. Phys.* **99**:177–195 (1985).

[29] L. Mora and M. Viana. Abundance of strange attractors. *Acta Math.* **171** (1): 1 – 71 (1993).

[30] E. Ott, J. C. Sommerer, J. C. Alexander, I. Kan and J. A. Yorke. Scaling behaviour of chaotic systems with riddled basins. *Phys. Rev. Lett.* **71**:4134–4137 (1993).

[31] E. Ott, J.C. Sommerer, J.C. Alexander, I. Kan and J.A. Yorke. A transition to chaotic attractors with riddled basins. *Physica D* **76**: 384–410 (1994).

[32] E. Ott, J.C. Sommerer. Blowout bifurcations: the occurrence of riddled basins and on-off intermittency. *Phys. Lett. A* **188**: 39–47 (1994).

[33] L.M. Pecora and T.L. Carroll. Synchronisation in chaotic systems. *Phys. Rev. Lett.*, **64**:821–824 (1990).

[34] A.S. Pikovsky and P. Grassberger. Symmetry breaking of coupled chaotic attractors. *J. Phys. A*, **24**:4587–4597 (1991).

[35] N. Platt, E.A. Spiegel and C. Tresser. On-off intermittency; a mechanism for bursting. *Phys. Rev. Lett.* **70**:279–282 (1993).

[36] N. Platt, S.M. Hammel and J.F. Heagy. Effects of additive noise on on-off intermittency. *Phys. Rev. Lett.* **72**:3498–3501 (1994).

[37] I.N. Stewart. Bifurcations with symmetry, In: *New Directions in Dynamical Systems*, L.M.S. Lecture Note series **127**, C.U.P. (1988).

[38] T. Yamada and H.Fujisaka. Stability theory of synchronised motion in coupled-oscillator systems II. *Prog. Theor. Phys.*, **70**:1240–1248 (1983).

BIFURCATIONS AND CHAOS IN AN OSCILLATOR WITH INERTIAL NONLINEARITY

V.S.ANISHCHENKO
and
V.V.ASTAKHOV
Laboratory of Nonlinear Dynamics,
Saratov State University, Saratov, Russia

ABSTRACT

In the work an original oscillator with chaotic behavior which is the generalization of classic Van-der-Pole model on case of three- dimensional phase space is presented. A detailed multiparametrical bifurcational analysis of its dynamics is made. Auto-oscillation regimes of autonomous oscillator in wide range of variation of control parameters and the corresponding transitions to chaos are investigated. The special attention is paid to compare the results of numerical simulation with the experimental data.

1. Introduction

Autonomous systems with a three-dimensional phase space are the most simple ones among the differential systems exhibiting chaotic behavior. This simplicity is relative and is to be understood in the sense that theoretical, numerical and experimental analyzes of the dynamics of three-dimensional systems are simpler as compared to the analysis of multi-dimensional systems to find characteristic homoclinic trajectories and to study typical bifurcations in their vicinity by using theoretical and experimental methods. At the same time, the dynamics of systems, realizing one or another homoclinics type, obeys general regularities.

In this work one of the most simple autonomous system with three-dimensional phase space is presented. This is an original oscillator with chaotic behavior which is the generalization of classic Van-der-Pole model on case of three-dimensional phase space. The suggested simple one-and-a-half- freedom-degree oscillator is a convenient system to study chaos which is characteristic for dynamical systems with non-robust homoclinic trajectory of a loop of saddle-focus separatrix type.

In the work a detailed investigations of auto-oscillation regimes of autonomous oscillator in wide range of variation of control parameters are made. The results are presented in terms of detailed comparisons of numerical and experimental bifurcation diagrams of oscillation regimes in parametric space of the investigated system. Radiophysical experiment is both a verification of roughness of typical auto-oscillation regimes and their bifurcations, established by numerical simulation, and an independent technique for the investigating of complicated dynamics of chaotic systems. Traditional methods and measuring set used in radiophysics allow to provide a visual physical interpretation of many quantitative regularities underlying of studied phenomena, and to estimate also the role of fluctuations in regimes of dynamical chaos.

2. The Diagram and Equations of Modified Oscillator with Inertial Nonlinearity

The classical oscillator with inertial non-linearity was proposed and described by Theodorchik.[1] Self-oscillations are provided in the system by introducing into an oscillatory circuit the thermoresistance $R(T)$ with the properties non-linearly and inertially depending on the current through it. The diagram of Theodorchik's oscillator with inertial non-linearity is presented in Fig.1. The equations for current $i(t)$ in the circuit have the form:

$$d^2i/dt^2 + [R(T)/L - MS_0/LC]di/dt + [(LC)^{-1} + L^{-1}\partial R(T)/\partial T]i = 0, \qquad (1)$$

where S_0 is the curve slope of the amplifier which is assumed to be linear, M is the mutual induction of the feedback circuit, $R(T)$ is thermo-resistance depending on the absolute temperature T, L and C are the induction and capacity of the oscillatory circuit.

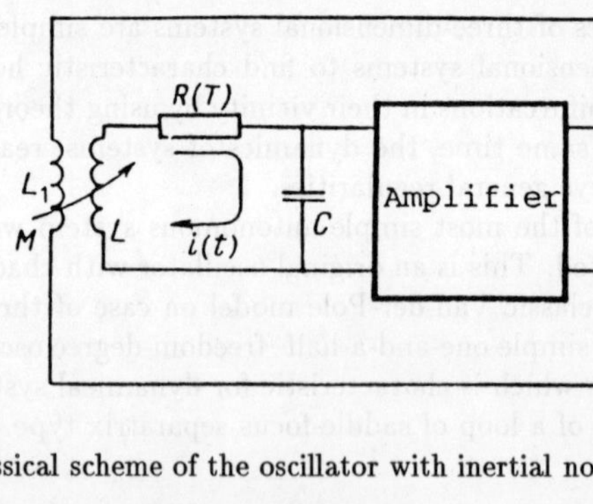

Fig.1. Classical scheme of the oscillator with inertial non-linearity.

Assuming the dependence $R(T)$ to be linear

$$R(T) = R_0 + LbT \qquad (2)$$

and setting the heat exchange process to obey the Newton's law

$$\rho q dT/dt + k_B T = R(T)i^2, \tag{3}$$

where q and ρ are the specific heat capacity of the thermistor filament and its mass, respectively, we obtain a closed system of equations in the form

$$\begin{aligned} d^2i/dt^2 + \omega_0^2 i &= (\mu - bT)di/dt - bidT/dt, \\ dT/dt + \gamma T &= \alpha(T)i^2. \end{aligned} \tag{4}$$

Here, the following notation is used:

$$\mu = \omega_0^2 S_0 M - R_0/L, \quad \omega_0^2 = 1/LC, \quad \gamma = k/\rho q, \\ \alpha(T) = \alpha_0 + bLT/\rho q, \quad \alpha_0 = R_0/\rho q. \tag{5}$$

The use of the dimensionless variables as follows:

$$x = ai, \quad \dot{y} = -x, \quad z = bT/\omega_0, \quad \tau = \omega_0 t, \quad a = (\alpha b \rho q/\omega_0 k)^{1/2} \tag{6}$$

provides Eq.4 and 5 in the form:

$$\dot{x} = mx + y - xz, \quad \dot{y} = -x, \quad \dot{z} = -gz + gx^2, \tag{7}$$

$$m = \mu/\omega_0 = \omega_0 S_0 M - R_0/\omega_0 L, \quad g = \gamma/\omega_0, \quad \dot{x} = dx/d\tau$$

In the three-dimensional two-parametric system (7), parameter m is proportional to the difference between energies introduced and scattered in the circuit, and g is a parameter defining the relative time of the thermistor relaxation. Parameters m and g will be called further the oscillator excitation and inertia parameters, respectively.

Consider the diagram shown in Fig.2. The depicted oscillatory circuit has no nonlinear in contrast to the classical case (Fig.1). Amplifier 1 is controlled by the additional feedback circuit comprising a linear amplifier and an inertial transformer. The differential equations for the oscillator may be explicitly written by particularizing the slope dependence of amplifier 1, $S(x, V)$, and specifying the equations of the inertial transformation $V(x)$.

Fit the function $S^1(x)$, i.e., the slope of amplifier 1 without reference to the effect of the additional feedback, by the following polynomial:

$$S^1(x) = S_0 - S_1 x^2, \tag{8}$$

where x is the voltage at the input of amplifier 1 and S_0 and S_1 are the fixed positive coefficients. Suppose the mechanism of the inertial feedback circuit influence to obey the regularity

$$S = S^1(x) - bV = S_0 - S_1 x^2 - bV, \tag{9}$$

where $V = V(x)$ is voltage at the output of the inertial transformer and b is a parameter. Let the inertial transformation be performed according to the equation

$$\dot{V} = -\gamma V + \Phi(x). \tag{10}$$

The following equation for current in the oscillator circuit (Fig.2)

$$Ldi/dt + Ri + C^{-1} \int (i - MSdi/dt)dt = 0, \tag{11}$$

along with Eq.9 and 10, give a closed system which is reduced, in the dimensionless variables, to the form

$$\dot{x} = mx + y - xz - dx^3, \quad \dot{y} = -x, \quad \dot{z} = -gz + g\Phi(x), \tag{12}$$

where m and g are the excitation and inertia parameters, $d = d(S_1)$ is a parameter corresponding to the degree of the influence of the curve slope non-linearity, and $\Phi(x)$ is a function describing the properties of the inertial transformer.

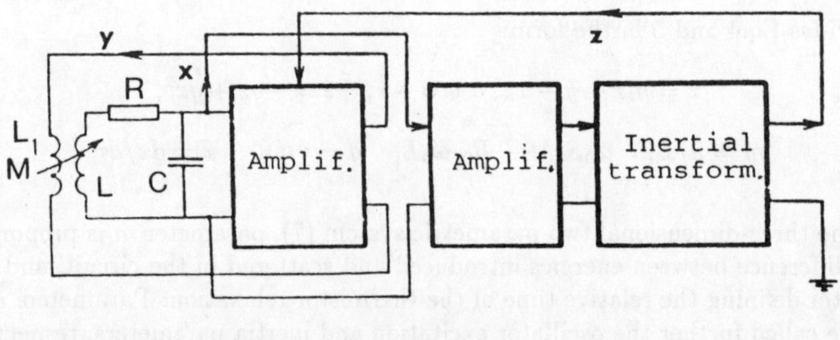

Fig.2. Modified scheme of the oscillator with inertial non-linearity

Two mechanisms of the non-linear oscillation amplitude restriction act in the oscillator (see Eq.9 and 10). The former is non-inertial and is associated with the non-linearity of the amplifier characteristics, and the latter is inertial and is due to the slope S as a function of voltage V. Let the amplifier operate on the linear slope segment ($s_1 = 0$). The inertial transformer is connected in the circuit of a two-half-wave square-law detector with RC-filter (Fig.3) and is described by the equation:

$$\dot{z} = -gz + gx^2. \tag{13}$$

The inertia parameter g is equal to the ratio of the circuit oscillation period T_0 to filter time constant $\tau_d = R_d C_d$. With the above assumptions, Eq.12 are transformed into the classical equations of the oscillator with inertial non-linearity (Eq.7). Thus, if amplifier 1 is linear and if the inertial transformer satisfies Eq.13, then the mathematical oscillator models in Fig.1 and 2 are identical. The circuit comprising an inertial detector is more convenient for leading the experiment since it allows the inertial properties of the oscillator to be changed by controlling the time constant of the filter what is impracticable when using a thermistor.[2,3]

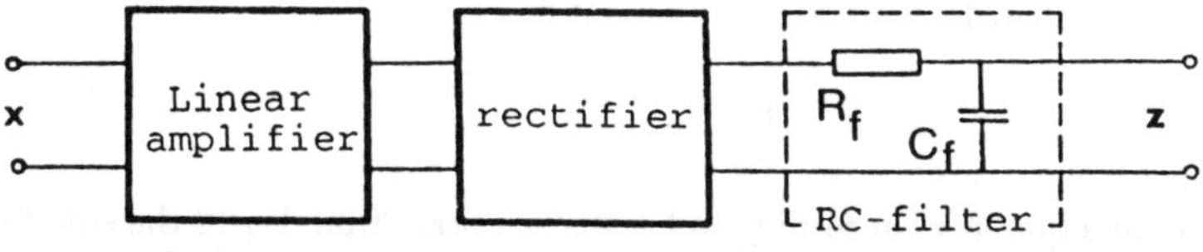

Fig.3. A scheme of a version of practical realization of inertial transformer.

The form of Eq.12 holds if an RC-chain in the form of Wien-bridge is employed as a selective element.[3] To provide for oscillation conditions in this case, two amplifier stages must be used, as it is shown in Fig.4. For a symmetrical Wien-bridge, the control parameters m and g in Eq.12 are represented simply and experimentally more convenient via the circuit parameters

$$m = K_0 - 3, \qquad g = R_0 C_0 / \tau_d, \tag{14}$$

where K_0 is the amplification factor of a two-stage amplifier and R_0C_0 and τ_d are the time constants of the Wien-bridge and the detector filter, respectively. Parameters m and g are easy to vary and to measure by varying the amplification factor and the time constant of the filter.

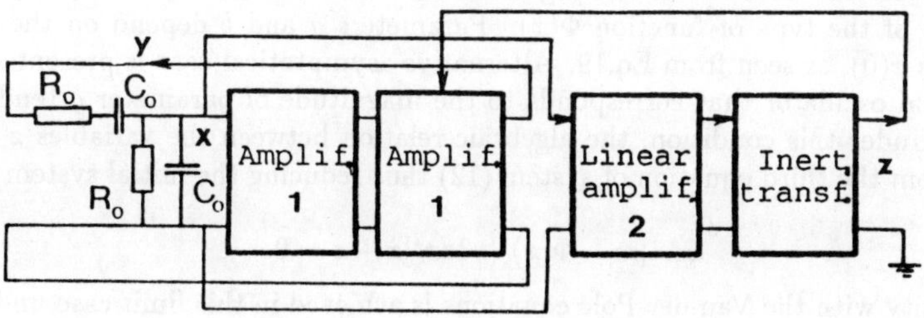

Fig.4. Scheme of RC-oscillator with inertial non-linearity.

The dynamics of system (12) depends in much on the specific form of function $\Phi(x)$ defining the inertial properties of transformer. A more abundant pattern of oscillation regimes, including chaotic ones, and of their bifurcations is observed in the case when a one-half- period detector is used.[4,5] Function $\Phi(x)$ may have the form

$$\Phi(x) = e^x - 1 \tag{15}$$

or, what is convenient for numerical investigation,

$$\Phi(x) = \frac{x + |x|}{2} x, \tag{16}$$

Defining the function $\Phi(x)$ according to Eq.16, we obtain from Eq.12 the equations of a modified oscillator with inertial non-linearity. They represent the following three-dimensional three-parametric non-linear dissipative system:

$$\dot{x} = mx + y - xz - dx^3, \quad \dot{y} = -x, \quad \dot{z} = -gz + g\frac{x + |x|}{2} x. \tag{17}$$

By eliminating the variable y, the equations for the oscillator with inertial non-linearity (Eq.17) are reduced to the form

$$\ddot{x} - (m - z - 3dx^2)\dot{x} + [1 - gz + g\Phi(x)]x = 0, \qquad \dot{z} = -gz + g\Phi(x). \tag{18}$$

An automatically controlled non-linear oscillator (Eq.18) is defined by the inertial dependence of dissipation and frequency on the variable x. In case of a strong system inertia ($\tau_d >> T_0$), as $g \to 0$, the initial system (17) degenerates into the two-dimensional one:

$$\ddot{x} - a(1 - bx^2)\dot{x} + x = 0, \qquad a = m - z_0, \quad b = 3d/(m - z_0), \quad z_0 = z(0), \tag{19}$$

and coincides in the form of writing with the Van-der-Pole oscillator's equations regardless of the type of function $\Phi(x)$. Parameters a and b depend on the initial conditions $z(0)$, as seen from Eq.19. Alternative asymptotical case is presented by a non-inertial oscillator that corresponds to the magnitude of parameter g tending to infinity. Under this condition, the algebraic relation between the variables z and x follows from the third equation of system (12) thus reducing the initial system to the form

$$\ddot{x} - [m - \Phi(x) - 3dx^2]\dot{x} + x = 0. \tag{20}$$

The identity with the Van-der-Pole equations is achieved in this limit case under the condition $\Phi(x) = x^2$. In a real oscillator with inertial non-linearity, the region of the inertial parameter g values, where the system behaves like the three-dimensional one, is bounded by a certain interval $g_1 \leq g \leq g_2$. Beyond it, the considered asymptotical equations in the phase plane can serve as an approximate description.

3. The Bifurcation of a Cycle Birth

Consider the mathematical model of oscillator only with inertial mechanism of auto-oscillation amplitude restriction ($d = 0$ in Eq.17)

$$\dot{x} = mx + y - xz, \quad \dot{y} = -x, \quad \dot{z} = -gz + g\frac{x + |x|}{2}x. \tag{21}$$

Eq.21 define the flow in R^3: $-\infty < x < \infty, \quad -\infty < y < \infty, \quad 0 \leq z < \infty$, where variable z is defined on a positive half-axis since, from physical point of view, it presents a detected voltage at the output of the filter. This system is featured by the only one singular point at the origin. The linearization of the system at this point gives the characteristic polynomial:

$$(g + s)(s^2 - ms + 1) = 0, \tag{22}$$

with the eigenvalues:

$$s_{1,2} = m/2 \pm (i/2)(4 - m^2)^{1/2}, \qquad s_3 = -g. \tag{23}$$

The real parts of all the eigenvalues are negative and the singular point is stable in the region of the parameter plane $g > 0$, $-2 < m < 0$. From the physical point of view, parameter g is always positive being the ratio of the characteristic times of the system (of the oscillation period to the relaxation time of the filter). Parameter m may be both smaller (oscillator is not excited) and greater than zero (in the generation regimes). The singular point in the interval of $0 < m < 2$ is a saddle-focus with two-dimensional unstable and one-dimensional stable manifolds. The line $m = 2$ is bifurcational and corresponds to the substitution of the saddle-focus by a saddle-node.

As evident from Eq.23, the eigenvalues $s_{1,2}$ cross the imaginary axis at the bifurcation point $m = 0$ with the non-zero velocity

$$\partial Res_{1,2}(m)/\partial m \mid_{m=0} = 1/2. \tag{24}$$

The third eigenvalue $s_3 = -g$ is separated from the imaginary axis. A classic Andronov-Hopf bifurcation is realized, the bifurcation of a cycle birth from the saddle-focus. Hence, the line $g > 0, \quad m = 0$ is a bifurcation line of the cycle birth in the region of the control parameters of the system that is physically realizable and presents a positive quadrant of the plane $m \geq 0, \quad g > 0$.

To solve the problem on the stability of a cycle being born, analyze the type of Andronov-Hopf bifurcation. Numerical calculations have shown the first Lyapunov quantity $L_1(g)$ to be negative at the singular point everywhere along the line of cycle birth. The system's limit cycle being born is stable (supercritical bifurcation). The first Lyapunov quantity may be approximately found by calculating analytically[1] with

---

[1]Problems with using the above algorithm appear due to a discontinuity of the second derivative $\Phi(x) = \frac{x+|x|}{2}x$ in Eq.21. If one approximates the $\Phi(x)$ by the exponential function $exp(x) - 1$ and restricts himself to the first three terms of its Taylor's expansion, then calculation can be completed and can be shown that $L_1(g) < 0$ for any $g > 0$.

the use of Bautin's algorithm.[6] Approximate analytical and numerical results agree qualitatively. Thus, in system (21) on the line $m = 0, g > 0$ a stable limit cycle is born with the radius growing in proportion to $\sqrt{m}$ and with the period, according to the theorem, being equal to

$$T_0 \approx 2\pi/|s_{1,2}(0)| = 2\pi. \tag{25}$$

It was found by numerically integrating the system (21) that, for values $0 < m < 2$ and $0 < g < 2$, the stable periodic oscillation with the amplitude $\simeq \sqrt{m}$ and period $T_0 \simeq 2\pi$ is the solution of the Cauchy task with the initial conditions close to the singular point at zero. For the values of $m < 0.5$, the agreement to the theorem of cycle birth, within the accuracy of calculation, is virtually adequate.

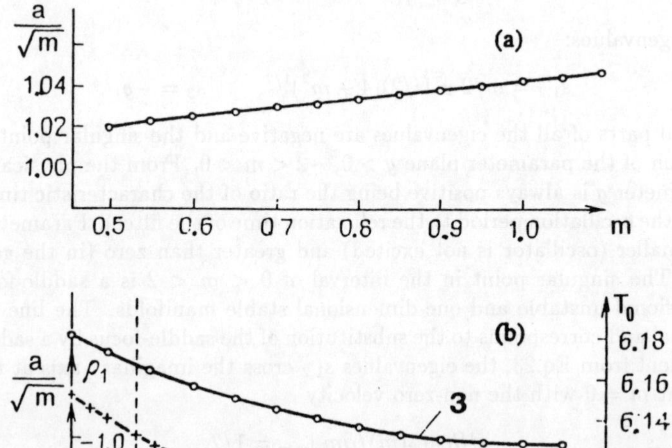

Fig.5. Normalized amplitude of stable cycle $\Gamma_0(m)$ (a), the amplitude $a/\sqrt{m}$, the largest multiplier ρ_1 and period T_0 of cycle $\Gamma_0(m)$ as a function of parameter m when passing through the bifurcation point of period doubling (b).

With the increase of $m > 0.5$, small deviations appear in the amplitude and the period of the cycle $\Gamma_0(m)$ dependencies from theoretical predictions.

In Fig.5a the data are indicated which were calculated for the size of the stable limit cycle $\Gamma_0(m)$ divided by $\sqrt{m}$ (normalized) for $g = 0.097$. The deviation of the dependence from the constant is visible in the region of $0.5 < m < 1.1$ and is not higher than 2.5-3.5%. The normalized size and period and the greatest in modulus cycle multiplier are given in Fig.5b as a function of parameter m for $g = 0.2$ when the parameter passes the point of period doubling bifurcation ($m^* = 0.966...$). As calculations demonstrate, the limit cycle evolution is described well enough by Andronov-Hopf bifurcation theorem not only in the small vicinity of the parameter value corresponding to the birth and not only of stable, but also of saddle cycle Γ_0. The deviation of the oscillation period from 2π and that of cycle size from $\simeq \sqrt{m}$ are here $\pm 3\%$, though the exceeding over the generation threshold is essential and corresponds to the regimes of non-linear oscillations.

4. The Transition to Chaos via a Cascade of Period Doubling Bifurcations

Let us investigate bifurcational transitions at the varying of system parameters which are observed on the base of cycle Γ_0 being born from the unstable saddle-focus at $m = 0$. These transitions are accompanied by change of only stable regimes. Numerical simulation and physical experiments show that in this case the sequence of being softly born period doubling cycles which is completed by the appearance of chaos is observed.

Carry out a numerical experiment on studying one-parametrically the sequence of bifurcational values of parameter μ_k corresponding to the points of period doubling bifurcations of cycles $2^k T_0$. Choose two directions of motion in parameter plane: the fixed $g = 0.3$ provides the variation of parameter m and, as $m = 1.45$, the variation of parameter g. The bifurcational values of parameters μ_k are to be calculated for which the multiplier of the cycle with period $2^k T_0$ becomes -1. The values of parameters m_k and g_k were defined[7] with the accuracy of $\epsilon = 10^{-6}$. They are indicated in Table 1 for $k \leq 4$.

Table 1. Bifurcational values of parameters m_k, g_k, Feigenbaum constant δ_k and critical points under period doubling in system (21).

	$\delta = 0.3$			$m = 1.45$		
k	m_k	δ_k	m^*	g_k	δ_k	g^*
0	0.7700	-	-	0.1200	-	-
1	1.0200	-	1.0880	0.16898	-	0.18233
2	1.0713	4.873	1.0853	0.18162	3.876	0.18506
3	1.08216	4.724	1.08511	0.18438	4.582	0.18513
4	1.08449	4.66896	1.08512	0.18497	4.66836	0.18513

The magnitude of Feigenbaum's constant δ_k and values of critical points m^*, g^* (Table 1) were calculated by using the following asymptotic relations:

$$\delta_k = \Delta_k/\Delta k+1 = (\mu_{k+1}-\mu_k)/(\mu_{k+2}-\mu_{k+1}), \qquad \mu^* = \frac{\mu_{k+1}\delta - \mu_k}{\delta - 1}, \tag{26}$$

The data obtained are the evidence that a converging sequence of period doubling bifurcations, obeying Feigenbaum's law, takes place when moving in the parameter plane, both along m and g.

The analysis of the system (21) dynamics, carried out during its evolution to stochasticity via a series of period doubling bifurcations of the main family of cycles, demonstrates a comparatively rare possibility to verify the conclusions, obtained numerically, in experiment not only qualitatively, but also quantitatively. Really, cycle Γ_0 is the only limit cycle of system (21) being softly born from a singular point at zero with small m and finite g. Every subsequent cycle with doubled period is softly born from the preceding one at the bifurcation points m_k, g_k and is stable.

If the intrinsic noise of the system is small enough, then the evolution of limit cycles can be followed in a radiophysical oscillator with the excitation parameter growth up to the critical point of the onset of chaos and of small threshold exceedings. At this point the tangency bifurcations of cycles with comparatively small periods do not influence the oscillation regimes (their multiplicity lines lie over the line of critical parameter values). Such an experiment is not related with an exact specification of initial conditions that is generally difficult to provide in a physical system. The physical experiment on quantitatively investigating the doubling bifurcations in inertial non-linearity oscillator becomes possible.

The RC-oscillator for a low radio-frequency range[3] ($f_0 = T_0^{-1} \simeq 7kHz$) was developed and fabricated for full-scale experiments according to Fig.4. The basic oscillator involved two amplifier stages having a sufficiently large linear segment of the curve and a selective element in the form of Wien-bridge. The complementary chain of negative inertial feedback consisted of the anode-like half-wave square-law detector and RC-filter. The highest possible correspondence between real and mathematical system models was experimentally achieved by thoroughly choosing the oscillator element parameters and characteristics of detector, amplifiers and filter.[3,8] A measuring set allowed one to independently vary and to quantitatively evaluate the magnitude of excitation parameter m by measuring the amplification factor of the basic amplifier K_0 and the inertia parameter g controlled by the magnitude of the filter capacity C_d. The accuracy of the measurement of control parameters m and g was better than $\pm 1\%$. Note an extreme feature of system (21) and thus of the oscillator, and namely: the independent variation of parameters m and g makes it feasible to control separately the degrees of contraction ($s_3 = s_3(g)$) and expansion ($s_{1,2} = s_{1,2}(m)$) in the vicinity of a saddle-focus in the course of experiments, i.e., to independently control the properties of stable and unstable manifolds of a singular point (see Eq.23).

The oscillation regimes and their bifurcations were studied with the help of a system of oscilloscopes, spectrum analyzer and selective micro-voltmeter involved in the measuring set.

We follow the evolution of oscillation regimes when increasing the parameter m for a number of discrete parameter g values. The experimental diagram of oscillation regimes in a portion of the parameter plane is indicated in Fig.6. As it was expected, when varying m in experiment, the accumulation of bifurcation doubling lines l_{k1} towards the line of critical values of parameters l_{cr} is observed. Doubling bifurcations are reliable recorded for $k = 0, 1, 2$, i.e., up to the transition $4T_0 \to 8T_0$. The transition $8T_0 \to 16T_0$ was observed experimentally as well, but it was unstable in some regions of the parameter plane due to natural and technical setting noise. The corresponding bifurcation line l_{31} is not shown in Fig.6.

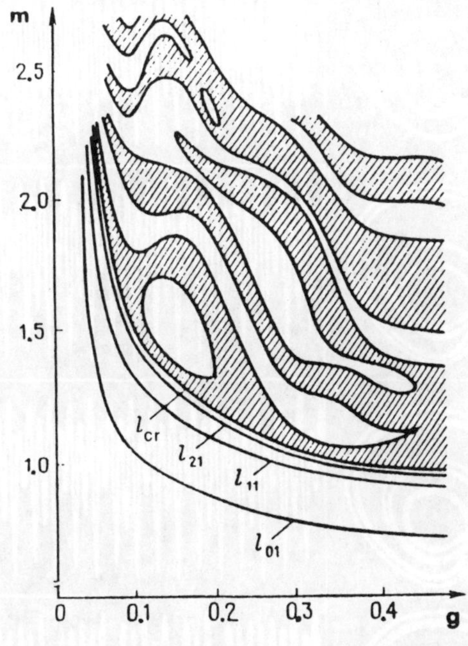

Fig.6. The bifurcation diagram of experimental oscillation regimes in oscillator. Hatched regions correspond to chaos.

The transition to chaos, including a small threshold exceeding, proceeded without hysteresis. The above bifurcation lines l_{k1} and l_{cr} coincided within the accuracy of measurement when both increasing and decreasing the parameter m. A noticeable threshold exceeding resulted in abrupt transitions to the regimes of oscillations with period $p2^kT_0$, $(p = 1, 2, 3, ..., \quad k = 1, 2, ...)$, which were alternated with the regions of strange attractors exhibiting another structure.[3] In this region the hysteresis phenomenon was typical and the data, shown in Fig.6, present the picture of regime change under increase of the parameter. The properties of the system under the considerable exceeding of the stochasticity threshold will be considered in the following sections. Here, our attention focuses on the transition to chaos via a series of doubling bifurcations.

Fig.7 shows a number of oscillation regime photographs illustrating doubling bifurcations $2T_0 \to 4T_0 \to 8T_0$ and the form of a strange attractor near the critical point. The experimental results correspond to the value $g = 0.2$ and to different m. The period-one cycle, preceding the first doubling, is not shown in the figure.

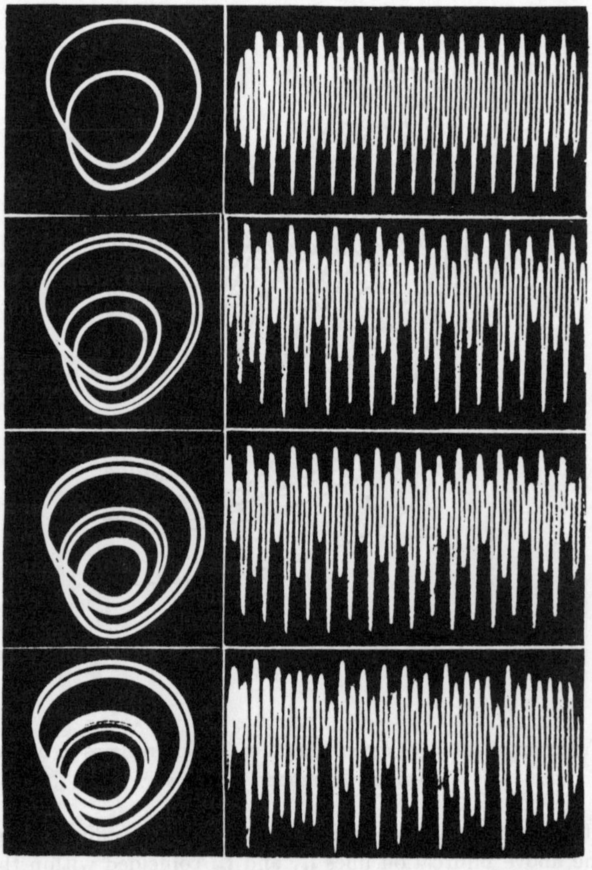

Fig.7. Phase trajectories' projections onto xy-plane and the corresponding parts of time series $x(t)$ for the transition to chaos via the sequence of period doubling bifurcations.

The photographs display the evolution of time realization $x(t)$ and the projections of phase trajectories onto the plane of variables x, y for different m. The qualitative picture of transitions is quite obvious. To ascertain quantitative regularities, we present the findings of numerical and physical experiments below.

The fix the parameter value $g = 0.2$ and construct, by using computation, a bifurcation tree of branching of the amplitudes of stable cycles when varying the parameter m. To do this, introduce in the system's phase space a secant plane $y = 0$ and record the coordinate x on the secant under variation of the parameter m. The calculation results are indicated by solid curves in Fig.8. Having set the value $g = 0.2$ in the course of physical experiment, measure the oscillation amplitudes of variable $x(t)$. The normalized results are shown for comparison in Fig.8 by dashed curves.

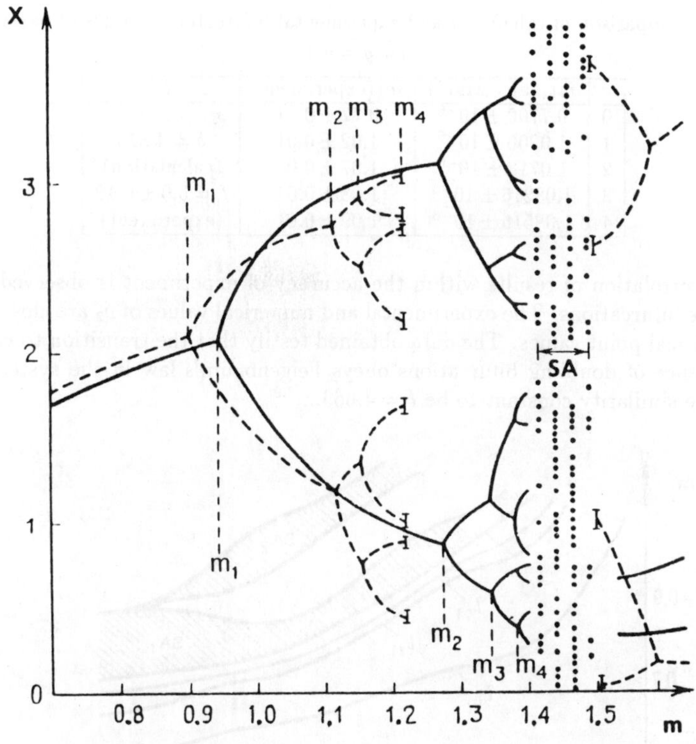

Fig.8. The bifurcation tree of amplitude branching during doubling (solid curves: calculation, dashed-and-doted curves: experiment).

The comparison of numerical and experimental data gives a good qualitative agreement. The calculated and measured parameter values m_k do not coincide, however, at doubling bifurcation points. The reason turned out to be hidden in the following. It is apparent from the model equations that the detector characteristic $\Phi(x)$ obeys the square law (Eq.16). The real characteristic $\Phi(x)$ can differ somewhat from approximation (16) in the oscillator, but it allows correction by varying the detector anode voltage of the inertial cascade of oscillator.

In Fig.9, the bifurcation diagram of oscillation regimes is presented which was obtained experimentally in the plane of parameters m and U_a, the latter being the detector anode voltage for the value $g = 0.3$. Notation in the figure is the same as in Fig.6. Special measurements have shown the best agreement between the experimental characteristics $\Phi(x)$ and theoretical approximation (16) to be achieved as $U_a^0 = 120V$. For $U_a = U_a^0$, the bifurcational values of parameter $m_k (k = 0, 1, 2, 3)$ and m_{cr} were measured. The results obtained are indicated in Table 2.

Table 2. Comparison of calculated and experimental bifurcational values of parameter m for $g = 0.3$.

k	m_k(calculation)	m_k(experiment)	δ_2
0	0.7700 ± 10^{-6}	0.77 ± 0.01	
1	1.0200 ± 10^{-6}	1.02 ± 0.01	$\delta = 4.873$
2	1.0713 ± 10^{-6}	1.07 ± 0.01	(calculation)
3	1.08216 ± 10^{-6}	1.08 ± 0.01	$\delta = 5.0 \pm 0.12$
4	1.08516 ± 10^{-6}	1.09 ± 0.01	(experiment)

A good correlation of results within the accuracy of experiment is observed for the first three bifurcations. The experimental and numerical values of δ_2 are close, as well as the critical point values. The data obtained testify that the transition to chaos via the sequence of doubling bifurcations obeys Feigenbaum's law in the system under study, the similarity constant to be $\delta = 4.669....$[7,9]

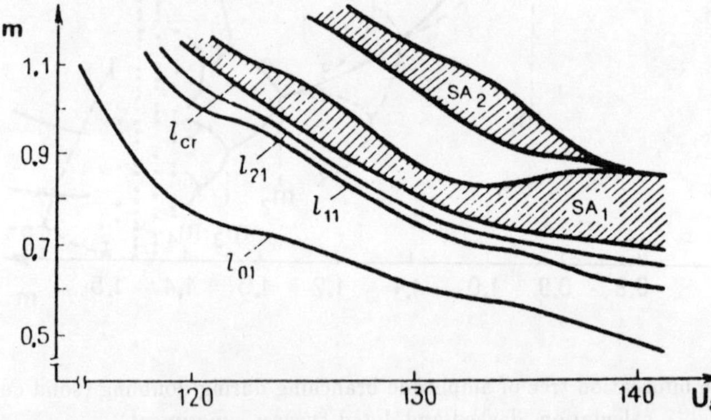

Fig.9. The experimental bifurcation diagram of regimes on the plane of parameters m, U_a for $g = 0.3$. SA_1 and SA_2 - the first and the second chaotic attractor regions.

Consider the evolution of the oscillation power spectrum on the way to chaos via a series of period doubling bifurcations. The sequence in discrete spectrum enrichment with subharmonic correspond in a one-to-one manner to a set of cycles with periods, being doubled while passing the bifurcation points. Subharmonic amplitudes satisfy the quantitative regularity proceeding from the scale-invariant properties of branching of the cycle amplitudes. In Fig.10, the experimental power spectra $S_x(f)$ of oscillations are shown. The corresponding time dependencies $x(t)$ and phase trajectory projections are given in Fig.7.

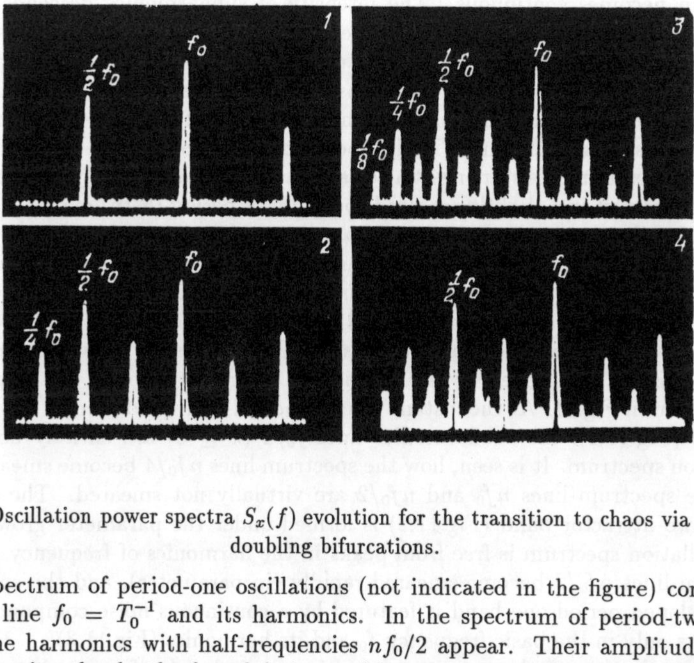

Fig.10. Oscillation power spectra $S_x(f)$ evolution for the transition to chaos via period doubling bifurcations.

The spectrum of period-one oscillations (not indicated in the figure) consists of the basic line $f_0 = T_0^{-1}$ and its harmonics. In the spectrum of period-two oscillations, the harmonics with half-frequencies $nf_0/2$ appear. Their amplitudes grow when increasing the depth of modulation and become saturated at the instant when the subsequent doubling bifurcation proceeds. The components of spectrum $nf_0/4$ appear and at the bifurcation point smoothly increase with the parameter growth corresponding to the soft birth of a cycle with period $4T_0$. The picture is the same for the transition $4T_0 \rightarrow 8T_0$ Thorough measurements with the help of the spectrum analyzer,comprising a memory, have demonstrated the ratio of the subharmonic intensity $f_0/2^k$ to $f_0/2^{k+1}$ to be 10-12 dB for $k = 0, 1$. As $k = 2$, it reaches the value of 13.0 ± 0.4 dB. The calculations of the power spectrum of realization $x(t)$, corresponding to the cycle with period $16T_0$ ($m = 1.084$, $g = 0.3$), provide a close result of 13.03 ± 0.3 dB. The calculations were performed by using the algorithm of Fast Fourier-transform for subharmonic intensity ratios $f_0/4$ to $f_0/8$. The theoretical value of $\simeq 13.5$ dB is assumed to be reached with sufficiently large k.

The process of spectrum enrichment by subharmonics on the way to chaos via doublings is defined by the scale-invariant properties. If the spectrum of period-two cycle is to be considered, then it comprises two lines with a lower intensity $f_0 \pm f_0/2$ in the vicinity of frequency f_0. $\Delta f_2 = f_0/4$ is the scale unit for a period-four cycle. In the vicinity of line $f_0/2$, the spectrum has as well two lateral components $f_0/2 \pm f_0/4$. Within the limit of accumulation of the period doubling bifurcations ($k \gg 1$) the scale-invariant structure of the spectrum becomes valid at the bifurcation points.

When achieving the critical point, the number of proceeding doublings tends to infinity, and the frequency interval between the nears subharmonics $\Delta f_k \rightarrow \infty$. The spectrum becomes continuous. The intensity of subharmonics descends, however, quickly with growing k and tends to zero as $k \rightarrow \infty$. Because of this, the real experimental spectrum will be discrete under the finite recording accuracy at the critical point. With exceeding the chaos threshold, the process of broadening the spectrum lines of the highest subharmonics sets in that leads to a higher intensity of continuous noise component in the spectrum. The continuous spectrum is really recorded in experiments starting at a certain parameter value (always over the critical level). Usually, it is a slightly smeared spectrum of period-(8-16) cycle.

Fig.10,4 illustrates the spectrum of a strange attractor under a small threshold exceeding which corresponds to the smeared spectrum of period-eight cycle. Such an attractor presents an period-eight band in the phase space (Fig.7).

With the growth of supercriticality, the successive merging of multi-periodic attractor bands to period-one band proceeds forming, as a result, a developed period-one band. Consider the process of the merging of a period-two band to the period-one band shown in Fig.11. At the instant where a period-four band merges to the period-two one (Fig.11,1), the subharmonics lines $f_0/2^k (k \geq 3)$ are entirely noise in the oscillation spectrum. It is seen, how the spectrum lines $nf_0/4$ become smeared. With this, the spectrum lines nf_0 and $nf_0/2$ are virtually not smeared. The developed period-two attractor band (Fig.11,2) is formed under the parameter growth. Here, the oscillation spectrum is free from peaks in the harmonics of frequency $f_0/4$. The spectrum lines $nf_0/2$ become smeared rapidly (in parameter), and the spectrum of the developed period-one band is featured by a continuous noise component having the peaks only in the basic frequency f_0 and its harmonics (Fig.11,3).

The mechanism of the merging of multi-periodic bands of an attractor with the growth of supercriticality, followed by successively smearing spectral lines of subharmonics $nf_0/2^k$ ($k = \infty, ..., 2\ \ 1$) is due to homoclinic effects. A dynamical system is characterized by the presence of the infinite number of saddle cycles with periods $2^k T_0$ at the critical point of transition. The parameter growth results in the intersection of two-dimensional stable and unstable cycle manifolds followed by robust homoclinic curves. In the Poincare map on the secant plane, the stable and unstable separatrixes of a saddle point, that make the prototypes of appropriate cycle manifolds, intersect robustly providing for the birth of a homoclinic point and, thus of a homoclinic structure. Clear succession in the appearance of such structures is observed being inverse to the regularity of the doubling bifurcation accumulation process. Under the supercriticality growth, the homoclinic structures of saddle cycles with periods $2^k T_0$

appear in the inverse order in k. In connection with this, the attractor bifurcations in a supercritical region are called sometimes the inverse, or connectivity, bifurcations.

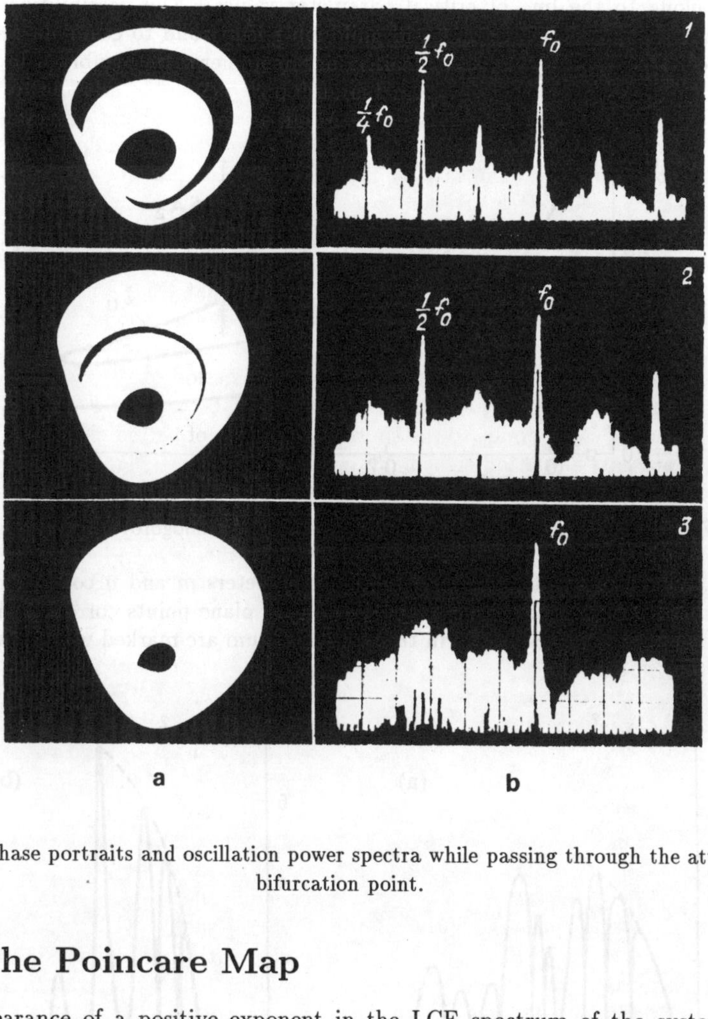

Fig.11. Phase portraits and oscillation power spectra while passing through the attractor bifurcation point.

5. The Poincare Map

The appearance of a positive exponent in the LCE spectrum of the system is a criterion of stochasticity. For systems in R^3 the signature of LCE spectrum may be, in general case, as follows: "–", "–", "–" for a stable steady-state point, "0", "–", "–" for a limit cycle, and "+", "0", "–" for a strange attractor. Calculate the whole LCE spectrum of system (21) while varying the parameters in those sections where the bifurcational values of m_k and g_k were calculated for doublings ($m = 1.45$ and

$g = 0.3$). The LCE spectrum signature is insensitive to regime rebuilding due to doublings up to the critical point where a positive exponent appears in the spectrum indicative of the birth of a strange attractor. The LCE spectrum bifurcation point must belong to the line of critical parameter values. The possibility is given to independently determine the critical parameter values and to compare these values to the data obtained by calculating the cycle multipliers with doublings.

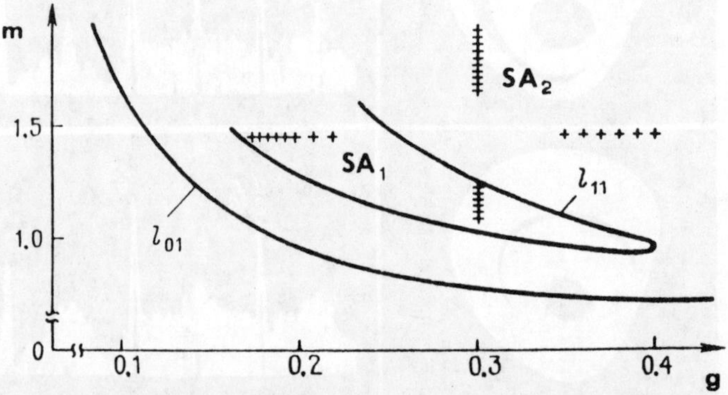

Fig.12. Bifurcation digram portion with positive Kolmogorov entropy points.

Fig.12 shows a portion of the plane of parameters m and g comprising the bifurcation lines of doublings l_{01} and l_{11}. Here, the plane points corresponding to the presence of a positive exponent in the LCE spectrum are marked with crosses.

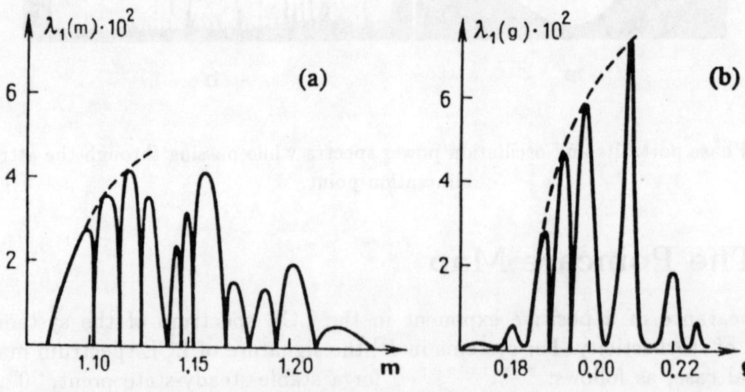

Fig.13. The results of $\lambda_1(m)$ and $\lambda_1(g)$ calculation; a: $g = 0.3$, b: $m = 1.45$.

The LCE spectrum is of the type "0", "–", "–" at other points of the straight lines indicated, testifying the presence of periodic oscillations in the system. When varying both m and g, two localized stochasticity regions are found. We call them the first and second regions, SA_1 and SA_2. Examination has shown every of the regions to be a complicated region where different types of both regular and strange attractors are realized replacing each other under variation of parameters and initial conditions. The calculation results for $\lambda_1(m)$ and $\lambda_1(g)$ are given in Fig.13.

The LCE spectrum calculations using relatively great averaging time intervals ($\tau \geq 10^4$) and the linear approximation of the positive exponent as a function of parameter in the immediate vicinity of the transition point have confirmed the results indicated in Table 1, i.e., $m^* = 1.0853 \pm 5 \cdot 10^{-4}$ for $g = 0.3$ and $g^* = 0.1854 \pm 5 \cdot 10^{-4}$ for $m = 1.45$. Thus, the critical points, within the given accuracy of calculation of the LCE spectrum, correspond to those found numerically by Feigenbaum's law. Note that one must be particularly careful when calculating in the vicinity of the critical point, where the maximum exponent of LCE spectrum undergoes the bifurcation "0" $\rightarrow$ " + ", due to a slow convergence of the exponent magnitudes in time. The averaging time decrease under determination of λ_1 leads to lowered critical point values as compared to the precise ones.

The knowledge of the whole LCE spectrum of the strange attractor of the system enables one to determine its Lyapunov dimension. In the two above stochasticity regions, $D_L = 2+d$, as $0 < d < 0.4$, which decreases with the growth of g. The dimension is 2.341 and 2.187 at the points of developed stochasticity $m = 1.45, \quad g = 0.21$ and $m = 1.16, \quad g = 0.3$, respectively. The fractional part of d near the critical line tends to zero, thus substantiating the approximate description of the transition dynamics with the help of one-dimensional map.[10,11] Numerical experiments give evidence in favor of such an approximation to be possible. A two- dimensional map on the secant plane $yz(x = 0)$ has the shape of a horseshoe being near to one-dimensional curve (Fig.14a).

The model map $y_{n+1} = \Phi(y_n)$ (Fig.14b), constructed numerically for variable y, is presented as a function belonging to the class of Feigenbaum map discontinuities are absent, and there is one smooth maximum. Cantor structure is not visible in the transverse section of the almost one-dimensional map in the scale of the figure (Fig.14a) due to comparatively strong dissipation ($g = 0.3$). However, if the value of parameter $g \leq 0.1$ is decreased, then the system map on the secant would appear as a system of nested, nearly one-dimensional curves of horseshoe-type demonstrating the Cantor structure features even without varying the pattern scale.[7]

The Poincare section can be visualized in a physical experiment.[12–14] To accomplish this, the electron beam of the oscilloscope is to be intensified at the time moments of a phase trajectory traveling through a given cross-section. For example, short intensifying pulses can be formed when realization $x(t)$ passes through zero values. In this case, the projection of the Poincare section by the plane $x = 0$ will be displayed by brighter points on oscilloscope screen simultaneously with the attractor projection onto the phase plane. The photographs of two types of chaotic attractor are presented in Fig.15 along with the corresponding Poincare sections. The oscillation regime shown in Fig.15a corresponds to the parameter values for which the section and return functions shown in Fig.14 were calculated.

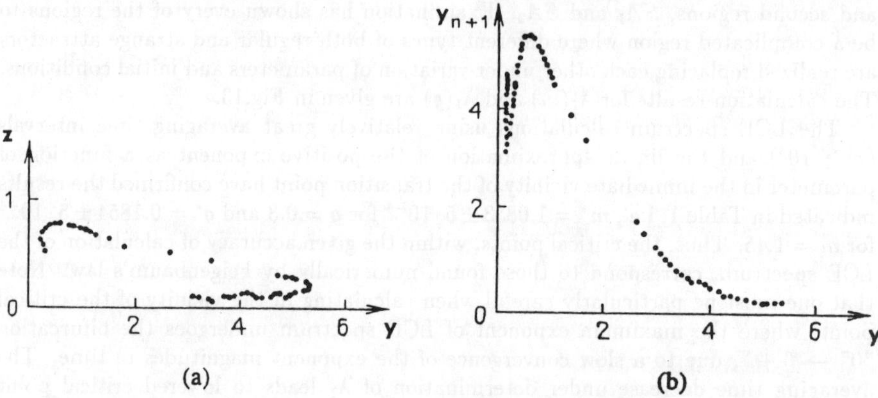

Fig.14. The Poincare section in plane $x = 0$ (a) and model return map (b), calculated for the chaotic regime with $m = 1.16$, $g = 0.3$.

The qualitative agreement of full-scale and numerical experiment results is obvious. Notice that a period-two band of an attractor with the Poincare map, comprising two non-crossing chaotic zones, is realized in the situations presented in Fig.14 and 15. In case of the developed chaos, these zones merge and the Poincare section takes the characteristic shape very similar to the one-dimensional square parabola (Fig.15b).

Fig.15. The Poincare section in physical experiments.

Let us verify quantitatively the regularities in the character of the positive LCE spectrum as a function of the supercriticality level. As seen from Fig.13, the nonlinear dependencies $\lambda_1(m)$ and $\lambda_1(g)$ are true near the threshold interrupted by sudden depressions when varying the parameter. The broken character of curves $\lambda_1(m)$ and $\lambda_1(g)$ corresponds to the appearance of periodic motion stability "windows" due to tangency bifurcations. The envelope of plots $\lambda_1(m)$ and $\lambda_1(g)$ near the threshold allows approximation by using the following relations:

$$\lambda_1(m) = c_1(m - m^*)^\gamma, \quad g = 0.3; \qquad \lambda_1(g) = c_2(g - g^*)^\gamma, \quad m = 1.45, \tag{27}$$

where $c_1 = 0.181, m^* = 1.085$, $c_2 = 0.395, g^* = 0.185$, and $\gamma = ln2/ln\delta \simeq 0.449$. The approximating curves are drawn in Fig.13 by dashed line. They corroborate quantitatively the theoretical dependence $\lambda_1(\mu)$ ascertained for Feigenbaum's transition on the base of one-dimensional maps.[9,15]

6. System Dynamics in the Supercritical Range of Parameter Values. Hysteresis and Transition to Chaos via Intermittency Induced by Fluctuations

We return to discussion of the results of calculating the maximum LCE spectrum exponent $\lambda_1(m)$ (Fig.13a). With exceeding the threshold, dependence $\lambda_1(m)$ involves discrete regions in parameter m where the positive exponent becomes zero. Initially, $\lambda_1(m) = 0$ in the narrow parameter value intervals. But the interval widths increase with the growth of supercriticality. This is a sufficient demonstration of the presence of stable periodic regimes which may be observed in the course of an experiment. The appearance of stability windows above the critical point is caused by two reasons. First, the stable and unstable cycles of different families are born from the trajectories' closeness as a result of saddle-node bifurcations. Secondly, initial cycles with periods $2^n T_0$, which are all saddle at the critical point, acquire their stabilities via inverse doubling bifurcations (the multiplier enters the unit circle through -1). The width of regions of stable periodic regime existence above the critical point can be defined by the corresponding cycle multiplier as a function of parameter and, thus, by location of multiplicity line in the diagrams.

As an example, consider the evolution of period-two cycle Γ_1 being born from cycle Γ_0 as a result of doubling. The analysis will be performed in the section $g = 0.3$ by varying the parameter m. At the bifurcation point $m = 0.77$, cycle Γ_1 is softly born and is stable further. Fig.16 presents the maximum multiplier ρ_1 of cycle Γ_1 as a function of parameter m obtained by calculation.

Curve $\rho_1(m)$ is not simple and has the shape of a loop with peculiarities at points B, D, F, C and E where the multiplier modulus is a unity. At point B ($m_B \simeq 1.267$), cycle Γ_1 undergoes the doubling bifurcation and further, up to point D ($m_D \simeq 1.257$), remains saddle. At point D, the multiplier enters the unit circle through -1, and cycle Γ_1 acquires its stability again. At point F ($m_F \simeq 1.267$) cycle Γ_1 vanishes by merging with the unstable cycle Γ_1'' as a result of bifurcation $\rho_1 = +1$. A window of cycle Γ_1 stability appears which has the width in parameter $m_F - m_D \simeq 0.01$. Then at point C ($m_C \simeq 1.101$), if one increases the parameter m, a pair of cycles is born, stable

Γ_1' and unstable Γ_1''. The second window of stability appears since cycle Γ_1' is stable within the interval in parameter $m_E - m_C \simeq 0.22$. Then, cycle Γ_1' loses its stability at point E ($m_E \simeq 1.32$) via doubling, and when increasing m, the cascade of Feigenbaum doubling bifurcations becomes possible as applied to the family of cycles Γ_1'. If we calculate the dependence $\rho_1(m)$ for cycle Γ_1', then it will qualitatively follow the loop shown in Fig.16.

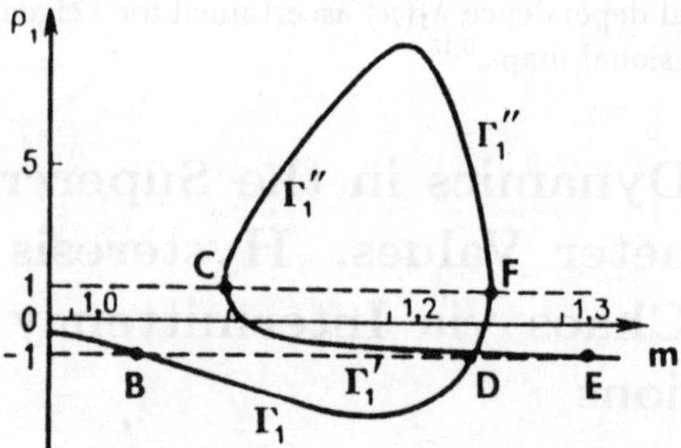

Fig.16. $2T_0$-cycle largest multiplier ρ_1 as a function of parameter m at $g = 0.3$.

Abrupt changes of oscillation regimes (catastrophes) and intricate hysteresis phenomena are observed at the boundaries of stability windows. Let, for example, the initial conditions correspond to cycle Γ_1 in the region of its stability between points D and F. When increasing m, the cycle disappears at the point F, and the jump of a representative point onto the stable cycle Γ_1' would be probable. With decreasing parameter m, the stable cycle Γ_1' exists everywhere up to the critical point C. Due to abrupt bifurcation it is difficult to predict what new regime will appear when the old one vanishes behind point C. The realization of one of the multi-periodic cycles of the system or of a strange attractor is not improbable since there are no stable cycles with periods of $\simeq 2T_0$ below the point C in the parameter m.

Investigate the effect of abrupt bifurcations near points C and F (Fig.16) in detail. At the point $m_C = 1.10112$, system (21) was integrated for times $\tau_0 \geq 10^3$, and the power spectrum, autocorrelation and distribution functions, the entire LCE spectrum and attractor dimension were calculated. Then, parameter m was slightly decreased while retaining a point on cycle Γ_1' as initial conditions, and the above calculations were performed again. The following was revealed. With a small offset from point C ($m = 1.1010 < m_C$), a long laminar oscillation phase was observed nearly repeating the motion on cycle Γ_1'. However, the system has gone over to the regime of a strange attractor localized in the system's phase space in such a manner that it comprised no vicinity of the fixed point of cycle Γ_1' in the Poincare map. The long-duration of the laminar phase was decreased when moving along the parameter away from point C, however, the phenomenon of intermittency, i.e., that of an accidentally repeating return of a trajectory to the vicinity of cycle Γ_1', was not recorded. The non-stationary character on the prolonged process of transition to chaos nearby the point C led to smearing the spectrum lines of initial cycle Γ_1', and a continuous spectrum of the Feigenbaum attractor appeared when moving away from the point of tangency bifurcation. The calculation results are illustrated in Fig.17. The spectrum lines nf_1' of stable cycle Γ_1' are gradually broadened under motion away from point C. As a consequence, new lines nf_1 appear which rigorously correspond to Γ_1, and the transition to an attractor is performed that appears as a result of doubling sequence at the critical point $m^* = 1.085...$. In the vicinity of point F, the picture is similar in the whole, but the switching is performed from the vanishing cycle Γ_1 to the stable cycle Γ_1'. Only abrupt transitions either to chaos or to the stable periodic motion proceed at the bifurcation points C and F with no intermittency. This is a radical finding which must be rigorously explained.

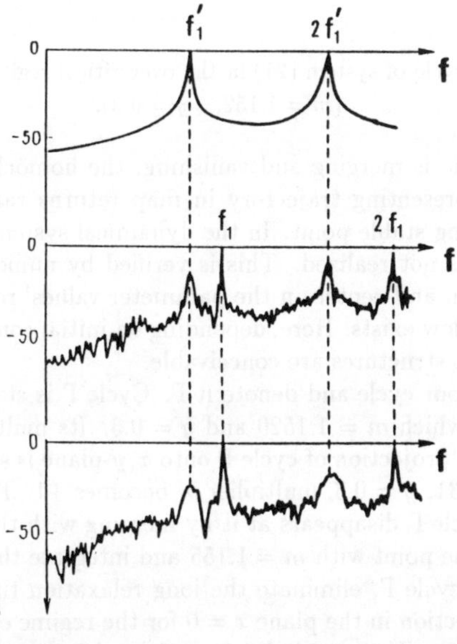

Fig.17. Oscillation power spectra with departing along the parameter from the point of tangency bifurcation.

How does a true intermittency appear? In the Poincare map, the fixed points of saddle and node (or focus) types correspond to the unstable cycle Γ_1'' and the stable cycle Γ_1, respectively. Before their merging unstable saddle separatrixes may not intersect with the stable ones. The intermittency is realized if the instant of saddle-node bifurcation is preceded by formation of a homoclinics in the vicinity of the above point.

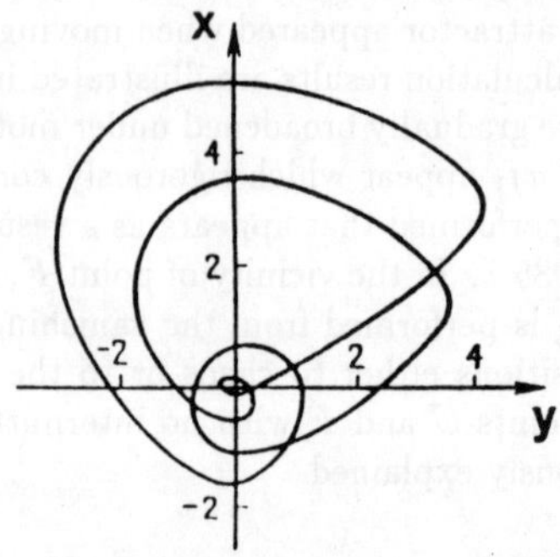

Fig.18. Stable $4T_0$-cycle of system (21) in the overcritical region of parameter values ($m = 1.152, \quad g = 0.3$).

As the saddle-node is merging and vanishing, the homoclinic structure becomes attractive, and a representing trajectory in map returns randomly in time to the vicinity of disappearing stable point. In the dynamical system (21), the above bifurcational mechanism is not realized. This is verified by numerical experiments. We apply to Fig.13a again and center on the parameter values' region close to $m = 1.15$ where a stability window exists. Here, depending on initial conditions, stable periodic regimes with different structures are conceivable.

Choose a period-four cycle and denote it Γ. Cycle Γ is stable at the point of the parameter plane for which $m = 1.1520$ and $g = 0.3$. Its multipliers are $\rho_1 = 0.3890$ and $\rho_2 = 0.0228$. The projection of cycle Γ onto x, y-plane is shown in Fig.18. At the point with $m = 1.15231$, $g = 0.3$, multiplier ρ_1 becomes $+1$. This point is identical to point F in Fig.16, cycle Γ disappears at it by merging with the unstable cycle. Vary parameter m up to the point with $m = 1.155$ and integrate the system (21) with the initial conditions on cycle Γ, eliminate the long relaxation time $\tau_y \geq 10^3$, and construct the Poincare section in the plane $x = 0$ for the regime established. The results are indicated in Fig.19a. They are indicative of an abrupt switching from cycle Γ to the strange attractor. There is no intermittency as demonstrated by the relative position of the attractor and the fixed point of cycle Γ denoted by Q. Point Q is isolated, i.e., its vicinity is not involved in the attractor. If one constructs the one-dimensional return function $y_{n+4} = \Phi(y_n)$ using the section calculation results (Fig.19a), then the plot of a model map turns out to be typical for tangency bifurcation in the local vicinity of point Q. The non-local map properties do not provide, however, the trajectory return to the point Q vicinity thereby eliminating the intermittency.

The effects of switching from one regime to another, that exhibit the character of a catastrophe, are manifested, under computation, with a high accuracy degree in the intervals of the parameter values $\Delta m \simeq 10^{-3}$.

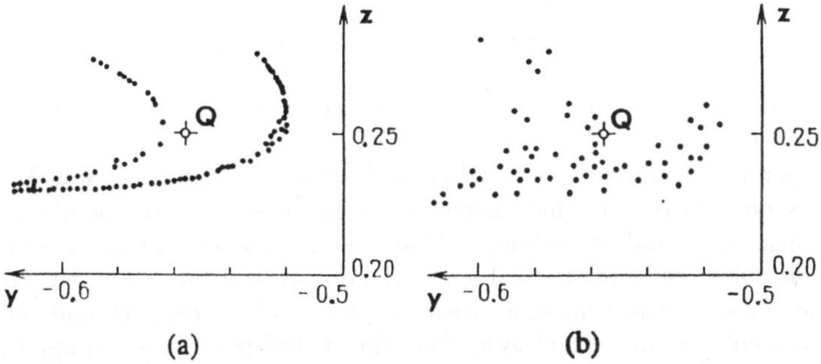

Fig.19. The Poincare sections at the abrupt switching to a chaotic attractor resulting from bifurcation of +1; a: fluctuations are absent, b: with fluctuation intensity $D = 10^{-3}$.

In a radiophysical oscillator, the controlled variation of parameter m is feasible with the accuracy of 1%, rougher by an order. Thus, the detailed investigation of switching effects is obviously excluded in the full-scale experiment. Nevertheless, the switching, hysteresis and intermittency phenomena are confidently observed in the physical experiment. How is this explain?

There are at least two reasons of possible disagreement between experimental and calculated data. The former resides in mathematical model's incompatibility with a real oscillator at the parameter values lying in the supercritical region. The numerical and full-scale experiments reject such a proposal and this is illustrated by the whole following description. The latter is the influence of natural and technical fluctuations on oscillation processes in the oscillator. It is of principle and deserves particular attention.

The effect of natural fluctuation is modeled by using the appropriate system of stochastic differential equations for the oscillator[16]:

$$\dot{x} = mx + y - xz + \xi_1(\tau), \qquad \dot{y} = -x + \xi_2(\tau), \qquad \dot{z} = -gz + g\frac{x+|x|}{2}x + \xi_3(\tau), \quad (28)$$

where $\xi_i(\tau)$ is a normally distributed noise satisfying the conditions

$$\langle \xi_i(\tau) \rangle = 0, \qquad \langle \xi_i \xi_j \rangle = \begin{cases} 0, & i \neq j, \\ D\delta(\tau - \tau'), & i = j \end{cases} \quad (29)$$

and D is the noise intensity.

We investigate oscillation processes in the system in the presence of additive noise on the assumption the external influence to be a small perturbation of initial equations (Eq.21). The dependence $\lambda_1(m)$ for $g = 0.3$ is to be calculated by varying noise intensity D. Fig.13 corresponds to the case of $D = 0$. The growth of the noise intensity leads to the successive vanishing of the smallest in parameter stability windows of the multi-periodic cycles. As $D \simeq 10^{-2}$, the system attractor becomes as if hyperbolic, i.e., it comprises no stable periodic trajectories in a certain finite region of parameter values Δm. The critical point of stochasticity birth is shifted towards lower parameter values due to the noise influence on cycles with period $2^k T_0$ starting at a certain $k > k_0$.

The system's response to the additive external normal noise was found to be principally non-linear in regimes being close in parameters to the points of tangency bifurcations. It depends on intensity D and on the distance in parameter from the bifurcation point. Near the critical point, the duration of the nearly periodic oscillations (the laminar phase) increases with the growth of intensity D, and the quantity λ_1 is decreased. The movement away from the critical point gives an opposite result, i.e., the positive exponent λ_1 grows with the increase of noise intensity D, and the time of switching to another stable regime decreases. The results of calculating the influence of additive fluctuations on the system dynamics are illustrated in Fig.20.

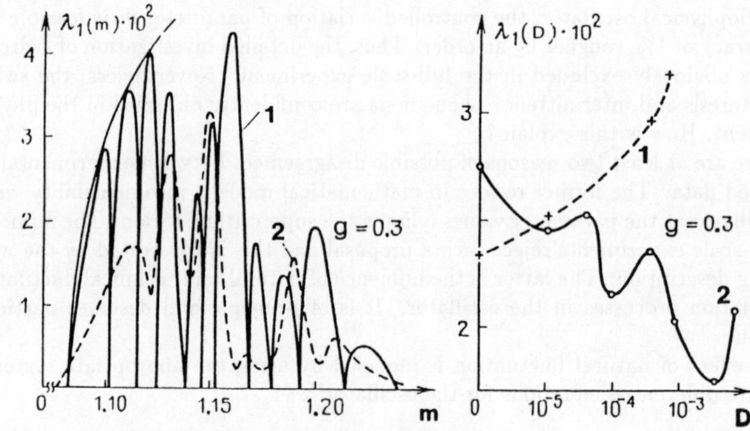

Fig.20. Noise effect on motion chaoticity degree; $a - \lambda_1(m)$ for $D = 0$ (curve 1) and $D = 10^{-2}$ (curve 2), $b - \lambda_1(D)$ for $m = 1.1005$ (curve 1) and $m = 1.1010$ (curve 2); $m^* = 1.10112$.

Thus, the effect of external noise on system (21) near the point of tangency bifurcation is manifested in that the characteristic relaxation times for a new stable regime depend complexly on the fluctuation intensity and parameter values. The intermittent stochasticity phenomenon is not induced, however, by the external additive noise.

We refer to Fig.19b where the Poincare section is shown for the parameter values given in Fig.19a, but as the additive noise with an intensity of $D =: 10^{-3}$ is introduced. Being close to the one-dimensional band in the absence of noise, the attractor band "swells" as affected by fluctuations, becomes principally two-dimensional. But the representative points fill the region near an unperturbed attractor essentially with the equal probability and do not concentrate in the vicinity of point Q.

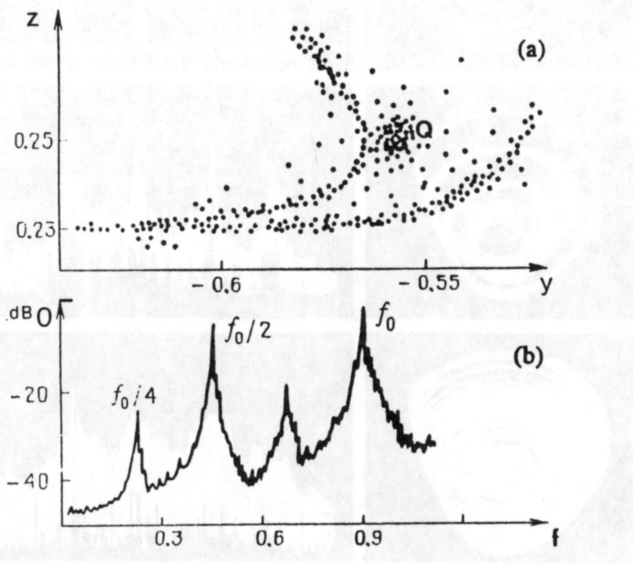

Fig.21.The Poincare section (a) and oscillation power spectra (b) for by modulation intermittency ($m = 1.1522$, $g = 0.3$, $D = 0.01$, $p = 0.01$).

Cycle Γ (Fig.18) is concerned again. For the parameter values $m \leq m^0 = 1.15231$, the cycle is stable. The increase of parameter $m \geq m^0$ implies the abrupt system's switching to the SA regime (Fig.19a). The system will generate either periodic or stochastic oscillations if parameter m is slowly varied as compared with the basic oscillation period of the oscillator $T_0 \simeq 2\pi$ within the small limits of $\pm(0.005 \div 0.010)$. Under the random parameter variations, the oscillations will exhibit the laminar phases (motion on the limit cycle Γ) that are irregularly interrupted in time by the turbulent phases (motion on the SA with the parameter values in excess of the critical one). The time realization of the oscillation process $x(\tau)$ will not differ from that being typical for intermittency. However, this process is not self-oscillating, but a modulating process, induced by the system parameter fluctuations. Unlike the systems of Morse-Smale type, where noise leads to a broader spectrum line of oscillation, the intermittent chaos can be induced by the multiplicative noise in the systems which exhibit stochastic behavior near the multiplicity lines.

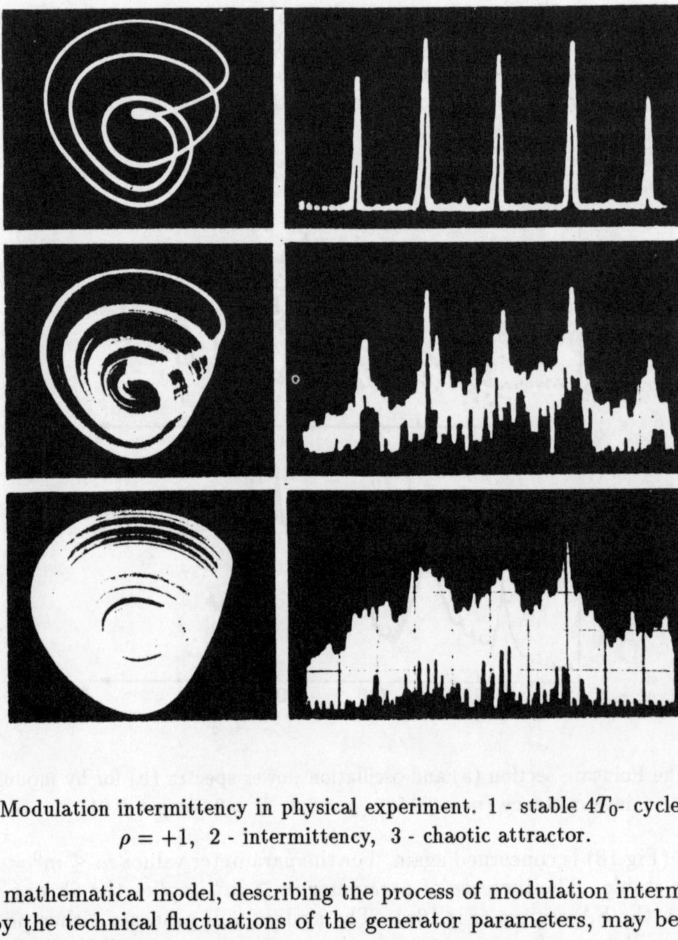

1

2

3

Fig.22. Modulation intermittency in physical experiment. 1 - stable $4T_0$- cycle near point $\rho = +1$, 2 - intermittency, 3 - chaotic attractor.

The mathematical model, describing the process of modulation intermittency induced by the technical fluctuations of the generator parameters, may be written in general form as follows

$$\dot{x} = [m^0 + \nu_1(\tau)]x + y - xz, \quad \dot{y} = -x, \quad \dot{z} = [g^0 + \nu_2(\tau)][\frac{x+|x|}{2}x - z], \tag{30}$$

where m^0 and g^0 are the parameter values corresponding to the periodic oscillation regime at the point of tangency bifurcation (points on the bifurcation multiplicity lines), and $\nu(\tau)$ are the functions of periodic perturbation type having random amplitudes and phases. The initial conditions to solve the Cauchy problem must belong to periodic oscillation regime of an unperturbed system.

Carry out the numerical experiment with system (30) by specifying values $\nu_1(\tau) = \xi(\tau)sin(p\tau)$, $\nu_2(\tau) = 0$, considering $\xi(\tau)$ to be a white noise with intensity D, and assuming $p = 0.01$. Fig.21 presents the calculation data for the Poincare section and power spectrum of oscillation regime under the above conditions. It is seen that the representative points in the section are concentrated in the vicinity of point Q testifying to the trajectory return (laminar oscillation phases). At the same time, in the Poincare section, the representative points are present in the vicinity of chaotic band (see Fig.19a as D=0) testifying to the "turbulent" motion phases. The spectrum of a regime, being discrete in the absence of noise, acquires the characteristic form which represents the process of modulation intermittency.

Now, look at the physical experiment. The stable periodic regime in the vicinity of the bifurcation point $\rho_1 = +1$, $m \simeq 1.15$ and $g = 0.3$ is shown in Fig.22a. In Fig.22c, a chaotic attractor is presented to which the system is switched as parameter m increases to a value of $\simeq 1.16$. In the Fig.22b, the phase trajectory projection onto xy-plane is presented in the regime of intermittency realized by specifying the intermediate parameter value of $1.15 < m < 1.16$. As seen from the comparison of calculated and experimental data, just the modulation intermittency is realized in the physical experiment with the regularities in the character of the process spectra being typical.

In the course of experiments, the phenomenon of modulation intermittency near the multiplicity line of a multi-periodic cycle was confidently recorded for the period-three, -four, and -five cycles. The analysis of truncated realizations of time processes by using a storage oscilloscope has proved the random character of appearance and duration of the laminar phases, of oscillations interrupted by turbulent motion on the chaotic attractor.

7. Interaction of Chaotic Attractors. Intermittency of "Chaos-Chaos" Type

The infinite number of attractors, with different structures is realized in quasi-hyperbolic systems with the chaotic dynamics depending on parameters and initial conditions. The recording of a control parameter does not change the picture: the infinite number of attractors holds and their basins of attraction are separated in the phase space by separatrix surfaces. The problem of describing the dynamics of a quasi-hyperbolic system becomes highly complicated due to many regular regimes coexisting along with the chaotic attractors in the phase space.

The modulation intermittency is a typical phenomenon for quasi-hyperbolic systems considered in section 4. Conditions for its realization are, first, the coexistence of only two stable (regular and chaotic) regimes close in parameters and in the phase space, and, second, parameter fluctuations. In the present section, consider just one more phenomenon, namely, the interaction between chaotic attractors leading to the intermittency of "chaos-chaos" type.[17–21]

We perform a numerical experiment using the dynamical system (21) in the section of the parameter plane $g = 0.097$ to study the oscillation regimes when varying the parameter m. The basic (initial) limit cycle of the system with period T_0 evolves, as a result of accumulation of period doubling bifurcations, to a strange attractor (denote it SA_1) that is softly born at the critical point $m^* = 2.39$. This attractor was considered in detail in sections $g = 0.2$ and 0.3 above. In the supercritical region $m \geq 2.39$, the attractor SA_1 evolves according to the regularities discussed earlier.

At the bifurcation point $m = 2.31$, a pair of period-three cycles, saddle and stable ones, is born abruptly as a result of bifurcation $\rho = +1$. The stable period-three cycle undergoes as well the cascade of period doubling bifurcations with the growth of parameter m which is completed by the birth of attractor SA_3 at the critical point $m_3^* = 2.35$. While the parameter m grew to the values being lower than a certain critical one $2.44 < m_0^* < 2.45$, both SA_3 and SA_1 demonstrate typical regularities, including the appearance of the stability windows of multi-periodic cycles of different families. Both strange attractors, SA_1 and SA_3, exist independently in the interval of parameter $2.40 < m < m_0^*$ with the basins of attraction in the phase space separated by a complicated two-dimensional separatrix surface.

The calculation results for the Poincare section of system (21) in plane $y = 0$ and for the appropriate spectra are presented in Fig.23. Sections (a) and (b) correspond to the value of parameter $m = 2.41$ with the initial conditions on attractors SA_1 and SA_3, respectively. It is seen from the location of points on the Poincare section that there exist two independent attractors localized in different phase space regions of system (21). Attractors SA_1 and SA_3 maintain their structures and statistical properties under a small variation of control parameters and the influence of small noise.

The attractors begin to interact as parameter m grows. The separatrix surface, separating SA_1 and SA_3 upon reaching a certain threshold m_0^*, is collapsed and the attractors merge. A merged attractor SA_0 is formed that comprises both SA_1 and SA_3. In Fig.23c, a stochastic set is shown in the secant plane for the parameter value of $m = 2.45 > m_0^*$ involving attractors SA_1 (a) and SA_3 (b) if one takes into account their minor evolution with the growth of parameter m. The merged attractor is realized independently of whether the initial conditions belong to the SA_1 or SA_3 basins of attraction, or not, showing the separatrix surface to collapse. The merging of the SA_1 and SA_3 basins of attraction seems to take place under formation of the merged attractor SA_0.

The detailed study of time realizations, autocorrelation, power spectrum and integral intensity of oscillations in the SA_1, SA_3 and SA_0 regimes shows that, while the parameter m passes the critical point of merging of attractors m_0^*, the system first spends a longer time on SA_1 continuously. The switching to SA_3 proceeds irregularly in time and has the character of "turbulent" splashes on the background of relatively more prolonged "laminar" motion phase on SA_1 with the energy being remarkably lower than that of SA_3. The switching to attractor SA_3 can be determined by the time realization $x(\tau)$ when a negative splash appears in dependence $x(\tau)$ with the amplitude exceeding the maximum possible ones when moving on SA_1.

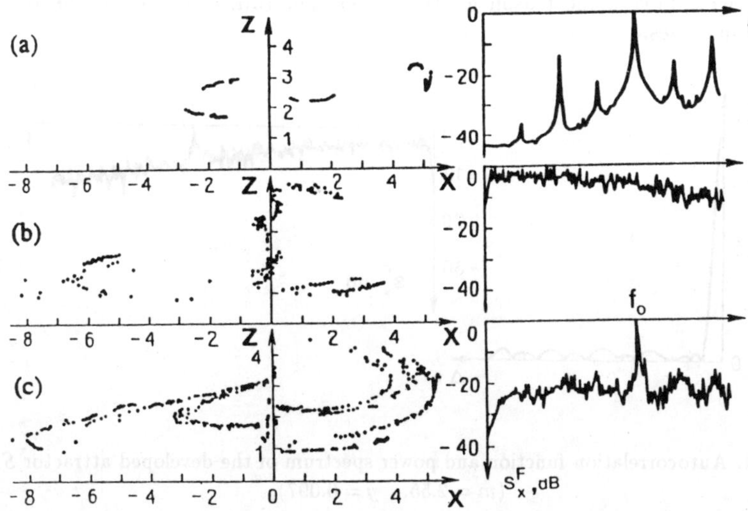

Fig.23. The Poincare section and power spectra of intermittency "chaos-chaos".

The statistical analysis of the relative residence time of the system in the states of "turbulent" and "laminar" phases in the merged attractor regime has shown that for small threshold exceeding, the dependence of the mean residence time of the system on SA_1 is evaluated approximately in the form of a ratio being valid for the "cycle-chaos" intermittency[18]:

$$< \tau_1 >= c(m - m_0^*)^{-1/2}, \qquad c = 18.0, \qquad m_0^* = 2.445. \tag{31}$$

The duration of laminar oscillation phase (here, it corresponds to a chaos regime which is structurally close to SA_1) decreases with the growth of supercriticality m having the critical index of -1/2 while the mean duration of turbulent splashes (motion on SA_3) increases, respectively. An obvious analogy to the "cycle-chaos" intermittency regime allows one to call the phenomenon described the intermittency of "chaos-chaos" type.[18]

As supercriticality grows, turbulent splashes occur more often leading to the increase of SA_0 mean energy. The intensity of self-oscillation is remarkably built up and its power spectrum becomes smoother evolving simultaneously to lower frequencies.

The autocorrelation function of oscillation process seeks to be δ-shaped demonstrating a strong intermixing.

The averaged power spectrum and autocorrelation function of the merged attractor SA_0 are indicated in Fig.24. They were calculated using realization $x(\tau)$ for $m = 2.55$. The turbulent splashes proceed within relatively larger time intervals on the average. This is the reason for the power spectrum to evolve to the region of lower frequencies.

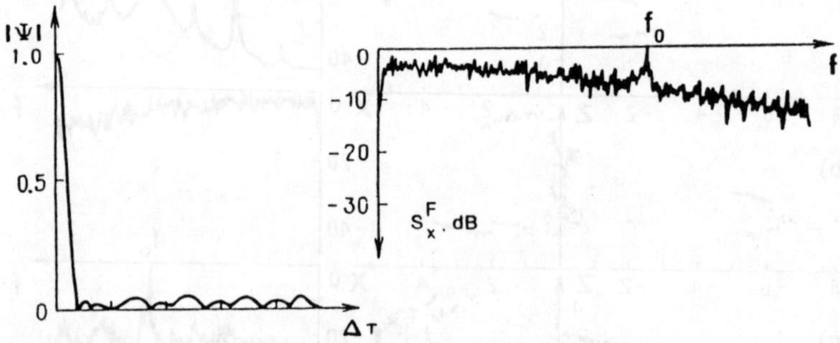

Fig.24. Autocorrelation function and power spectrum of the developed attractor SA_0 ($m = 2.55$, $g = 0.097$).

The merged attractor does not collapse as affected by an additive δ-correlated noise. The noise effect resulted, under the numerical simulation of a process by stochastic equations (Eq.28), in variation of the particular time moments of the system's switching from SA_1 to SA_3. However, the mean statistical characteristics of SA_0 were practically not altered. A more intricate type of "chaos-cycle-chaos" intermittency can be induced depending on the intensity of noise influence on SA_0. Such a self-oscillation process randomly includes the realization segments belonging to SA_1, SA_3 and to one or several multi-periodic cycles.

The results presented here may be treated, particularly, as an experimental proof of the fact that the branching of chaotic solutions of a system of non-linear differential equations is possible when varying the control parameters.

8. Conclusion

Modified oscillator with inertial nonlinearity, presented in this work, one can consider as the one of base models of dynamical chaos theory of finite-dimensional systems. It has a least dimension of phase space at which the realization of strange attractor regime is possible. Its equations have enough simple form. This oscillator demonstrates all known typical mechanisms of the transition to chaos, which are characteristic for systems with quasi-hyperbolic attractor. When the control parameters' values in supercritical region are varied, the evolution of chaos obeys also well-known universal regularities. Radiotechnical model of modified oscillator with inertial nonlinearity is easily realized in experiment and precisely corresponds to the mathematical model.

Unfortunately, the results of investigations for some important and interesting questions, having a general character, were not presented due to limited volume of the article. For example, about an influence of non-inertial dissipative nonlinearity on chaotic dynamics of the system; about dynamics of oscillator in the neighborhood of homoclinic trajectory of a loop of saddle-focus separatrix type in the excited mathematical model of oscillator; about an influence of external fluctuations on bifurcations of attractors; about breakdown of two-frequency oscillations in non-autonomous oscillator with inertial nonlinearity.

One can read widely with materials on mentioned above and other questions, which concern dynamics of modified oscillator with inertial nonlinearity, in monographs[4,5,9,22] of V.S. Anishchenko.

9. Acknowledgements

We thank International Science Foundation of Soros (Grant RNO 000) and the State Committee of Russia on Science and Higher School (Grant 93-8.2-10).

We also thank G. Strelkova for the help in preparing the manuscript for publication.

References

1. K.F. Theodorchik, *Zh. Tekhn. Fiz.* **16** (1946) 845.
2. P.S. Landa, *Auto-Oscillations in Systems with Finite Number of Degrees of Freedom* (Nauka Publishers, Moscow, 1980).
3. V.S. Anishchenko, V.V. Astakhov, *Radiotekhnika i elektronika* **28** (1983) 1109.
4. V.S. Anishchenko, *Complicated oscillations in simple systems: Appearance, routes, structure and properties of dynamical chaos on radiophysical systems* (Nauka Publishers, Moscow, 1990).
5. V.S. Anishchenko, *Dynamical chaos - models and experiments* (World Scientific, Singapore, 1995) (to be published).
6. N.N. Bautin, *Zh. Exp. i Teor. Fiz.* **8** (1938) 648.
7. V.S. Anishchenko, V.V. Astakhov, T.E. Letchford, M.A. Safonova, *Izv. Vuzov - Radiofizika* **26** (1983) 169.
8. V.S. Anishchenko, V.V. Astakhov, T.E. Letchford, *Zh. Tekhn. Fiz.* **53** (1983) 152.

9. V.S. Anishchenko, *Stochastic Oscillations in Radiophysical Systems. Part 2. Typical Bifurcations and Quasi-attractors in Nonlinear Systems with Low Number of Degrees of Freedom* (Saratov University Publisher, Saratov, 1986).

10. F.M. Izrailev, M.I. Rabinovich, A.D. Ugodnikov, *Phys. Lett.* **A86** (1981) 321.

11. F.M. Izrailev, M.I. Rabinovich, A.D. Ugodnikov, in Proc.: *IX Intern. Conf. on the Nonlinear Oscillations* **3** (Naukova Dumka, Kiev, 1984) .396.

12. J. Testa, J. Perez, C.Jeffries, *Phys. Rev. Lett.* **48** (1982) 714.

13. J. Testa, J. Perez, C. Jeffries, *Evidence for Bifurcation and Universal Chaotic Behavior in Nonlinear Semiconducting Devices* (University of California, Berkeley, 1982) 32 p.

14. R. Buskirk, C. Jeffries, *Phys. Rev.* **31A** (1985) 3332.

15. B.A. Huberman, J. Rudnick, *Phys. Rev. Lett.* **45** (1980) 154.

16. V.V. Anishchenko, V.V. Astakhov, T.E. Letchford, M.A. Safonova, *Izv. Vuzov - Radiofizika* **26** (1983) 832.

17. F.T. Arecchi, F. Lisi, *Phys. Rev. Lett.* **49** (1982) 94.

18. V.S. Anishchenko,*Sov. Tech. Phys. Lett.* **10** (1984) 266.

19. V.S. Afraimovich, M.I. Rabinovich, A.D. Ugodnikov, *Pisma v Zh. Eksp. i Teor. Fiz.* **38** (1983) 64.

20. A.S. Dmitriev, V.Ya. Kislov, S.O. Starkov, *Zh. Tekhn. Fiz.* **55** (1985) 2417.

21. V.I. Oseledez, *Works of Mosk. Math. Sci.* **19** (1968) 179.

22. V.S. Anishchenko, *Dynamical Chaos – Basic Concepts.* (Teubner-Texte zur Physik, Leipzig, 1987).

BIFURCATION AND CHAOTIC STATE IN FORCED OSCILLATORY CIRCUITS CONTAINING SATURABLE INDUCTORS

Tetsuya YOSHINAGA
School of Medical Sciences, The University of Tokushima
Kuramoto, Tokushima 770, Japan
yosinaga@medsci.tokushima-u.ac.jp
and
Hiroshi KAWAKAMI
Department of Electrical and Electronic Engineering
The University of Tokushima
Minamijosanjima, Tokushima 770, Japan
kawakami@ee.tokushima-u.ac.jp

ABSTRACT

Bifurcations of periodic, quasi-periodic and chaotic oscillations are most frequently observed in forced nonlinear circuits containing saturable inductors, whose dynamics are described by nonautonomous differential equations with periodic external forcing terms. A Poincaré mapping is commonly applied to the study of the bifurcation phenomena. We present some elementary discussion of the codimension one and two bifurcations of periodic solution and give a practical method to obtain various bifurcation diagrams. As illustrated examples, we show numerical results for typical circuits.

1 Introduction: Circuit Model and its Dynamics

An electric circuit containing a saturable inductor exhibits various nonlinear phenomena[1], such as the coexistence of several periodic oscillations which are correlated with the jump and hysteresis behaviors, the fundamental, higher-harmonic and sub-harmonic resonances, the appearance of quasi-periodic and chaotic states of oscillations, etc. The state change due to the variation of system parameters in dynamical systems is referred to as bifurcation. The generic codimension one bifurcations are

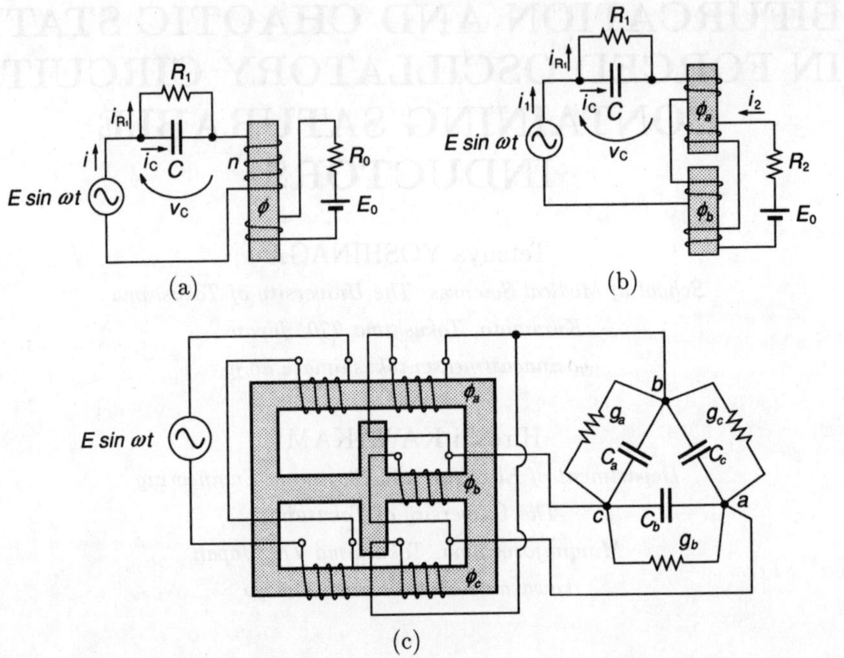

Figure 1: Electric circuits containing saturable inductors: (a) resonant circuit, (b) resonant circuit with two saturable inductors and (c) phase converter circuit.

classified into three types: tangent, period-doubling and the Hopf bifurcations, each of which corresponds to generation of a pair of periodic solutions, branching of periodic solution and appearance of quasi-periodic solutions, respectively. For analyzing such nonlinear phenomena, we present a numerical method to obtain the codimension one bifurcation sets. The method is easily applicable to the case of bifurcations with codimension two. In this chapter, to see various nonlinear phenomena, we consider simple but typical three circuits[1, 2] shown in Fig. 1.

In the rest of this section, we introduce fundamental equations derived from the circuits. Section 2 is devoted to summarize mathematical concepts and technical procedures for analyzing topological properties of periodic solutions and their bifurcations. Our main results of analysis are shown in Sec. 3. In Sec. 4 are given complementing remarks and problems related to the subjects.

As the first example, let us consider the electric circuit shown in Fig. 1 (a), see also Chap. 5 in Ref. 1. In the circuit, nonlinear oscillations occur due to the presence of saturable-core coil under the impression of a voltage source $E\sin\omega t$. Following the notation in the figure, we have

$$
\begin{aligned}
n\frac{d\phi}{dt} + v_C &= E\sin\omega t \\
i_C &= C\frac{dv_C}{dt}
\end{aligned}
\tag{1}
$$

with

$$
v_C = R_1 i_{R_1}, \qquad i = i_{R_1} + i_C \tag{2}
$$

where n is the number of turns of the inductor coil and ϕ is the magnetic flux in the core. Note that the saturable reactor has a secondary coil through which the biasing direct current flows. Then, neglecting hysteresis, we assume the saturation curves of the core be characterized by

$$
ni = a_1\phi + a_2\phi^2 + a_3\phi^3 \tag{3}
$$

where a_1, a_2 and a_3 are constants. We introduce nondimensional variables u and v, defined by

$$
i = I_n u, \qquad \phi = \Phi_n v \tag{4}
$$

where I_n and Φ_n are appropriate base quantities of the current and the flux, respectively. Then Eq. (3) may be rewritten as

$$
\begin{aligned}
u &= \frac{1}{nI_n}\left(a_1\Phi_n v + a_2{\Phi_n}^2 v^2 + a_3{\Phi_n}^3 v^3\right) \\
&= b_1 v + b_2 v^2 + b_3 v^3
\end{aligned}
\tag{5}
$$

Although the base quantities I_n and Φ_n are arbitrarily, we choose them as

$$
n\omega^2 C\Phi_n = I_n, \qquad {b_2}^2 + 3(1 - b_1 - b_3)b_3 = 0 \tag{6}
$$

Then, eliminating i_{R_1} and i_C in Eqs. (1) and using Eqs. (4) and (6), we obtain

$$
\frac{d^2v}{d\tau^2} + k\frac{dv}{d\tau} + b_1 v + b_2 v^2 + b_3 v^3 = B\cos\tau \tag{7}
$$

where

$$
\begin{aligned}
&\tau = \omega t - \tan^{-1}k, \qquad k = \frac{1}{\omega C R_1} \\
&B = \frac{E}{n\omega\Phi_n}\sqrt{1+k^2}
\end{aligned}
\tag{8}
$$

Equation (7) can be transformed to the alternative form

$$\frac{d^2v'}{d\tau^2} + k\frac{dv'}{d\tau} + c_1 v' + c_3 v'^3 = B_0 + B\cos\tau \tag{9}$$

where $v' = v + b_2/(3b_3)$ and c_1, c_3, B_0 are constants determined by b_1, b_2 and b_3, and, from the second equation of Eqs. (6), c_1 and c_3 are restricted by the condition

$$c_1 + c_3 = 1 \tag{10}$$

The second example is an electric circuit as shown in Fig. 1 (b), see also Chap. 11 in Ref. 1. Following the notation in the figure, the circuit equations are written as follows:

$$\begin{aligned} n\frac{d}{dt}(\phi_a + \phi_b) + v_C &= E\sin\omega t \\ n\frac{d}{dt}(\phi_a - \phi_b) + R_2 i_2 &= E_0 \\ i_C &= C\frac{dv_C}{dt} \end{aligned} \tag{11}$$

with

$$v_C = R_1 i_{R_1}, \qquad i_1 = i_{R_1} + i_C \tag{12}$$

where ϕ_a and ϕ_b are the magnetic fluxes in the cores, and n is the number of turns of each coil wound around the cores. The nonlinear characteristics of the cores are assumed to be

$$c_3{\phi_a}^3 = ni_1 + ni_2, \quad c_3{\phi_b}^3 = ni_1 - ni_2 \tag{13}$$

where c_3 is a constant dependents on the nature of the cores. We introduce the nondimensional variables defined by

$$i_1 = I_n u_1, \qquad i_2 = I_n u_2, \qquad \phi_a = \Phi_n v_1, \qquad \phi_b = \Phi_n v_2 \tag{14}$$

and fix the base quantities I_n and Φ_n by

$$n\omega^2 C\Phi_n = I_n, \qquad c_3{\Phi_n}^3 = nI_n \tag{15}$$

Then Eqs. (11) may be transformed to

$$\begin{aligned} \frac{d^2}{d\tau^2}(v_1 + v_2) + k_1\frac{d}{d\tau}(v_1 + v_2) + u_1 &= B\cos\tau \\ \frac{d}{d\tau}(v_1 - v_2) + k_2 u_2 &= B_0 \end{aligned} \tag{16}$$

where

$$k_1 = \frac{1}{\omega C R_1}, \qquad k_2 = \omega C R_2, \qquad \tau = \omega t - \tan^{-1} k_1$$
$$B = \frac{E}{n\omega\Phi_n}\sqrt{1+{k_1}^2}, \qquad B_0 = \frac{E_0}{n\omega\Phi_n} \tag{17}$$

The nonlinear characteristics (13) also may be expressed by

$${v_1}^3 = u_1 + u_2, \qquad {v_2}^3 = u_1 - u_2 \tag{18}$$

Introducing new variables

$$a = v_1 + v_2, \qquad b = v_1 - v_2 \tag{19}$$

yields, from Eqs. (16) and (18), the following simultaneous equations:

$$\begin{aligned} \frac{d^2a}{d\tau^2} + k_1\frac{da}{d\tau} + \tfrac{1}{8}\left(a^2 + 3b^2\right)a &= B\cos\tau \\ \frac{db}{d\tau} + \tfrac{1}{8}k_2\left(3a^2 + b^2\right)b &= B_0 \end{aligned} \tag{20}$$

Note that if the secondary circuit is removed, i.e., $b \equiv 0$, then Eqs. (20) become Duffing's equation with symmetrical nonlinear characteristics. Hence we call Eqs. (20) a third order system of Duffing's type.

At this moment let us derive an autonomous equation correlated with Eqs. (20), whose equilibria and limit cycles correspond to harmonic and quasi-periodic solutions of Eqs. (20), respectively. Following Hayashi's formal averaging procedure[1], the variables a and b may be assumed to take the form

$$a = x(\tau)\cos\tau + y(\tau)\sin\tau, \qquad b = z(\tau) \tag{21}$$

where $x(\tau)$, $y(\tau)$ and $z(\tau)$ are slowly varying functions of the time τ. Substituting Eqs. (21) into (20) and equating the coefficients of the terms containing $\cos\tau$ and $\sin\tau$ and the nonoscillatory terms separately to zero gives the averaged equations:

$$\begin{aligned} \frac{dx}{d\tau} &= \tfrac{1}{2}\left(-k_1 x - Ay\right) \\ \frac{dy}{d\tau} &= \tfrac{1}{2}\left(Ax - k_1 y + B\right) \\ \frac{dz}{d\tau} &= B_0 - \tfrac{1}{16}k_2\left(3r^2 + 2z^2\right)z \end{aligned} \tag{22}$$

with

$$A = 1 - \tfrac{3}{32}\left(r^2 + 4z^2\right), \qquad r^2 = x^2 + y^2 \tag{23}$$

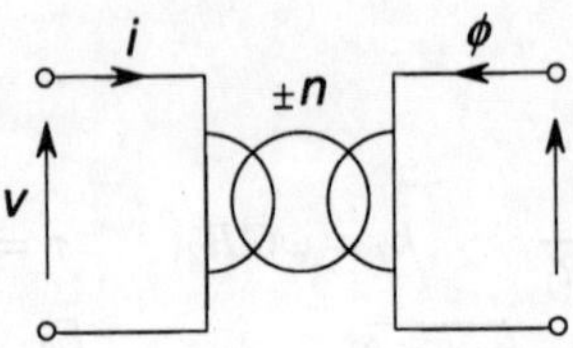

Figure 2: The dynamic gyrator, a translator between the electric (v, i)-characteristics and the magnetic (f, ϕ)-characteristics.

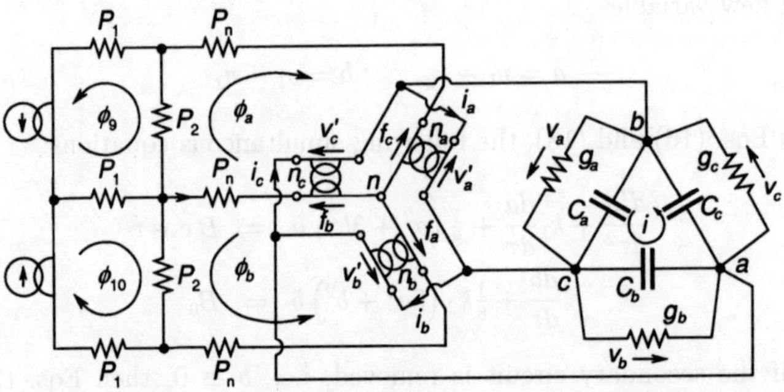

Figure 3: An equivalent circuit of the phase converter shown in Fig. 1 (c).

under the assumptions that $dx/d\tau$, $dy/d\tau$, $dz/d\tau$ and k_1 are small.

We consider the third example of circuit shown in Fig. 1 (c), which is a single-phase to three-phase converter of ferro-resonance type[2]. The electric-magnetic circuit is considered as a composed system such that electric and magnetic sub-circuits are connected through *dynamic gyrators.* The dynamic gyrator and an equivalent circuit of the phase converter are illustrated in Figs. 2 and 3, respectively. The characteristics of the dynamic gyrator is expressed by

$$\begin{bmatrix} f \\ v \end{bmatrix} = \begin{bmatrix} 0 & \pm n \\ \mp n\frac{d}{dt} & 0 \end{bmatrix} \begin{bmatrix} \phi \\ i \end{bmatrix} \tag{24}$$

where f, ϕ, v and i are the magnetomotive force, flux, voltage and current, respectively, and n is the number of turns of inductor coil.

In the equivalent circuit for the phase converter, we assume that the magnetic reluctance P_n, whose characteristics of the magnetomotive force $f(\phi)$ be a nonlinear function of the flux ϕ:

$$f(\phi) = a_1\phi + a_3\phi^3 \tag{25}$$

We also assume the other elements be linear. Following the notation in Fig. 3, the circuit equations are derived, from the magnetic sub-circuit,

$$\begin{aligned} f_a - f_c - f(\phi_c) + f(\phi_a) + P_2(\phi_a + \phi_9) &= 0 \\ f_b - f_c - f(\phi_c) + f(\phi_b) + P_2(\phi_b + \phi_{10}) &= 0 \end{aligned} \tag{26}$$

and, from the electrical sub-circuit,

$$\begin{aligned} C_a\frac{dv_a}{dt} + g_a v_a + i_c &= C_c\frac{dv_c}{dt} + g_c v_c + i_a \\ C_b\frac{dv_b}{dt} + g_b v_b + i_c &= C_c\frac{dv_c}{dt} + g_c v_c + i_b \end{aligned} \tag{27}$$

where the variables are restricted by the relation

$$\phi_a + \phi_b + \phi_c = 0 \tag{28}$$

$$v_a + v_b + v_c = 0 \tag{29}$$

In the following, we consider a symmetrical case, that is,

$$\begin{aligned} &C_a = C_b = C_c = C, \qquad g_a = g_b = g_c = g, \qquad N_1 = N_2 = N \\ &n_a = n_b = n_c = n, \qquad \phi_a = \phi_b = \Phi(t) = \Phi_m \cos\omega t \end{aligned} \tag{30}$$

so that the characteristics of elements become

$$\begin{aligned} &f_a = ni_a, \qquad f_b = ni_b, \qquad f_c = ni_c \\ &v_a = -v_a' = n\frac{d\phi_a}{dt}, \qquad v_b = -v_b' = n\frac{d\phi_b}{dt}, \qquad v_c = -v_c' = n\frac{d\phi_c}{dt} \end{aligned} \tag{31}$$

Then, from Eqs. (26), (27), (28) and (29), introducing new variables defined by $x = \phi = \phi_a - \phi_b$, $u = \phi_c$ and $\tau = \omega t$ yields

$$\begin{aligned} \frac{d^2x}{d\tau^2} + k\frac{dx}{d\tau} + (c_0 + c_1)x + \tfrac{1}{4}c_3(x^2 + 3u^2)x &= 0 \\ \frac{d^2u}{d\tau^2} + k\frac{du}{d\tau} + (\tfrac{1}{3}c_0 + c_1)u + \tfrac{1}{4}c_3(x^2 + 3u^2)u &= B\cos\tau \end{aligned} \tag{32}$$

where

$$\begin{aligned} &k = \frac{g}{\omega C}, \qquad c_0 = \frac{P_2}{\omega^2 n^2 C}, \qquad c_1 = \frac{a_1}{\omega^2 n^2 C} \\ &c_3 = \frac{a_3}{\omega^2 n^2 C}, \qquad B = \frac{2P_2\Phi_m}{3\omega^2 n^2 C} \end{aligned} \tag{33}$$

The system of Eqs. (32) can be considered as a four dimensional system of Duffing's type from their form. Because the system includes Duffing's equation as a subset, we may appear various complicated behavior such as chaotic state due to a period-doubling process of periodic solutions showing parametric excitations, codimension two bifurcations and so on.

2 Bifurcation of Periodic States

In this section, we first summarize mathematical concepts and techniques for analyzing topological properties of periodic solutions and their bifurcation conditions. Then we will show a computational method for finding codimension one and two bifurcation sets.

2.1 Poincaré Mapping and Classification of Hyperbolic Fixed Points

Let us consider a nonautonomous ordinary differential equation:

$$\frac{dx}{dt} = f(t, x, \lambda) \tag{34}$$

where $x \in R^n$ is the state and $\lambda \in R^m$ is the system parameter. Let $f : R \times R^n \times R^m \to R^n$ be sufficiently differentiable for all arguments and periodic in time with period 2π, i.e., $f(t + 2\pi, x, \lambda) = f(t, x, \lambda)$. We also assume that the solution of Eq. (34) with initial condition $u := x(0)$, denoted by $\varphi(t, u, \lambda)$, exists for all t. Since f has the period 2π, we can naturally define a diffeomorphism T_λ, called the Poincaré map, from the state space R^n into itself:

$$T_\lambda : R^n \to R^n \ ; \ u \mapsto T_\lambda(u) = \varphi(2\pi, u, \lambda) \tag{35}$$

If a solution $x(t) = \varphi(t, u, \lambda)$ is periodic with period 2π, then the point u becomes a fixed point of T_λ:

$$u - T_\lambda(u) = 0 \tag{36}$$

Hence the study of a periodic solution with period 2π, that is, $x(t) = \varphi(t, u, \lambda)$ of Eq. (34) is topologically equivalent to the study of a fixed point $u \in R^n$ satisfying Eq. (36). Note that a periodic solution with period $2\pi k$, can be studied by replacing T_λ with ${T_\lambda}^k$, k-th iterates of T_λ, in Eq. (36). Therefore in the following we consider only the property of fixed point of T_λ and its bifurcations. Similar argument can be applied to the periodic point of T_λ.

Assume $u \in R^n$ be a fixed point of T_λ. Then the characteristic equation of the fixed point u is defined by

$$\chi(u, \lambda, \mu) := \det(\mu I - DT_\lambda(u)) = \mu^n + a_1\mu^{n-1} + \cdots + a_{n-1}\mu + a_n = 0 \tag{37}$$

where I is the $n \times n$ identity matrix, and DT_λ denotes the derivative of T_λ. We call u is hyperbolic, if all the absolute values of the eigenvalues of T_λ are different from unity[3, 4]. The topological type of a hyperbolic fixed point is determined by the $\dim E^u$ and $\det L^u$, where E^u is the intersection of R^n and the direct sum of the generalized eigenspaces of $DT_\lambda(u)$ corresponding to the eigenvalues μ such that $|\mu| > 1$, and $L^u = DT_\lambda(u)|_{E^u}$.

A hyperbolic fixed point is called D-type, if $\det L^u > 0$, and I-type if $\det L^u < 0$. By this definition we have $2n$ topologically different types of hyperbolic fixed points. These types are:

$$_kD \quad (k = 0, 1, \cdots, n), \qquad _kI \quad (k = 1, \cdots, n-1) \tag{38}$$

where D and I denote the type of the fixed point and the subscript integer indicates the dimension of the unstable subspace: $k = \dim E^u$. This classification is also obtained from the distribution of characteristic multipliers of Eq. (37). That is, D and I correspond to the even and odd number of characteristic multipliers on the real axis $(-\infty, -1)$, and k indicates the number of characteristic multiplier outside the unit circle in the complex plane. The distribution can be checked by the coefficients of Eq. (37). For more detailed information see Refs. 5 and 6. Note that for two dimensional case: $n = 2$, we have four types of hyperbolic fixed points: $_0D$ (sink), $_1D$, $_1I$ (saddle), and $_2D$ (source), see Ref. 7. In the following we use the notation $_kD_l^m$ (resp. $_kI_l^m$) denoting a D-type (resp. I-type) hyperbolic m-periodic point of T_λ, and l denotes the number to distinguish several same sets, if necessary. If $m = 1$ or $l = 1$, it will be omitted.

2.2 Codimension One Bifurcations

Bifurcation occurs when the topological type of a fixed point is changed by the variation of system parameter λ. For the codimension one bifurcation, we have three different types of bifurcations: tangent, period-doubling and the Hopf bifurcations[3]. These bifurcations are observed when the hyperbolicity is destroyed, which corresponds to the critical distribution of the characteristic multiplier: $\mu = +1$, $\mu = -1$, or $\mu = e^{j\theta}$ where $j = \sqrt{-1}$.

I-a: Tangent bifurcation (ab. T-bifurcation)

Under the change of parameter λ, the generation or extinction of a couple of fixed points occurs. The types of bifurcation are

$$\phi \leftrightarrow {}_{k-1}D + {}_kD \tag{39}$$

$$\phi \leftrightarrow {}_{k-1}I + {}_kI \tag{40}$$

where the symbol $\leftrightarrow$ indicates the relation before and after the bifurcation and ϕ denotes the extinction of fixed points. This type of bifurcation is observed if one of the eigenvalues of Eq. (37) satisfies the condition $\mu = 1$, or equivalently

$$\chi(u, \lambda, 1) = 0 \tag{41}$$

and the remainder of the eigenvalues lies off the unit circle.

We note that a **D-type of branching** may appear in systems that possess some symmetrical properties. This type of bifurcation occurs when a real eigenvalue passes through the point $(1,0)$ in the complex plane. Then the bifurcation condition is a degenerate case of the T-bifurcation. The types of bifurcation are

$$_kD \rightarrow {}_{k+1}D + 2\,{}_kD \tag{42}$$
$$_kD \rightarrow {}_{k-1}D + 2\,{}_kD \tag{43}$$
$$_kI \rightarrow {}_{k+1}I + 2\,{}_kI \tag{44}$$
$$_kI \rightarrow {}_{k-1}I + 2\,{}_kI \tag{45}$$

where $2\,{}_kD$ (resp. $2\,{}_kI$) indicates two numbers of fixed points of type $_kD$ (resp. $_kI$).

I-b: Period-doubling bifurcation (ab. P-bifurcation)

The types of bifurcation are

$$_kD \rightarrow {}_{k+1}I + 2\,{}_kD^2 \tag{46}$$
$$_kD \rightarrow {}_{k-1}I + 2\,{}_kD^2 \tag{47}$$
$$_kI \rightarrow {}_{k+1}D + 2\,{}_kD^2 \tag{48}$$
$$_kI \rightarrow {}_{k-1}D + 2\,{}_kD^2 \tag{49}$$

This type of bifurcation is observed if $\mu = -1$, or equivalently

$$\chi(u, \lambda, -1) = 0 \tag{50}$$

and there are no other eigenvalues on the unit circle.

I-c: the Hopf bifurcation (ab. H-bifurcation)

The types of bifurcation are

$$_kD \rightarrow {}_{k+2}D + ICC \tag{51}$$
$$_kD \rightarrow {}_{k-2}D + ICC \tag{52}$$
$$_kI \rightarrow {}_{k+2}I + ICC \tag{53}$$
$$_kI \rightarrow {}_{k-2}I + ICC \tag{54}$$

where ICC indicates an *invariant closed curve* of T_λ, which generally corresponds to quasi-periodic solutions of Eq. (34). This type of bifurcation is observed if a simple pair of complex conjugate roots of Eq. (37) transverses the unit circle of the complex plane. The condition for this type of bifurcation is given by:

$$\chi(u, \lambda, e^{j\theta}) = 0 \tag{55}$$

Hence, eliminating θ in Eq. (55), we obtain a bifurcation condition

$$\chi_H(u, \lambda) = 0 \tag{56}$$

Note that in this case we need an additional inequality satisfying the condition: $|\cos\theta| < 1$.

In bifurcation diagram, we use notations:

G_l^m for the T-bifurcation set,
D_l^m for bifurcation set of the D-type of branching,
I_l^m for the P-bifurcation set, and
H_l^m for the H-bifurcation set

where $(\ )^m$ indicates a bifurcation set for m-periodic point, and l denotes the number to distinguish several same sets of $(\ )^m$, if they exist. If $m = 1$, it will be omitted.

The numerical determination of generic bifurcation values is accomplished by solving the system of equations that represent the relation of fixed point, i.e., Eq. (36), and a bifurcation condition for T-, P- or H-bifurcation. For this purpose Newton's method is used, see Ref. 5. On the other hand, because of the degeneracy of conditions for the D-type of branching, for finding the bifurcation values, we must use symmetrical properties of the system and its solution[5].

2.3 Codimension Two Bifurcations

Codimension two bifurcation generally takes place under two different kinds of bifurcation conditions. The codimension two bifurcation occurs naturally if system parameters are changed as two-dimensionally in parameter space. Hence the bifurcation value is generally obtained as an isolated point in the planar bifurcation diagram. Note that in a neighborhood of the codimension two bifurcation point, the dynamical behavior exhibits complicated features, and some types of codimension two bifurcations relate to the generation of chaotic states[8, 9, 10].

There are six types of codimension two bifurcations combining two conditions of three different codimension one bifurcations. While, in a bifurcation diagram of the parameter plane, we observe codimension two bifurcation at the intersection of several curves representing codimension one bifurcations. The possible types of planer bifurcation diagrams including codimension two bifurcation points are sketched in Fig. 4. In each diagram, the condition on codimension two bifurcation is different from each other. Intuitively, we consider codimension two bifurcation as the transversal intersection of two codimension one bifurcation curves on a fixed point manifold in the product spaces of states and parameters. According to each diagram shown in Fig. 4, we call all types of codimension two bifurcations as:

II-a: Double tangent bifurcation (ab. T^2-bifurcation)

II-b: Double period-doubling bifurcation (ab. P^2-bifurcation)

II-c: Double Hopf bifurcation (ab. H^2-bifurcation)

II-d: Tangent-Period-doubling bifurcation (ab. TP-bifurcation)

II-e: Tangent-Hopf bifurcation (ab. TH-bifurcation)

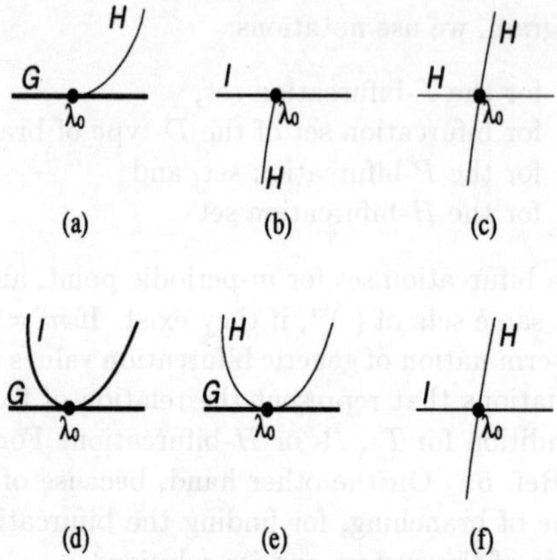

Figure 4: Classification of codimension two bifurcations by combinatorial intersection of codimension one bifurcation curves in parameter plane. The circled point λ_0 indicates the parameter values of (a) T^2-, (b) P^2-, (c) H^2-, (d) TP-, (e) TH- or (f) HP-bifurcation. The curves labeled by symbols G, I and H denote the T-, P- and H-bifurcation sets for fixed points, respectively.

II-f: Hopf-Period-doubling bifurcation (ab. HP-bifurcation)

The bifurcation value can be calculated by solving the system equations constructed by the fixed-point relation, i.e., Eq. (36), and two conditions for the codimension two bifurcation. In order to apply Newton's method, we must choose the bifurcation conditions so that the Jacobian matrix of the system equations is nonsingular. See Refs. 8 and 9, for the detail.

3 Results of Analysis

We now show numerical results obtained from the analysis of the dynamical systems that describe the circuits shown in Fig. 1.

3.1 *Duffing's Equations*

We rewrite Duffing's equation (9) as

$$\frac{d^2x}{dt^2} + k\frac{dx}{dt} + c_1x + c_3x^3 = B_0 + B\cos t \tag{57}$$

or equivalently

$$\begin{aligned}\frac{dx_1}{dt} &= x_2 \\ \frac{dx_2}{dt} &= -kx_2 - c_1x_1 - c_3{x_1}^3 + B_0 + B\cos t\end{aligned} \tag{58}$$

We consider the bifurcation sets of periodic solutions observed in the equations with

$$c_1 = 0.2, \qquad c_3 = 0.8 \tag{59}$$

Assuming the damping coefficient k is positive, we have the following properties:

1. Equations (58) are dissipative. Hence they have a maximum invariant set Δ, see Ref. 7.

2. The set Δ has zero area and cannot contain a simple closed curve which is invariant under the map T_λ.

We first fix $B_0 = 0$ and investigate the bifurcation sets in the parameter plane $\lambda = (B, k) \in R^2$. The bifurcation diagram shown in Fig. 5 exhibits well known nonlinear phenomena as jump and hysteresis behaviors. For example, by varying the value of B for fixed $k = 0.05$, we observe qualitative transition of fixed points as illustrated in Fig. 6. We see that jump phenomena occur at B_1 and B_2, and between (B_1, B_2) we have the hysteresis behavior of fixed points ${}_0D_1$ and ${}_0D_2$, which are, respectively, correlated with resonant and nonresonant periodic solutions[1]. The curves G^2 and G^3 in Fig. 5 indicate the tangent bifurcation sets of 1/2- and 1/3-harmonic oscillations, respectively.

Second, we fix $k = 0.05$ and consider the (B, B_0)-plane bifurcation problem, see Fig. 7 for the bifurcation diagram in relatively small values of B and B_0. Note that all bifurcations appeared in the diagram of the (B, B_0)-plane are symmetric with respect to the B-axis. The property is easily seen by the coordinate transformation $(x_1, x_2, t) \mapsto (-x_1, -x_2, t + \pi)$ in Eqs. (58). In Fig. 7, we see several cusp points exist, for example, at the points where the curves G_1 and G_2 meet tangentially. For relatively large values of B and B_0, the bifurcation sets of fixed and 2-periodic points are obtained in Fig. 8. The thick and thin curves indicate the tangent and period-doubling bifurcations, respectively. The curve by which the shaded region is bounded indicates the period-doubling bifurcation set of 2-periodic point. An enlarged diagram of Fig. 8 is shown in Fig. 9. On the curve aIb, a bifurcation ${}_0D \to {}_1I + 2\,{}_0D^2$ takes place, whereas on aIc, ${}_1I \to {}_0D + 2\,{}_1D^2$ is observed. Figure 10 shows a successive occurrence of period-doubling bifurcations

$${}_0D^{2^m} \to {}_1I^{2^m} + 2\,{}_0D^{2^{m+1}} \quad (\text{on } I^{2^m}) \qquad m = 0, 1, \cdots \tag{60}$$

The period-doubling bifurcations proceed to infinity and we observe a chaotic state of the solution.

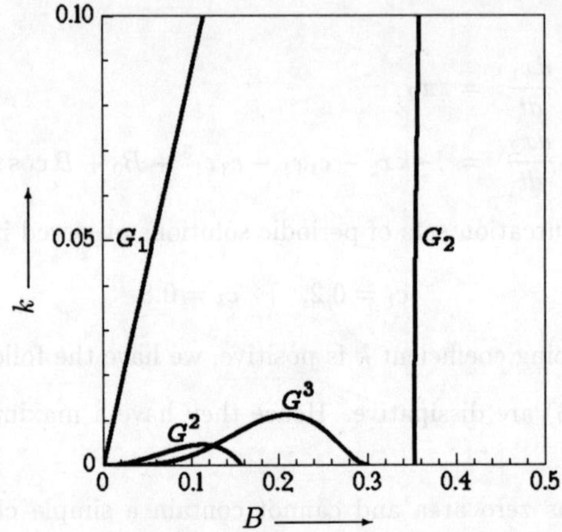

Figure 5: Bifurcation diagram of Duffing's equation (57).

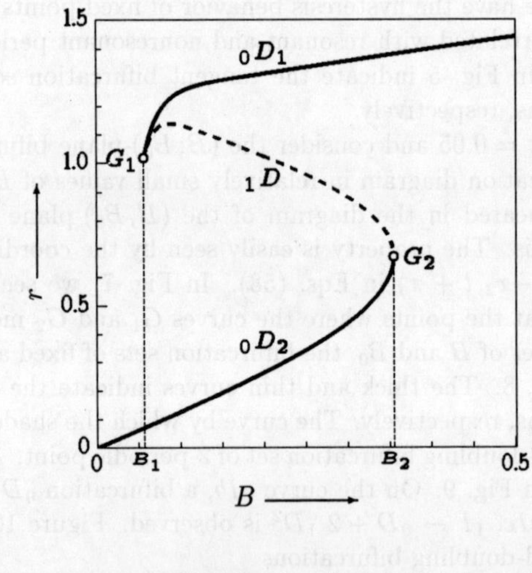

Figure 6: A norm r of fixed points by varying the value of B.

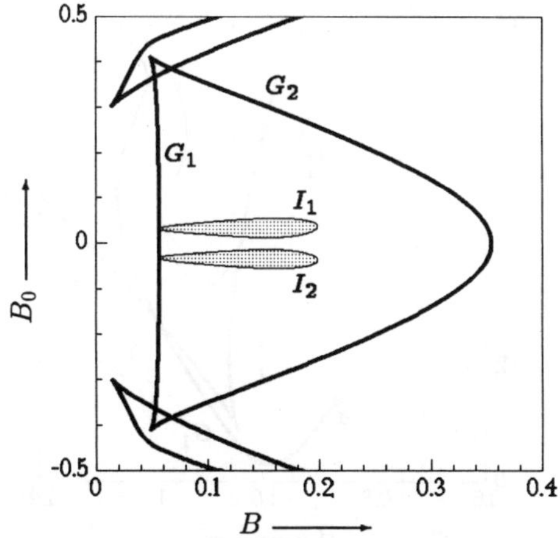

Figure 7: Bifurcation diagram in (B, B_0)-plane

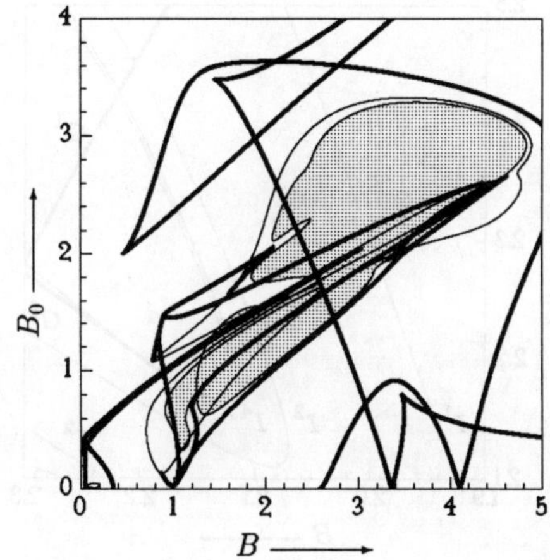

Figure 8: Bifurcation diagram of fixed and 2-periodic points.

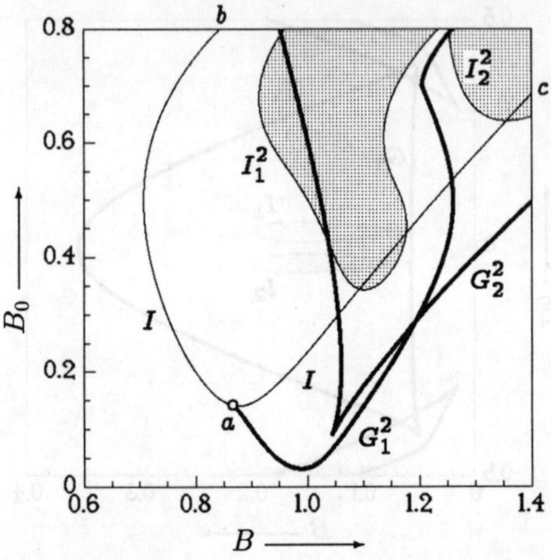

Figure 9: Enlarged bifurcation diagram of Fig. 8.

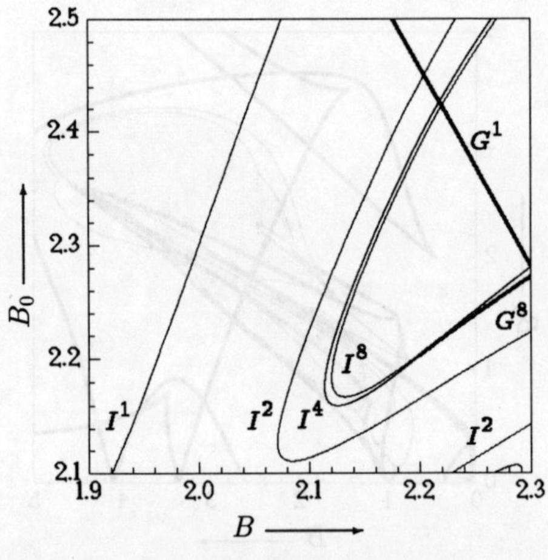

Figure 10: Enlarged bifurcation diagram of Fig. 8. The cascade of period-doubling bifurcations is observed.

3.2 Third Order System of Duffing's Type

As the second example, the bifurcation of quasi-periodic solution and the generation of chaotic state from the quasi-periodic oscillation are discussed in the system of Eqs. (20), see Refs. 11 and 12. We rewrite the equivalent differential equations of the first order form:

$$
\begin{aligned}
\frac{dx_1}{dt} &= x_2 \\
\frac{dx_2}{dt} &= -k_1x_2 - \tfrac{1}{8}\left({x_1}^2 + 3{x_3}^2\right)x_1 + B\cos t \\
\frac{dx_3}{dt} &= -\tfrac{1}{8}k_2\left(3{x_1}^2 + {x_3}^2\right)x_3 + B_0
\end{aligned}
\tag{61}
$$

and their averaged equations (22):

$$
\begin{aligned}
\frac{dy_1}{dt} &= \tfrac{1}{2}\left(-k_1y_1 - Ay_2\right) \\
\frac{dy_2}{dt} &= \tfrac{1}{2}\left(Ay_1 - k_1y_2 + B\right) \\
\frac{dy_3}{dt} &= B_0 - \tfrac{1}{16}k_2\left(3r^2 + 2{y_3}^2\right)y_3
\end{aligned}
\tag{62}
$$

with

$$
A = 1 - \tfrac{3}{32}\left(r^2 + 4{y_3}^2\right), \qquad r^2 = {y_1}^2 + {y_2}^2 \tag{63}
$$

In the following we fix

$$
B_0 = 0.03, \qquad k_2 = 0.05 \tag{64}
$$

and consider bifurcation problems for $\lambda = (B, k_1)$.

Figure 11 shows a bifurcation diagram of fixed points observed in the Poincaré map for Eqs. (61). In the diagram, dotted and solid curves denote the Hopf and tangent bifurcations, respectively, and the region in which a stable fixed point exists is shading. When the values of system parameters vary across the Hopf bifurcation curve in the direction from inside to outside of the stable region, we observe the appearance of an *ICC* of the Poincaré map, which corresponds to a quasi-periodic solution of Eqs. (61). Some phase portraits of *ICC*'s are shown in Fig. 12. The parameter values at which the oscillations exist are denoted by circled points with symbols a, b, c and d in Fig. 11. It is well known that there exists a cascade of period-doubling bifurcations of periodic solutions in nonlinear dynamical systems. Similar bifurcation process may occur in *ICC*'s of the Poincaré map: a stable *ICC* becomes unstable and there appears stable but twice winding *ICC*. In this case the winding number of the *ICC* is doubling. Figures 12 (a), (b) and (c) show this process observed in Eqs. (61) for the variation of the parameter k_1. Hence, after an infinite of the process, we observe a quasi-periodic-like chaotic solution, e.g., see Fig. 12 (d) for iterates of the Poincaré map. Because of the mathematical difficulty, little is known about this type of bifurcation for ICC. But we can deal with a quasi-periodic solution through an investigation of properties of limit cycles observed in the averaged system (62). Note that limit cycles of Eqs. (62) correspond to quasi-periodic solutions of Eqs. (61). Therefore the stability and bifurcation of quasi-periodic solutions can be discussed by results of an investigation of limit cycles.

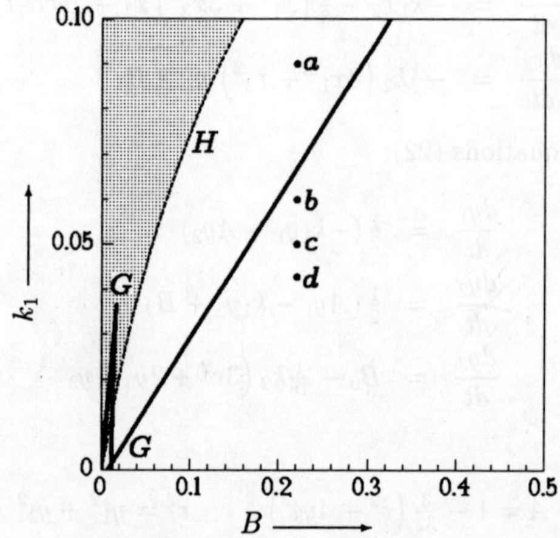

Figure 11: Bifurcation diagram of Eqs. (61).

To investigate qualitative properties of limit cycles observed in the autonomous system, we study fixed points of a Poincaré map defined by the local cross section or Poincaré section Π. Hence we define the Poincaré map as the first return map from Π to Π by the solution. Our method for calculating bifurcation sets, stated in Sec. 2, is applicable to the bifurcation problems of autonomous systems with a minor modification.

Before showing results, we summarize notations about bifurcations of equilibria and limit cycles observed in the autonomous system. We use symbols h and g_l ($l = 1, 2$) for the Hopf and tangent bifurcation sets of equilibria, respectively, and the symbol I^m for period-doubling bifurcation set of m-periodic limit cycle.

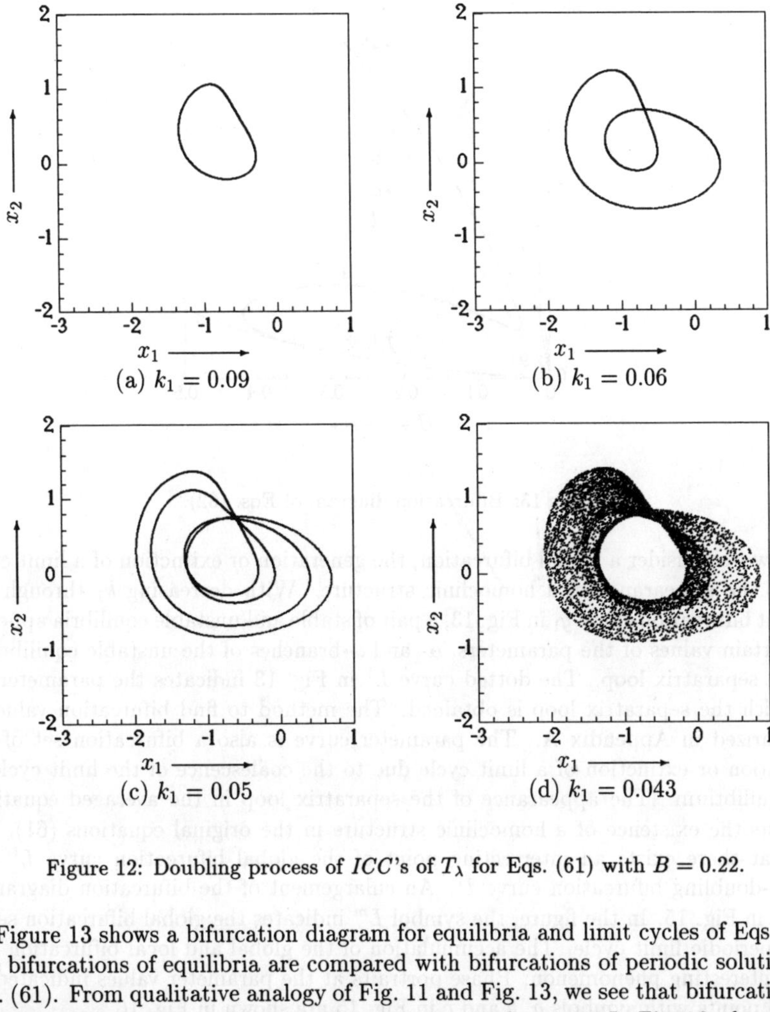

Figure 12: Doubling process of ICC's of T_λ for Eqs. (61) with $B = 0.22$.

Figure 13 shows a bifurcation diagram for equilibria and limit cycles of Eqs. (62). The bifurcations of equilibria are compared with bifurcations of periodic solutions of Eqs. (61). From qualitative analogy of Fig. 11 and Fig. 13, we see that bifurcations in the original system are well explained by the averaged system. In Fig. 13, the shaded region is a parameter area in which there exists a stable limit cycle generated by the Hopf bifurcation h of an equilibrium. Limit sets observed in Eqs. (62) with specified by circled points with symbols a, b, c and d in Fig. 13 are illustrated in Fig. 14. The doubling process of winding numbers of limit cycles occurs and entirely a chaotic attractor appears. From Fig. 12, we see that every limit set is well corresponding to the invariant set with the same parameter values.

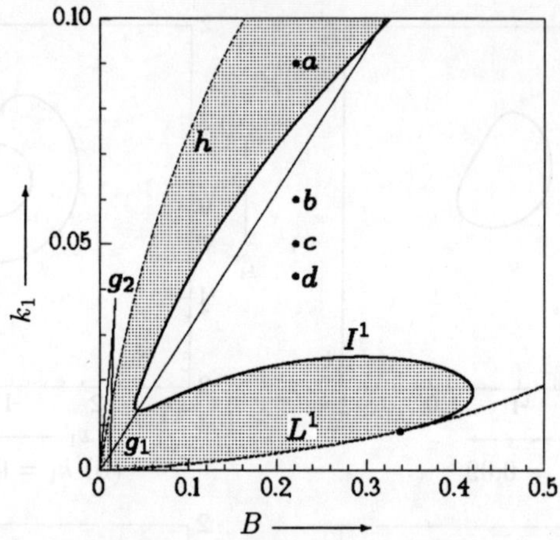

Figure 13: Bifurcation diagram of Eqs. (62).

Now we consider a global bifurcation, the generation or extinction of a limit cycle due to the appearance of a homoclinic structure. With decreasing k_1 through the tangent bifurcation curve g_1 in Fig. 13, a pair of stable and unstable equilibria appears. For certain values of the parameters, α- and ω-branches of the unstable equilibrium form a separatrix loop. The dotted curve L^1 in Fig. 13 indicates the parameter set on which the separatrix loop is obtained. The method to find bifurcation values is summarized in Appendix A. The parameter curve is also a bifurcation set of the generation or extinction of a limit cycle due to the coalescence of the limit cycle to the equilibrium. The appearance of the separatrix loop in the averaged equations suggests the existence of a homoclinic structure in the original equations (61). We see that there exists an intersecting point of the global bifurcation curve L^1 and period-doubling bifurcation curve I^1. An enlargement of the bifurcation diagram is shown in Fig. 15. In the figure, the symbol L^m indicates the global bifurcation set of an m-periodic limit cycle. The accumulation of the global and local bifurcation sets is an interesting phenomenon. Phase portraits at the parameter values indicated by circled points with symbols a, b and c in Fig. 15 are shown in Fig. 16.

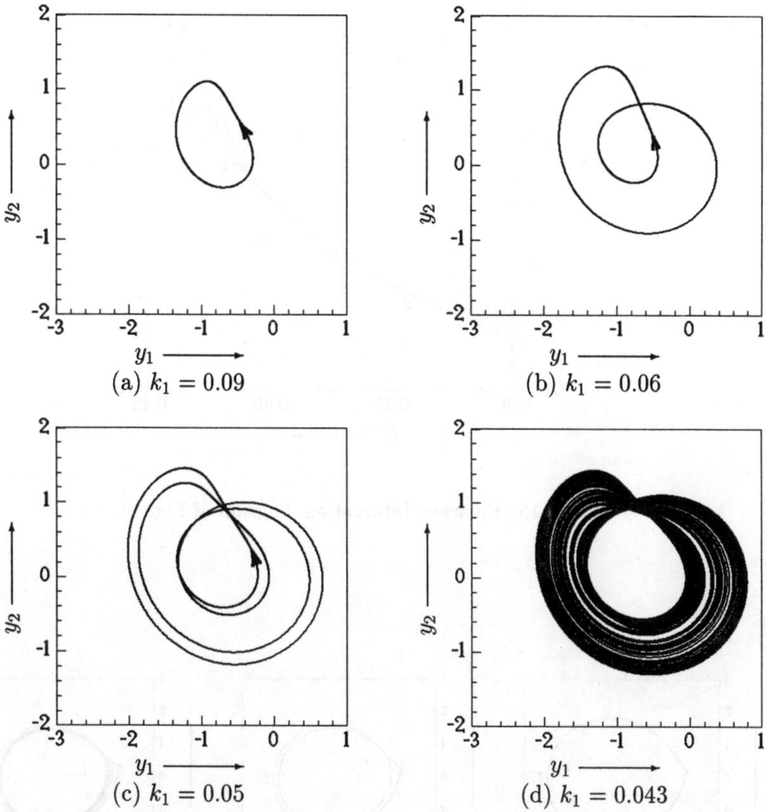

(a) $k_1 = 0.09$ (b) $k_1 = 0.06$ (c) $k_1 = 0.05$ (d) $k_1 = 0.043$

Figure 14: Phase portraits of limit cycles with (a) first, (b) second, and (c) forth order of periods, and (d) chaos observed in Eqs. (62) with $B = 0.22$. The arrow on each trajectory indicates the direction of trajectory.

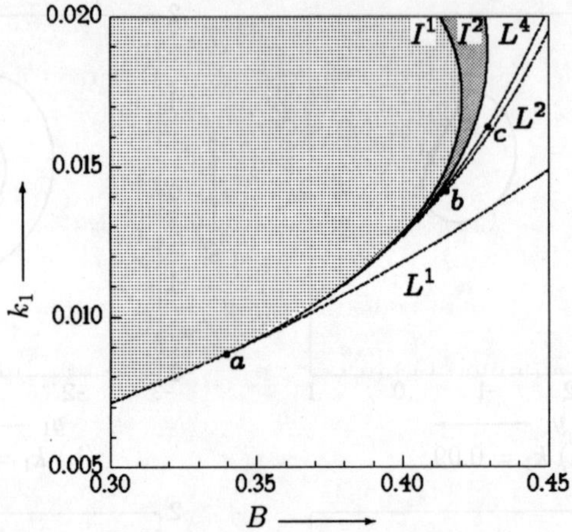

Figure 15: Enlarged bifurcation diagram of Fig. 13.

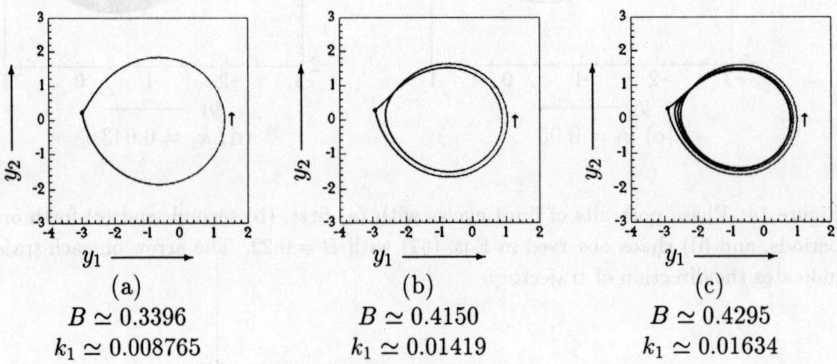

Figure 16: Phase portraits for separatrix loops observed in Eqs. (62) with parameter values denoted by circled points with symbols (a) a, (b) b and (c) c in Fig. 15. The circled point shows an equilibrium, and the arrow denotes the direction of the trajectory.

3.3 Phase Converter

Now let us consider the third example of Eqs. (32), or their first order form

$$\begin{aligned}
\frac{dx_1}{dt} &= x_2 \\
\frac{dx_2}{dt} &= -kx_2 - (c_0 + c_1)\, x_1 - \tfrac{1}{4}c_3 \left({x_1}^2 + 3{x_3}^2\right) x_1 \\
\frac{dx_3}{dt} &= x_4 \\
\frac{dx_4}{dt} &= -kx_4 - \left(\tfrac{1}{3}c_0 + c_1\right) x_3 - \tfrac{1}{4}c_3 \left({x_1}^2 + 3{x_3}^2\right) x_3 + B\cos t
\end{aligned} \tag{65}$$

with fixed system parameters

$$c_0 = c_1 = 0, \qquad c_3 = 1 \tag{66}$$

When we write the system of Eqs. (65) as $\frac{dx}{dt} = f(t,x)$, $x \in R^4$, we see that it has three types of symmetrical properties:

Prop. 1.

$$f(t, P_1 x) = P_1 f(t,x)$$

where

$$P_1 = \begin{bmatrix} -1 & 0 & 0 & 0 \\ 0 & -1 & 0 & 0 \\ 0 & 0 & 1 & 0 \\ 0 & 0 & 0 & 1 \end{bmatrix}$$

Prop. 2.

$$f(t+\pi, P_2 x) = P_2 f(t,x)$$

where

$$P_2 = \begin{bmatrix} -1 & 0 & 0 & 0 \\ 0 & -1 & 0 & 0 \\ 0 & 0 & -1 & 0 \\ 0 & 0 & 0 & -1 \end{bmatrix}$$

Prop. 3.

$$f(t+\pi, P_3 x) = P_3 f(t,x)$$

where

$$P_3 = \begin{bmatrix} 1 & 0 & 0 & 0 \\ 0 & 1 & 0 & 0 \\ 0 & 0 & -1 & 0 \\ 0 & 0 & 0 & -1 \end{bmatrix}$$

We call a solution $\varphi(t)$ as P_i*-symmetrical solution* if $\varphi(t) = P_i\varphi(t)$ holds for $i = 1, 2$ or 3. We note that the system of Eqs. (65) with $x_1 \equiv x_2 \equiv 0$ reduces to Duffing's equation of x_3 and x_4. Therefore there exist P_1-symmetrical solutions and their bifurcations. Figure 17 shows bifurcation diagrams of fixed points observed in the Poincaré map T_λ for Eqs. (65). In the diagrams, symbols D_1 and D_3 (resp. D_2 and D_4) show bifurcation sets of D-type of branchings of periodic points that correspond to the P_1-symmetrical (resp. P_2-symmetrical) solutions. If the parameter k decreases along the line ℓ in Fig. 17 (a), we observe the D-type of branching: ${}_0D \to {}_1I + 2\,{}_0D$ Due to the parameter variation, the stable P_1-symmetrical solution bifurcates to a pair of stable periodic solutions such that

$$-\psi_1(\pi) = x_1^0, \qquad -\psi_2(\pi) = x_2^0, \qquad \psi_3(0) = x_3^0, \qquad \psi_4(0) = x_4^0 \tag{67}$$

where $\psi(t) = (\psi_1(t), \psi_2(t), \psi_3(t), \psi_4(t))$ is the periodic solution passing through the fixed point $u = (x_1^0, x_2^0, x_3^0, x_4^0)$ at $t = 0$. Figure 18 shows an example of phase portrait of $\psi(t)$ and u at the parameter values denoted by the circled point a in Fig. 17 (a). The shaded region in Fig. 17 (a) shows a parameter area in which the solutions u and P_1u exist.

Figure 17 (b) shows the bifurcation diagram including the period-doubling bifurcation set I_1. There is an infinite doubling process of period-doubling bifurcations, and consequently we observe a chaotic state, see Fig. 19. Note that the period-doubling bifurcation corresponds to the parametric excitation phenomenon in the original circuit.

The circled point labeled by b in Fig. 17 (b) shows the TP-bifurcation set, which is an intersection of curves G_5 and I_2. Figure 20 illustrates an enlargement for the neighborhood of the TP-bifurcation parameter. In Fig. 20, the point a indicates another type of codimension two bifurcation, that is, T^2-bifurcation as an intersecting point of curves G^2 and H^2, which denote tangent and Hopf bifurcation sets of a 2-periodic point, respectively. The 2-periodic point is generated by the period-doubling bifurcation of a stable fixed point which corresponds to the P_1-symmetrical solution, or satisfies $P_1u = u$, where u is the state of the fixed point. For the 2-periodic point, say v, the condition $T_\lambda(v) = P_1v$ holds. The parameter regions in which stable fixed and 2-periodic points exist are, respectively, denoted by shading and in Fig. 20. We observe an *ICC* caused by the variation of system parameters through the Hopf bifurcation curve H^2 from inside to outside of the shaded region. Figure 21 (a) shows a (x_1, x_2)-plane phase portrait of the *ICC* of T_λ with parameter values denoted by the point c in Fig. 20. We see that two closed curves are symmetric with respect to the coordinate transformation of P_1. On the other hand, we observe a single closed curve of T_λ at the point d, see Fig. 21 (b). Therefore for certain value of parameters between the points c and d, the α- and ω-branches of the fixed point of type ${}_1I$ intersect each other and consequently form a homoclinic structure. The appearance of the homoclinic structure is also suggested by the observation of an attractor shown in Fig. 21 (c) and its enlarged diagram (d). When the value of B

continuously decreases along the line ℓ in Fig. 20, on the period-doubling bifurcation set I, the attractor suddenly disappears with the extinction of α-branch of the fixed point.

We finally mention that some kind of TP-bifurcation is related to a transition from a chaotic attractor with phase-drift-type to an almost-phase-locked chaotic attractor, see Ref. 8.

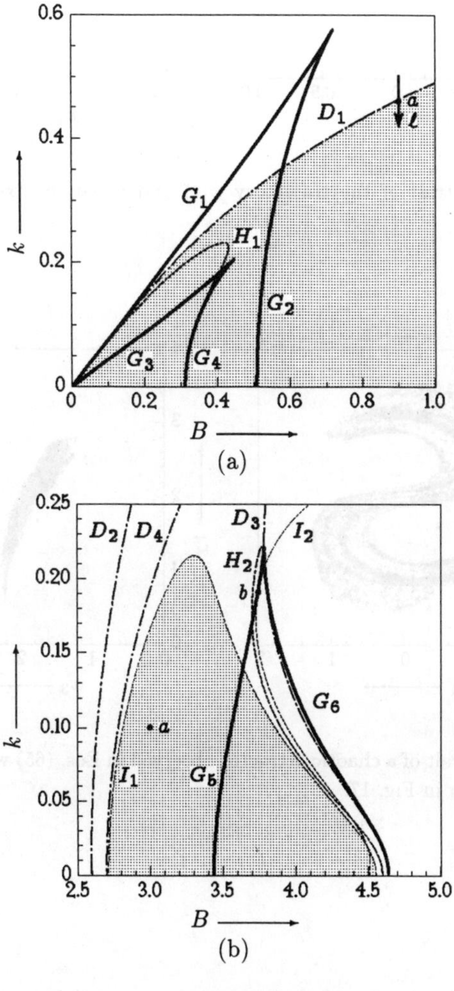

Figure 17: Bifurcation diagram of Eqs. (65).

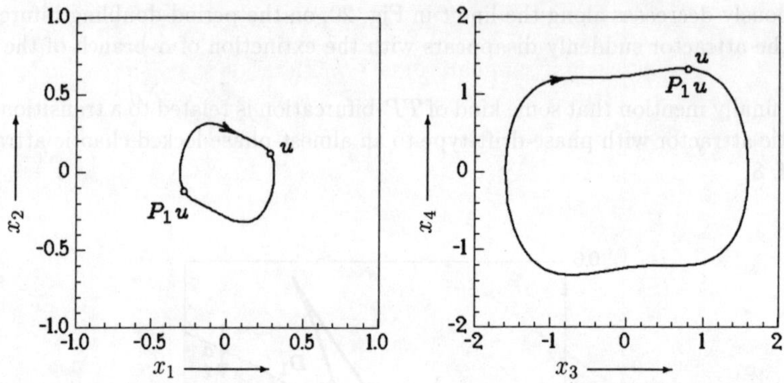

Figure 18: Phase portrait of the trajectory $\psi(t)$ and a pair of fixed points u and $P_1 u$ observed in Eqs. (65).

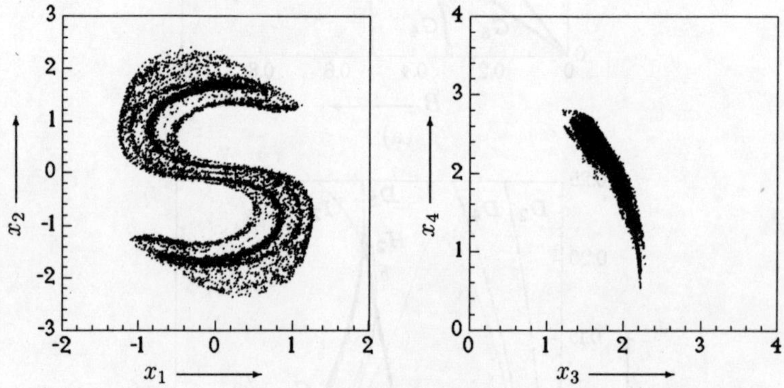

Figure 19: Phase portrait of a chaotic attractor observed in Eqs. (65) with parameter values denoted by the point a in Fig. 17.

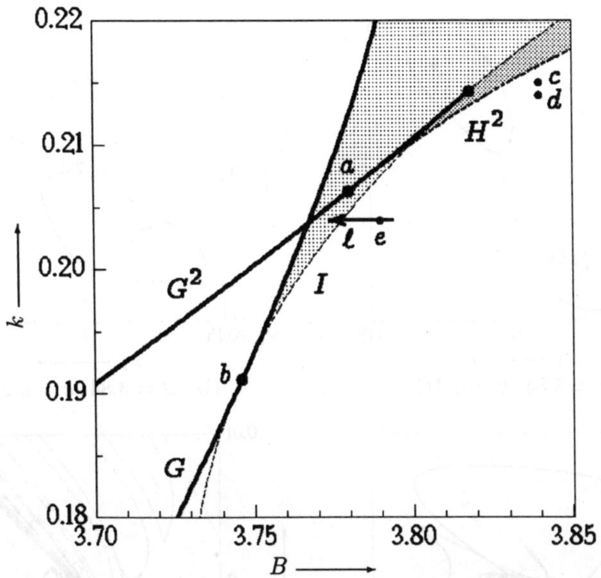

Figure 20: Enlarged bifurcation diagram of Fig. 17 (b). The bifurcation curves G and I correspond to G_5 and I_2 in Fig. 17 (b), respectively. A few bifurcation sets are omitted for the simplicity.

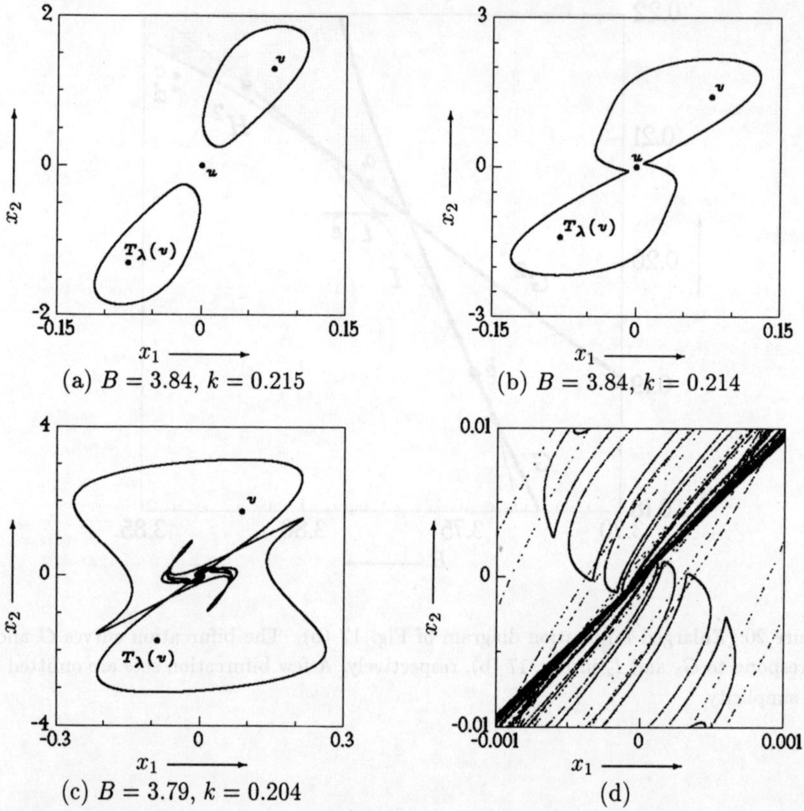

Figure 21: Phase portraits of T_λ for Eqs. (65) with parameter values denoted by circled points with symbols (a) c, (b) d and (c) e in Fig. 20. (d) Enlargement of (c). The points u and v are fixed and 2-periodic points of T_λ, respectively.

4 Remarks and Further Problems

Some nonlinear phenomena in dynamical systems that describe three electrical circuits have been investigated from the bifurcational viewpoint. For finding bifurcation values of parameters, we have shown the computational algorithm using a shooting method. Our method enables us to obtain directly bifurcation values from the original equations without special coordinate transformation. Hence we can easily trace out various bifurcation sets in an appropriate parameter plane.

Bifurcation of the invariant closed curve (*ICC*) of the Poincaré map, i.e., bifurcation of quasi-periodic solution of differential equation, is a global bifurcation and is very difficult for us to find directly bifurcation conditions. This kind of bifurcation appears naturally in our second and third circuit models, and is an interesting future problem to be considered.

References

1. C. Hayashi, *Nonlinear Oscillations in Physical Systems* (McGraw-Hill, 1964).
2. K. Bessho, "Theory and Performance of Single-Phase to Three-Phase Converter," *Trans. IEE of Japan*, **86**, 935 (1966) 1322–1331.
3. R. Abraham and J.E. Marsden, *Foundation of Mechanics* (Reading, MA: Benjamin/Cummings, 1974).
4. T.S. Parker and L.O. Chua, "Chaos: A Tutorial for Engineers," *Proc. IEEE*, **75** (1987) 982–1008.
5. H. Kawakami, "Bifurcation of Periodic Responses in Forced Dynamic Nonlinear Circuits: Computation of Bifurcation Values of the System Parameters," *IEEE Trans. on Circuits and Systs*, **CAS-31** (1984) 248–260.
6. K. Shiraiwa, "A Generalization of the Levinson-Massera's Equalities," *Nagoya Math. J.*, **67** (1977) 121–138.
7. N. Levinson, "Transformation Theory of Nonlinear Differential Equations of the Second Order," *Ann. Math.*, **45** (1944) 723–737.
8. H. Kawakami and T. Yoshinaga, "Codimension Two Bifurcation and its Computational Algorithm" in *Bifurcation and Chaos: Theory and Application*," ed. J. Awrejcewicz (Springer-Verlag, 1995).
9. T. Yoshinaga and H. Kawakami, "Codimension Two Bifurcation in Nonlinear Circuits with Periodically Forcing Term," *Trans. IEICE of Japan*, **J72-A**, 11 (1989) 1821–1828.

10. T. Yoshinaga and H. Kawakami, "A Property of Mean Value Defined on Period Doubling Cascade," in *Bifurcation Phenomena in Nonlinear Systems and Theory of Dynamical Systems*, ed. H. Kawakami (World Scientific, 1989) 183–195.

11. C. Hayashi and H. Kawakami, "Bifurcation and the Generation of Chaotic States in the Solutions of Nonlinear Differential Equations," *Fourth National Congress of Theoretical and Applied Mechanics* (1981) 1–6.

12. H. Kawakami and Y. Katsuta, "The Hopf Bifurcation and Chaotic States of Solutions in a Third Order Equation of Duffing's Type," *Trans. IECE of Japan*, **J64-A**, 11 (1981) 940–947.

Appendix

A Analysis of a Separatrix Loop

In this appendix, a numerical method is presented for calculating a separatrix loop or homoclinic bifurcation set of autonomous systems.

We consider an n-dimensional autonomous differential equation

$$\frac{dx}{dt} = f(x, \lambda), \qquad x \in R^n,\ \lambda \in R \tag{68}$$

Let f be sufficiently differentiable for all arguments. We also assume that the solution of Eq. (68) with initial condition $u := x(0)$, denoted by $x(t) = \varphi(t, u, \lambda)$ exists for all t. We consider an equilibrium of Eq. (68), say x_0, such that the eigenvalues of f at x_0 satisfy $\mu_1 > 0$ and $\mathrm{Re}(\mu_i) < 0$ for $i = 2, \cdots, n$. Then the α- and ω-branches with respect to the equilibrium x_0 are defined as

$$\alpha(x_0) = \left\{ x \in R^n \mid \lim_{t \to -\infty} x(t) = x_0 \right\} \quad \text{for } \alpha\text{-branch} \tag{69}$$

$$\omega(x_0) = \left\{ x \in R^n \mid \lim_{t \to \infty} x(t) = x_0 \right\} \quad \text{for } \omega\text{-branch} \tag{70}$$

Let

$$\ell := (\alpha(x_0) \cap \omega(x_0)) \cup x_0 \tag{71}$$

If $\ell \neq x_0$, then ℓ is called the separatrix loop.

When a limit cycle approaches to the equilibrium by varying λ, the α- and ω-branches are connected each other and then an orbit generates. The union of the orbit and the equilibrium becomes the separatrix loop. Hence conditions for the occurrence of separatrix loop is described as follows: there are points x_α and x_ω on the α- and ω-branchs, respectively, and T such that

$$\varphi(T, x_\alpha, \lambda_0) - \varphi(-T, x_\omega, \lambda_0) = 0 \tag{72}$$

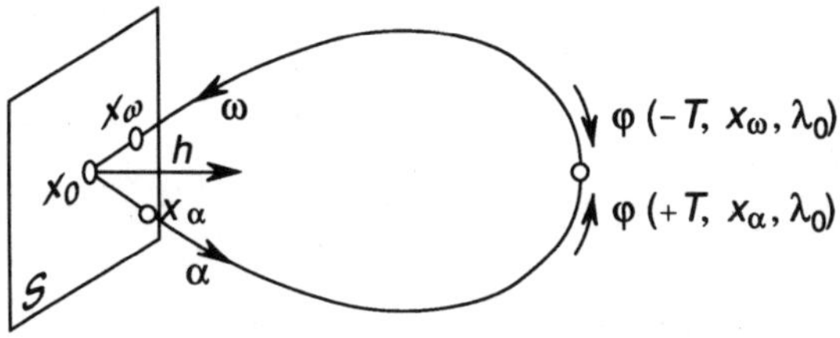

Figure 22: A separatrix of x_0 in R^3.

where λ_0 is the bifurcation value of the system parameter.

Figure 22 shows a schematic diagram for a separatrix of x_0 in R^3. In the figure, S is a plane that tangentially contacts to the stable manifold of the equilibrium x_0, and h is a vector that intersects orthogonally to S. Given $\delta_1 \ll 1$ and $\delta_2 \ll 1$, we have the following conditions:

$$f(x_0, \lambda_0) = 0 \tag{73}$$
$$[f_x(x_0, \lambda_0) - \mu_1 I]\,[x_\alpha - x_0] = 0 \tag{74}$$
$$[x_\alpha - x_0]^T\,[x_\alpha - x_0] - {\delta_1}^2 = 0 \tag{75}$$
$$\left[f_x^T(x_0, \lambda_0) - \mu_1^* I\right] h = 0 \tag{76}$$
$$h^T h - 1 = 0 \tag{77}$$
$$h^T\,[x_\omega - x_0] = 0 \tag{78}$$
$$[x_\omega - x_0]^T\,[x_\omega - x_0] - {\delta_2}^2 = 0 \tag{79}$$
$$\varphi(T, x_\alpha, \lambda_0) - \varphi(-T, x_\omega, \lambda_0) = 0 \tag{80}$$

where f_x is the Jacobian of f with respect to x at x_0, and $(\ )^T$ denotes the transposition of column vector or matrix.

Hence the problem is reduced to determine $(4n{+}4)$ variables $(x_0, x_\alpha, \mu_1, h, \mu_1^*, x_\omega, T, \lambda_0)$ for $(4n + 4)$ number of equations. If the Jacobian of the system equations is nonsingular, then we can apply Newton's method for solving Eqs. (73)–(80). In this case, the solution satisfying $\mu_1 = \mu_1^*$ gives the bifurcation set of the separatrix loop.

A BiCMOS Binary Hysteresis Chaos Generator

S. Ahmadi and R. W. Newcomb
Microsystems Laboratory
Electrical Engineering Department
University of Maryland
College Park, MD 20742 USA
email: newcomb@eng.umd.edu

Abstract:

A previous op-amp RC circuit which was proven to give chaotic signals is converted to a BiCMOS design more suitable to integrated circuit realization. The structure results from a degree two differential equation which includes binary hysteresis as its nonlinearity. The circuit is realized by differential (voltage to current) pairs feeding two capacitors, which carry the dynamics, with the key component being a (voltage to current) binary hysteresis circuit due to Linares.

I. Introduction

One of the fascinating application areas of chaos is in coding for secure communication in which chaos is synchronized between sender and receiver [1]. Hardware implementation of this coding rests upon having hardware that consists of accurately reproducible chaos generators [2][3]. And although our experience is that it is rather easy to modify oscillator and related circuits to obtain chaos it is more difficult to obtain good design formulations. Thus there has recently been an interest in circuits for generating chaos and this has led to a sequence of special issues of the IEEE devoted to the topic [4]. Much of the circuit results rest upon a study of the circuit due to Chua and first presented by Matsumoto [5]. However, there have been other philosophies, promising ones among which are those for which designs are based upon hysteresis [6][7][8]. A primary one among this class is our previous circuit using binary hysteresis [9][10]. This circuit has the advantages that it is only degree two, readily proven to give chaos, with its principles of operation conceptually easy to grasp. Previous realizations used operational amplifiers and could be considered voltage-mode. But for VLSI realization current-mode structures are usually more convenient, so here we give a different realization that is essentially current-mode, being dependent upon differential pair current outputs. Key to this is the design of the binary hysteresis with a current-mode device; fortunately for our designs such a device has recently been introduced by Linares [11].

In the following we review the principle of operation of the generic degree two binary hysteresis chaos generator, review the reason it yields chaos and then give circuit realizations in the form suitable for VLSI construction using the most recent fabrication technology of BiCMOS (Bipolar-Complementary Metal Oxide Silicon). For this latter we first introduce an ideal normalized realization of the describing equations and then denormalize to values that correspond to devices constructable by the standard MOSIS VLSI analog circuit BiCMOS fabrication process. The circuit diagrams and curves presented were all obtained using the Design Center CAD package associated with PSpice [12] and, consequently, adhere to its drawing conventions as well as to the standard SPICE nomenclature for labeling circuit components (in particular D represents a diode, F a bidirectional current mirror, G a voltage controlled current source; M and Q denote MOS and bipolar transistors, respectively).

II. The Basic Binary Hysteresis Chaos Generator

The basic philosophy of the binary hysteresis chaos generator can be seen by observing Fig. 1 where there are two half-planes that overlap and are separated by binary

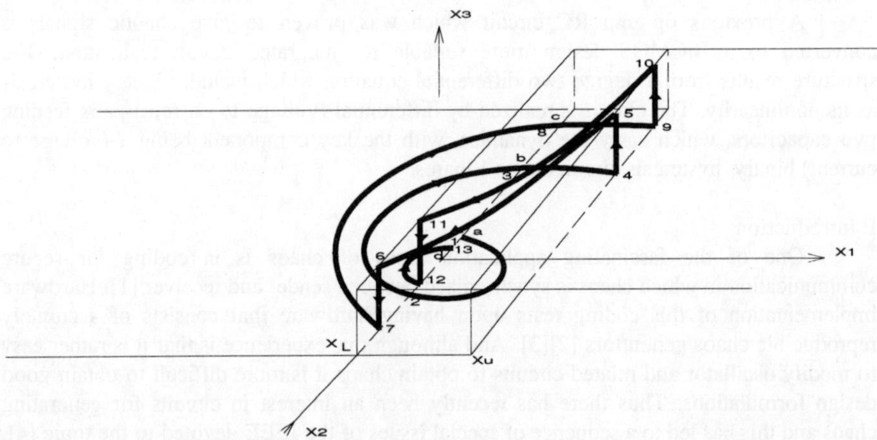

Figure 1

Trajectories in the Hysteresis Planes Illustrating a Set of Returns Satisfying d<a<b<c

hysteresis. In each plane are a set of spiral trajectories arising from a degree two unstable linear system. These two sets of trajectories are connected through the binary hysteresis to form the chaos generator's full set of trajectories. As an example of a tracing we can start at point 1, which is also labeled a, spiral in the lower plane through the points 2 and 3 (also labeled c) to 4 at which point a jump via the hysteresis is made to 5 and along an upper plane trajectory to 6 and then jump down to 7, spiral to 8=c to 9, jump up to 10, spiral to 11, jump down to 12 and on to 13=d, continuing on and on forever. A very important point is that the two sets of trajectories are, rather easily, designed to force $d<a<b<c$ so that there is a return on the upper half of the x_2 axis inside of the start point in such a way that a period three return map is obtained for the maximum of the x_2 axis projection of the trajectory. The system stems from an idea mentioned by Rössler [13] and was earlier proven to be chaotic [10, p. 330] by direct application of the basic Li-Yorke theorem on "period three implies chaos" [14]. An important practical, as well as theoretical, aspect of the system is that the same degree two subsystem can be used to

generate the spirals in the upper and the lower hysteresis half-planes with the binary hysteresis parameter serving to change the origin of the spirals in the upper half-plane from those in the lower. Thus, rather than using degree four, or even three, this system can be realized as a degree two system!

Although there are uncountably many choices of parameters which will yield chaos, in normalized form the following two differential equations have been rigorously proven to give chaos.

$$\frac{dx_1}{dt} = x_2 + a_1 h(x_1) \tag{1a}$$

$$\frac{dx_2}{dt} = -x_1 - 2\sigma x_2 + a_2 h(x_1) \tag{1b}$$

where the binary hysteresis is characterized by

$$h(x) = \begin{cases} 1 & \text{for } x_L < x \\ 0 & \text{for } x < x_U \end{cases} \tag{1c}$$

and the constants are chosen to be

$$x_L = 0, \quad x_U = 0.3, \quad \sigma = -0.2, \quad a_1 = -1, \quad a_2 = -1.34996673232527\ldots \tag{1d}$$

In Eqs. (1) x_1 and x_2 are the two dynamicel (state-like) variables and x_L and x_U the hysteresis jump points (to the lower and to the upper half lines, respectively). And, although Eq. (1c) indicates a double valued function, we consider true hysteresis which is single valued by requiring that h(x(t+)) retains the value of h(x(t)) unless x hits a jump point (practically in electronic circuits this is usually a consequence of the small parasitic capacitors [15]). The strange value of a_2 is chosen to give a continuous return map (the map of next crossings of the $x_2>0$ axis) in order that the Li-Yorke Theorem (which requires continuity) is satisfied; however continuity is not really necessary and the system is rather robust and the choice of $a_2=-1.35$ works well.

As was proven in [10] using the Li-Yorke Theorem, given appropriate initial conditions the system of Eqs. (1) will yield chaos.

III. Ideal DVCCS Realization

In this section we give a realization of Eqs. (1) using ideal components based upon voltage controlled current devices rather than the op-amps previously used. This will be done in such a manner as to conveniently allow denormalizations to practical VLSI circuits. Especially we will use the differential voltage controlled current source (DVCCS), using it to obtain all of the terms on the right hand side of Eqs. (1), including the hysteresis, while realizing the left had sides as currents in unit capacitors. The key subcircuit which makes this possible is that for the hysteresis for which we use the circuit presented by Linares in his dissertation.

Since the left hand sides of Eqs. (1a,b) are capacitor currents, the variables x_1 and x_2 are voltages. These are turned into currents for the right hand side of these equations, this being done by the use of a DVCCS for each nonhysteresis term and by a hysteresis DVCCS with a current mirror on the output to feed the same current, $h(x_1)$, with two different weights, a_1 and a_2, to the current summing nodes used to obtain Eqs. (1) by Kirchhoff's Current Law (KCL).

First we present the hysteresis circuit of Linares but as a preliminary, in Fig. 2(a), define the notation and symbolism for the ideal DVCCS, of SPICE symbol G, and in Fig. 2(b) the ideal current mirror. The output of the DVCCS is a current, i_o, whose value

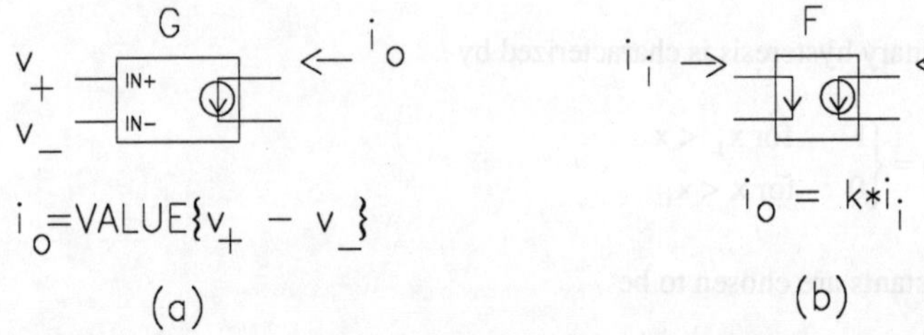

Figure 2
PSpice Symbols for Ideal Controlled Sources
(a) DVCCS, labeled G in SPICE (b) Current Mirror, labeled F in SPICE

depends upon the difference of the two input voltages, v_+ and v_- both measured with respect to ground. This output may be nonlinear (as needed here for hysteresis), in which case it is read into PSpice as a function through the VALUE{·} statement; if this is a linear function, $i_o=g_m(v_+-v_-)$, only the (mutual) transconductance g_m is read in. The ideal current mirror, of SPICE symbol F, is assumed linear with current gain k, $i_o=ki_i$, where i_o is the output and i_i the input currents.

With the DVCCS in hand we present in Fig. 3(a) the (slightly modified) basic ideal DVCCS binary hysteresis circuit of Linares [11, p, 76] with a printout of the binary hysteresis obtained in its DC input-output characteristic shown in Fig. 3(b). The operation of the circuit is as follows. The ideal diodes D1 and D2 conduct the output current, io, of G1 in i_o's positive and negative directions, respectively, to obtain the two branches of the hysteresis and necessitating that the GVALUE function of G1 takes on positive and negative values; let these latter be constant and designated I_o and $-I_o$, respectively. Then when i_o is postive D1 is forward biased and $v_+=X_U$ and $v_+-v>0$, that is $v<X_U$ while when io is negative then $v_+=X_L$ and $v_+-v<0$, that is $X_L<v$. Thus if $X_L<X_U$ we have a double valued response given by

$$i_o(v) = \begin{cases} I_o & v < X_U \\ -I_o & X_L < v \end{cases} \qquad (2)$$

In Fig. 3(a) this current is mirrored in the current mirrors, F1 and F2, to give the currents to be fed to the circuit, after a shift of the curve, by current sources I1 and I2, to bring

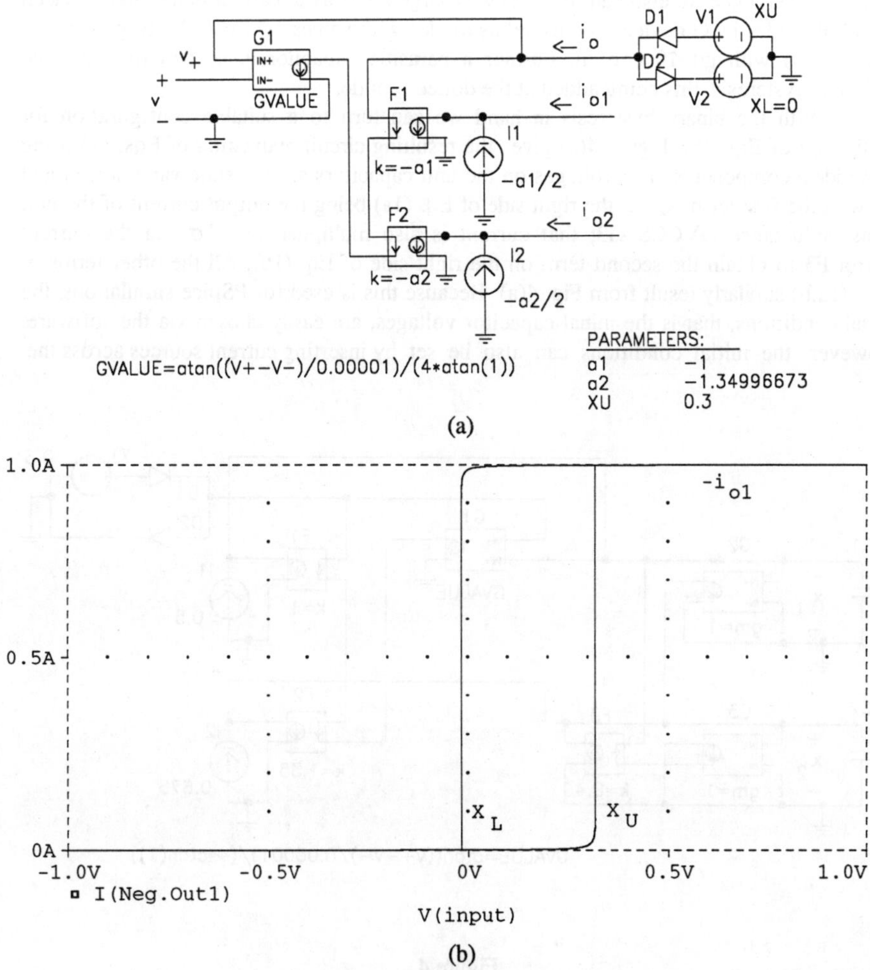

Figure 3
DVCCS Binary Hysteresis Following Linares
(a) Circuit Diagram (b) DC Characteristics

the lower branch to zero for $v<X_U$. For realization of G1 by a transistorized DVCCS we have chosen $I_o=\underline{I}_o$ and to obtain the (normalized) binary value we have chosen $I_o=1/2$ which yields the binary valued hysteresis after the shift by the current source I1 to give $i_{o1}=-h(v)$. A PSpice print out of $-i_{o1}$ is given in Fig. 3(b) where it is seen that the binary hysteresis has been well realized. To obtain Fig. 3(b) we used the approximation to a step function needed for G1 by way of the arctangent function, as indicated in Fig. 3(a); to obtain convergence we also had to insert very large resistors at certain nodes and replaced the ideal diodes by G devices connected as diodes with values set to VALUE=gd*((v_+-v_-) +abs(v_+-v_-)) with gd=10 chosen. Further a parasitic capacitor is needed to guarantee action as hysteresis, this being added at the dotted i_o node.

With the binary hysteresis in hand we can turn to a suitable configuration for realization of Eqs. (1). Figure 4(a) gives the resulting circuit realization of Eqs. (1) using these ideal components. The voltages on the unit capacitors are the state variables, x_1 and x_2, with the first term, x_2, on the right side of Eq. (1a) being the output current of the unit transconductance DVCCS G3; that current is also multiplied by -2σ via the current mirror F3 to obtain the second term on the right side of Eq. (1b). All the other terms of Eqs. (1a,b) similarly result from Fig. 4(a). Because this is used for PSpice simulations, the initial conditions, that is the initial capacitor voltages, are easily chosen via the software. However, the initial conditions can also be set by inserting current sources across the

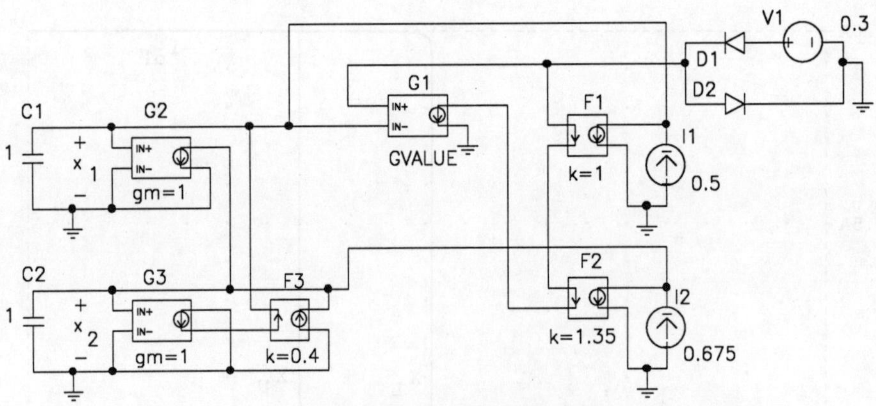

Figure 4
PSpice Ideal DVCCS Realization of Binary Hysteresis Chaos Generator
(a) Circuit Diagram

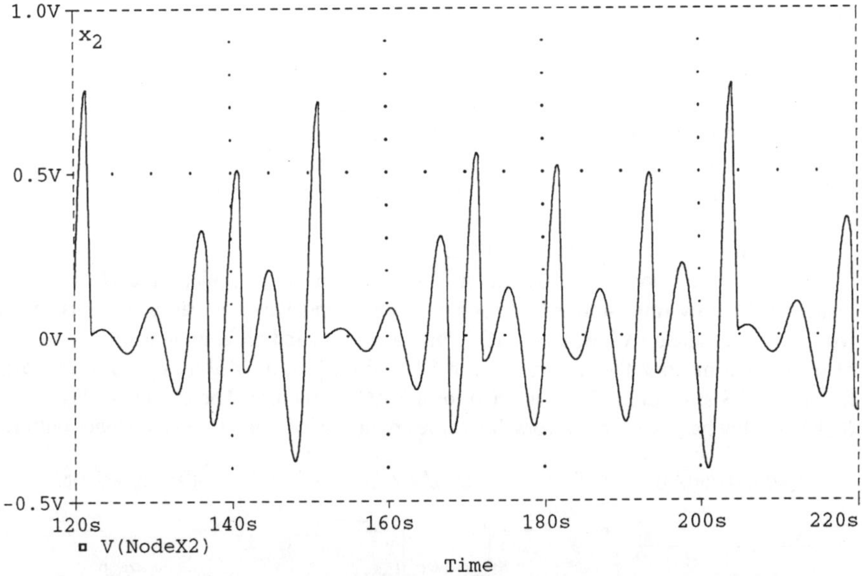

Figure 4 (b) Chaotic Response, Zero Initial Conditions

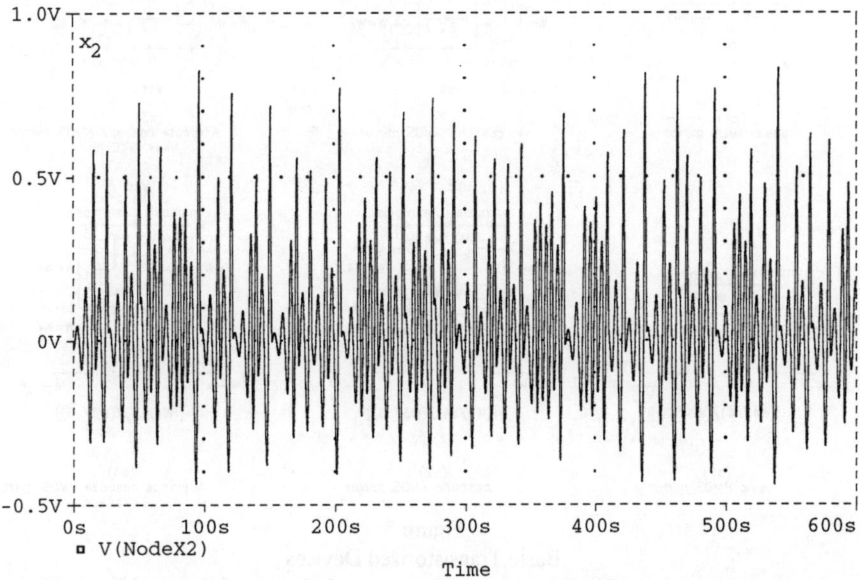

Figure 4 (c) Longer Time Span of Trajectory of (b)

capacitors with their currents being large but short pulses that drop to zero (to act as impulses); DVCCSs can also be used for realizing these initial condition current sources with these DVCCS inputs being pulses of voltage (which practically are easier to obtain than the pulses of current). However, in our case we have been able to achieve a design that uses zero initial conditions, by proper placement of the spirals in Fig. 1. In parts (b) & (c) of the figure are given a short time span and a long time span, respectively of a typical trajectory for x_1, this resulting from our choice of zero initial conditions.

IV. Practical DVCCS Realization for VLSI

Having the normalized circuit of Fig. 4 it remains to replace the ideal devices indicated there by transistorized components, something which is easily done using the work horses of analog VLSI, the CMOS current mirrors and differential pairs. In the case of the current mirrors, they are realized in VLSI form [16, p. 346] as shown in Fig. 5(a) for an NMOS structure and in Fig. 5(b) for a PMOS structure. The circuits of Fig. 5(a1) & (b1) are the basic current mirrors, but these are improved upon to obtain higher output

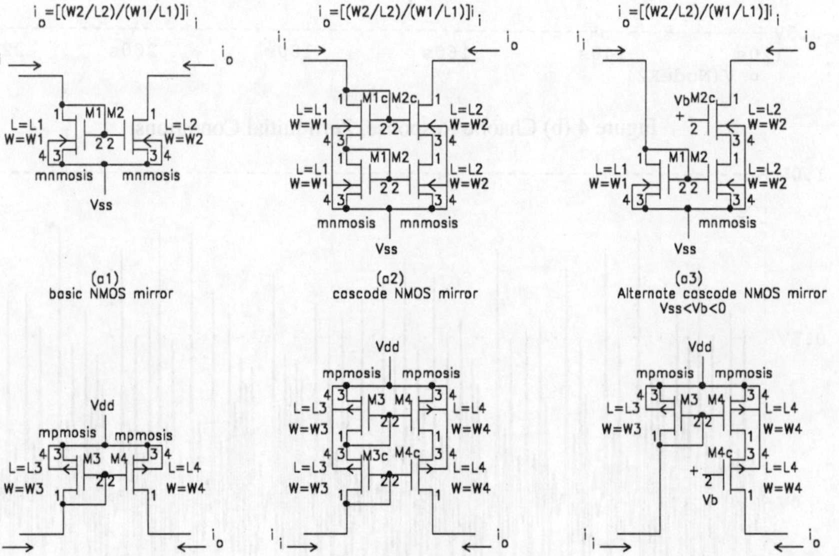

Figure 5
Basic Transistorized Devices
(a) NMOS & (b) PMOS Basic and Two Cascade Forms of Current Mirrors

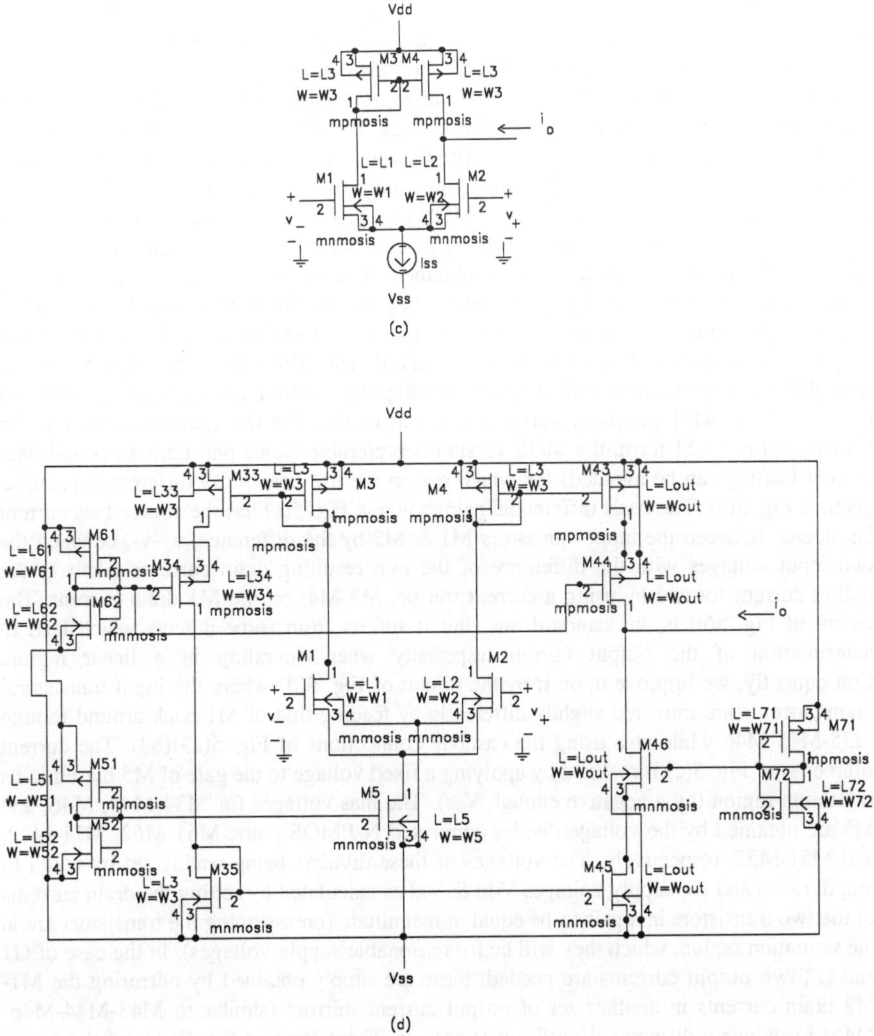

Figure 5
Basic Transistorized Devices
(c) Basic and (d) Practical Differential Pairs

impedance by use of the standard cascode connection of Fig. 5(a2) & (b2), this being important for the design of circuits sensitive to parameter changes, as is the case in chaos circuits. Figures 5(a3)(b3) show an alternate cascode connection [17, p. 404] which in our case is more effective since the output drive transistors, M2c & M4c, are held in saturation for almost all loads via the bias voltage Vb; although these are better for very sensitive structures, they do require the extra bias source. The current mirrors of Fig. 5 are actually uni-directional, the NMOS one giving a nonzero output current, i_o, only when the input current, i_i, is positive while the PMOS one yields only a nonzero output current when its input current is negative. But by placing the two types of current mirrors in parallel, a bidirectional current mirror is obtained. Both types of current mirrors work by transferring the voltage of the gate-source, V_{GS}, of the input transistor to the output transistor to force the drain currents of the two transistors to be equal (when $|V_{GS}|>|V_{threshold}|$ so the output transistor is turned on). The DVCCSs are readily constructed from differential pairs, often called differential amplifiers, with both MOS [16, p. 436] and bipolar [16, p. 446] transistor versions being available. For the chaos generator to be constructed in VLSI form, the MOS version is preferable (since pnp transistors and base current loading can be avoided) for which reason we use here MOS differential pairs, as given in Fig. 5(c). The basic differential pair shown in Fig. 5(c) has the source bias current Iss steered between the input transistors M1 & M2 by the difference, v+-v-, between the two input voltages with the difference of the two resulting drain currents taken as the output current formed by using a current mirror, M3-M4, on the M1 drain current. The circuit of Fig. 5(c) is the standard one, but it suffers from some defects which lead to deterioration of the output current especially when operating in a linear region. Consequently, we improve upon it by the circuit of Fig. 5(d) where the input transistors' drain currents are mirrored slightly differently by feeding that of M1 back around though M35-M45-M46 while also using the cascode connections of Fig. 5(a3)(b3). The current source Iss of Fig. 5(c) is realized by applying a fixed voltage to the gate of M5 biased in its saturation region (by a negative enough Vss). The bias voltages for M34-M44, M46, and M5 are obtained by the voltage divider connected N-PMOS pairs, M61-M62, M71-M72, and M51-M52, respectively. The voltages of these dividers being readily set by width to length ratios and the supply voltages Vdd & Vss as calculated by setting the drain currents of the two transistors in a pair to be equal in magnitude (on assuming the transistors are in the saturation region, which they will be for reasonable supply voltages). In the case of G1 and G2 two output currents are needed; these are simply obtained by mirroring the M1-M2 drain currents in another set of output current mirrors (similar to M43-M44-M46-M45) for which a different Wout/Lout is chosen. There remains the diodes of the binary hysteresis circuit. These can be realized as diode connected MOS transistors, in which the gates are tied to the drains. However, better diode characteristics result from junction diodes and, since our hysteresis needs sharp corners, we, consequently, construct the diodes from bipolar npn transistors, as are available in the BiCMOS process, by shorting bases to the collectors.

Before designing the fully transistorized circuit it is necessary to carry out a denormalization so that practical values of components can be used, in particular for gm of the differential pairs, capacitances, and hysteresis parameters. Since our anticipated fabrication is through the MOSIS facility we will use the MOSIS BiCMOS transistor parameters as given in the Appendix. For our designs, to be discussed below, we found by simulations of the circuits outlined above, a base gm=5.3 micro Amp/Volt, an upper hysteresis jump point of X_U=0.81 Volts and an hysteresis upper amplitude of H^+=15.7 micro Amp (along with a lower amplitude of 0).

To match these practical values obtainable by physical components we perform a denormalization of Eqs. (1) as follows. To adjust for the transconductance value we multiply Eqs. (1a) and (1b) by a constant K1. Then we adjust for the hysteresis amplitude by multiplying and dividing h(x1) by K2. To adjust the upper jump point of the hysteresis we scale x_1 to z_1 by setting $x_1=Xz_1$ and to keep similar labeling set $x_2=z_2$. Finally to adjust the capacitors, the time scale is next changed by naming $t=t_{old}$, replacing t_{old} by $t_{old}=t_{new}/C$ and renaming $t_{new}=t$ to get

$$CK_1X\frac{dz_1}{dt} = K_1z_2 + \frac{K_1}{K_2}K_2a_1h(Xz_1) \tag{3a}$$

$$CK_1\frac{dz_2}{dt} = -K_1Xz_1 - 2\sigma K_1z_2 + \frac{K_1}{K_2}K_2a_2h(Xz_1) \tag{3b}$$

Consequently, we choose K_1= gm=5.3 microA/V, $K_2=H^+$=15.7 microA, X=0.3/0.81=0.37 and $C=0.25\times10^{-3}$ to give capacitors C1=0.49 nanoFd and C2=1.325 nanoFd. Most width to length ratios are chosen to be unity for convenience, for which we used L=10 micron=W. But in the case of the linear DVCCS's we used a ratio of W/L=2/24 for their input transistors to achieve a larger region of linearity. Also the width to length ratios were appropriately adjusted in the case of the nonunity current ratios of F2 and F3.

Putting everything together Figure 6 results to show the total circuit. Although this looks rather formidable it is formed in a straightforward manner described above from Fig. 4(a). In Fig. 6 we have used a numbering consistent with the ideal circuit of Fig. 4(a) by using the numbers in Fig. 4(a) in the hundreds position of the transistors of Fig. 6. Also in Fig. 6 there are three ideal current sources to compensate for offset currents of the differential pairs; these can be realized in the same manner as Iss in Fig. 5(d) but are kept as current sources to prevent further clutter of the figure. Similarly there are two voltage sources to set the hysteresis jump points, which can be realized by voltage dividers similar to those setting the biases on the cascode current mirrors. In the figure we also include two copies of the supply sources, of values Vdd=+5 and Vss=-5, in order to avoid further crossing of lines, but only one of each is actually to be used.

The substitutions between the ideal devices of Fig. 4(a) and the transistorized circuit of Fig. 6 are catalogued in Table 1. Figure 7 shows the zero initial condition trajectory obtained from Fig. 6 with part (a) and (b) giving the time traces of the capacitor voltages and part (c) giving the phase plane plot of z_2 versus z_1. It is clearly seen when and where z_1 hits the hysteresis jump point to the upper half plane at 0.81 Volt. By capturing

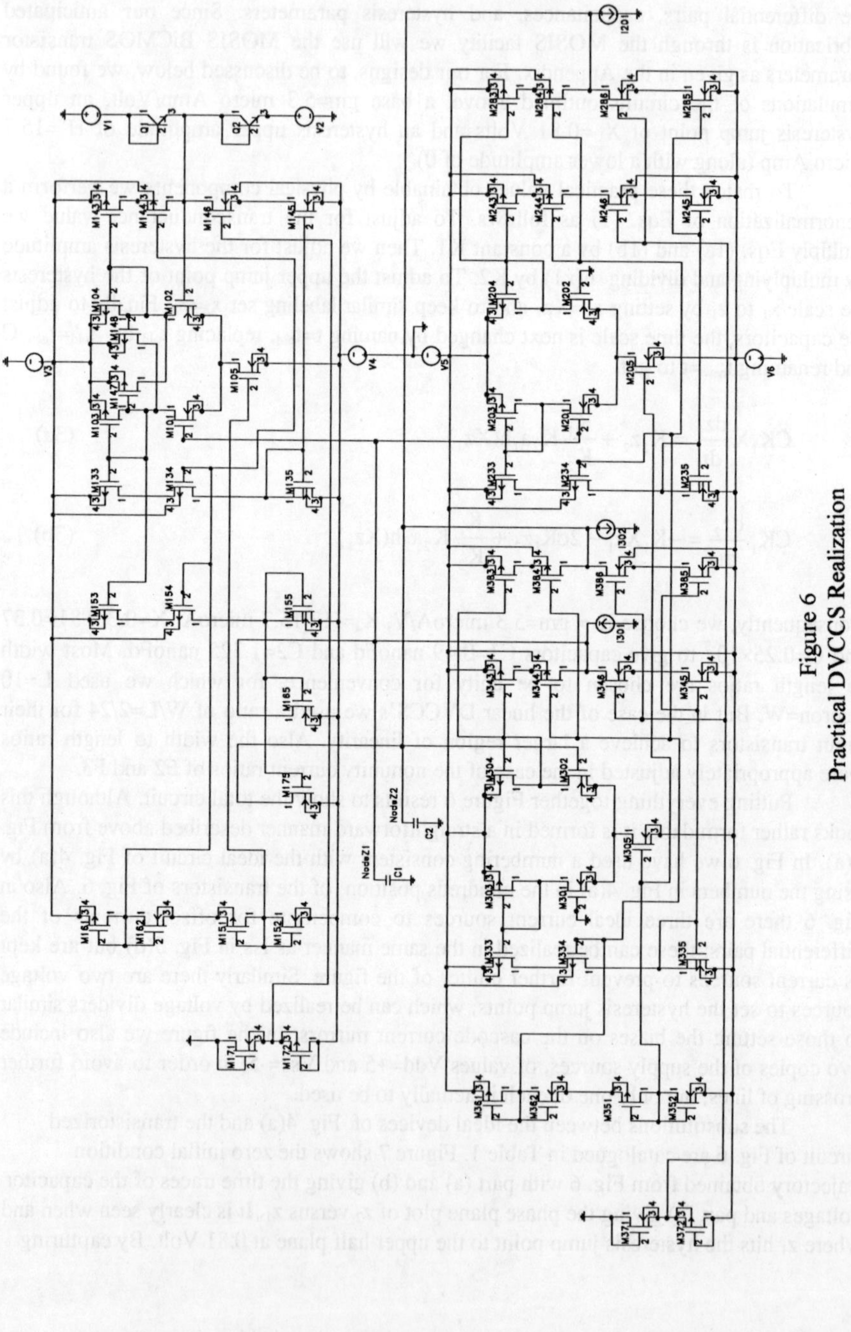

Figure 6
Practical DVCCS Realization

Table 1
Correspondences of Components of Figure 4(a) with Figure 6
all non listed L=10U, W=10U; x=0 through 9; U=10E-6

Fig. 4(a) Device	Fig. 6 Device	Purpose	Parameter	Comments
C1, C2	C1, C2	dynamics	C1=0.49n C2=1.325n	improvewith capacitor mult
D1-V1	Q1-V1	diode connected npn, X_U	V1=-0.6	Q1=diode connected npn
D2	Q2-V2	diode connected npn, X_L	V2=-0.2	Q2=diode connected npn V2 handles offsets
F1-I1	M165, 15x	for a_1 term	W165=3U	I1=0 due to mirror diode action
F2-I2	M175, 15x	for a_2 term	W175=4U	I2=0 due to mirror diode action
F3	M38x	for σ term	W383=4U W385=4U	
G1	M101, 102	input to differential pair for hysteresis		
	M103, 104 M13x, 14x	current mirrors		
	M16x & 17x	Vb	+2 & -2	
	M105, 15x	Iss of Diff. Pair	Iss gate vg=-3	
G2	M201, 202	input to linear differential pair	L=24U W=2U	
	M28x	current mirrors for z_1 term in (3b)	W283=3.7U W285=3.7U	
	M203, 204 M23x, 24x	current mirrors		
	M36x & 37x	Vb	+2 & -2	
	M205, 35x	Iss of Diff. Pair	Iss gate vg=-3	
	I201	offset current src		1 output
G3	M301, 302	input to linear differential pair	L=24U W=2U	
	M303, 304 33x, 34x	current mirrors M34x for z_2 term in (3a)		
	M36x & 37x	Vb	+2 & -2	
	M305, 35x	Iss of Diff. Pair	Iss gate Vg=-3	
	I301, 302	offset current srcs	I301=0.4U I302=0.6U	2 output currents
Vdd Vss	V3, V4, V5, V6	V3=V5=Vdd V4=V6=Vss	Vdd=+5 Vss=-5	Power Supply

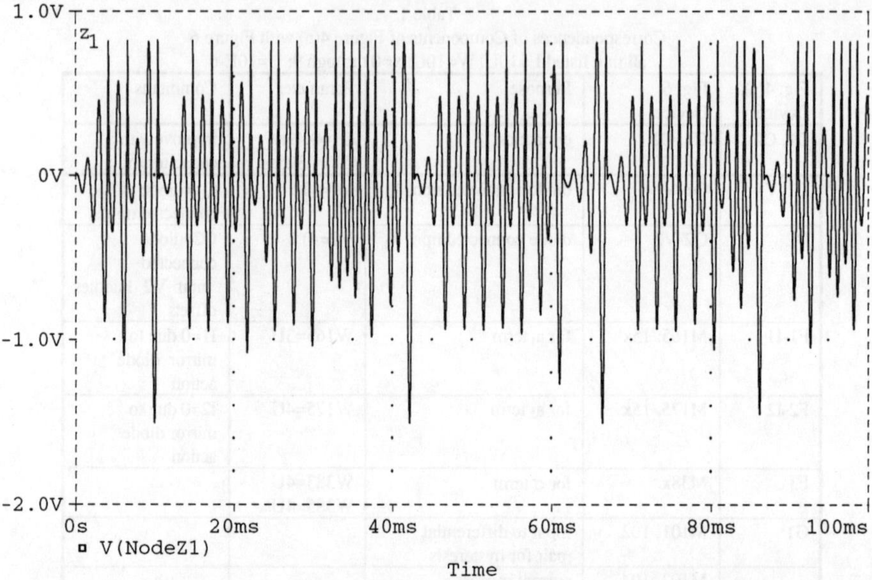

Figure 7 (a) Zero Initial Condition Responses of Practical Circuit, z_1 versus time

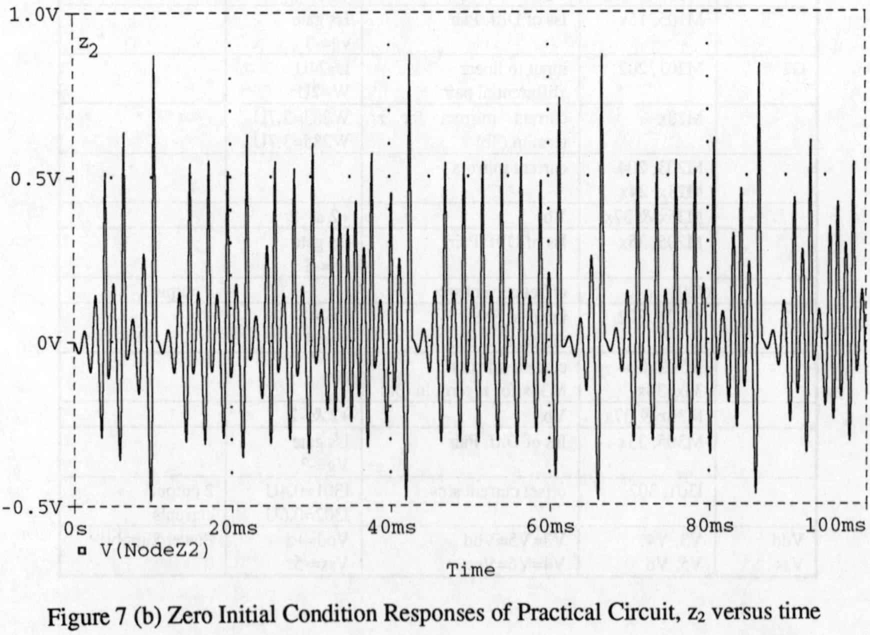

Figure 7 (b) Zero Initial Condition Responses of Practical Circuit, z_2 versus time

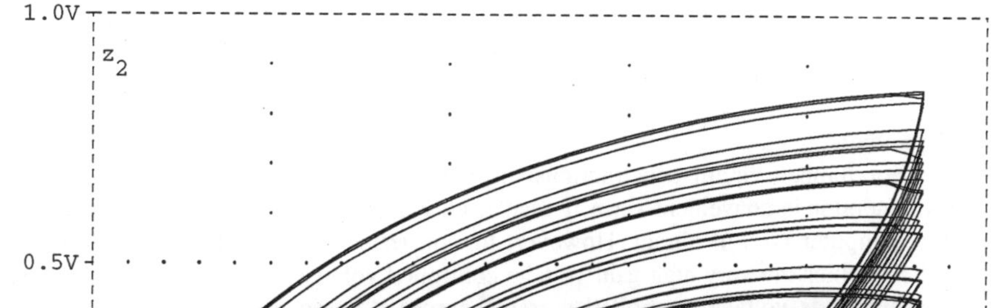

Figure 7 (c) Zero Initial Condition Responses of Practical Circuit, z_2 versus z_1 Plot

the times when these jumps occur one may be able to obtain some useful properties of this type of chaos.

The nature of responses of the circuit are quite dependent upon initial conditions. Although we have designed the circuit such that it gives a chaotic response with zero initial conditions, it may be desirable to allow for nonzero initial conditions. Such is easily accomplished by feeding the capacitors with a pulse of current which effectively acts as an impulse. Thus, by placing an MOS transistor's drain-source across a capacitor and feeding the gate-source with a short voltage pulse of appropriately large amplitude, desired nonzero initial conditions can be set.

On observation of the circuit of Fig. 6 it is seen that many components are present. Overall, however, the circuit presents no problem for integrated circuit fabrication. But the capacitors are larger than we would like and we feel that some optimization could be done to cut down on the number of transistors. To handle the capacitors, the sizes of which are needed to swamp parasitics, we would like to incorporate capacitor multipliers, with suitable ones possibly being available in the literature [18]. As for the number of transistors, by optimal choices of W/L ratios, it appears that the current mirrors of Fig. 5(a1)(b1) or (a2)(b2) could be used. In any event the circuit of Fig. 6 is a working circuit suitable for small chip fabrication, though we do feel with time it can be improved upon.

V. Discussion

For lumped circuit constructions the previous realization of a binary hysteresis chaos generator is convenient since it used readily available operational amplifiers in an analog computer type of circuit. However, for integrated circuit realizations operational amplifiers, which are very high gain voltage controlled voltage sources, are nowhere near as convenient as voltage controlled current sources. Consequently, we have presented here a redesign which is ready to be put into layout for VLSI fabrication.

The binary hysteresis chaos generator is particularly appealing since it is only of degree two and uses hysteresis which is readily made using standard available transistors. It should be noted that we are able to get away with a degree two system since the describing differential equations contain discontinuous functions. Because the electronic realization of the hysteresis relies upon capacitive parasitics it might be fair to believe that really this is a degree three system, the parasitic capacitor giving the third derivative. However, this parasitic can be vanishingly small so in practice we ignore it in the design. Further, if one were to use other means of realizing the binary hysteresis, for example by magnetic amplifier cores, then one need not rely on the parasitics to insure operation; all that is needed is to obtain the operation represented by Fig. 1 and Eqs. (1).

Chaos generating circuits are presently becoming a dime a dozen since by introducing appropriate return feedback almost any degree two or higher circuit can be made to create chaotic signals. However, obtaining desired characteristics of the chaos is another story since, as yet, there is not available a satisfactory design theory which takes into account the nature of the chaos. But it does seem that the binary hysteresis chaos generator is something like the basic element from which arbitrary chaos can be built by concatenations and other combinations, just as arbitrary vectors are constructed from basis vectors.

Finally we comment that the speed of hardware particularly shows up in working with chaos generators. Whereas it takes hours to obtain results with software the same is obtained in milliseconds with hardware.

Acknowledgments:

The authors would like to acknowledge the encouragement of Drs. T. Carroll and L. Pecora, without which this paper would not have been written, as well as the interest of students and colleagues from high school through post doctoral levels.

References:

[1]. W. L. Ditto and L. M. Pecora, "Mastering Chaos," Scientific American, Vol. 269, No. 2, August 1993, pp. 78 - 84.

[2]. T. L. Carroll and L. M. Pecora, "Synchronizing Chaotic Circuits," IEEE Transactions on Circuit and Systems, Vol. 38, No. 4, April, 1991, pp. 453 - 456.

[3]. J. Neff and T. L. Carroll, "Circuits that Get Chaos in Sync," Scientific American, Vol. 269, No. 2, August 1993, pp. 120 - 122.

[4]. L. O. Chua and M. Hasler, Editors, "Special Issue on Chaos in Nonlinear Electronic Circuits," Parts A, B, C, IEEE Transactions on Circuits and Systems, Vol. 40, No. 10, I and II, October 1993 and Vol. 40, No. 11, November 1993.

[5]. T. Matsumoto, "A Chaotic Attractor from Chua's Circuit," IEEE Transactions on Circuits and Systems, Vol. CAS-31, No. 12, December 1984, pp. 1055 - 1058.

[6]. R. Newcomb and S. Sathyan, "An RC-Op-Amp Chaos Generator," IEEE Transactions on Circuits and Systems, Vol. CAS-30, No. 1, January 1983, pp. 54 - 56.

[7]. T. Saito, "An Approach toward Higher Dimensional Hysteresis Chaos Generators," IEEE Transactions on Circuits and Systems, Vol. CAS-37, No. 3, March 1990, pp. 399 - 409.

[8]. T. Saito, "Reality of Chaos In Four Dimensional Hysteretic Circuits," IEEE Transactions on Circuits and Systems, Vol. CAS-38, No. 12, December 1991, pp. 1517 - 1524.

[9]. R. W. Newcomb and N. El-Leithy, "A Binary Hysteresis Chaos Generator," Proceedings of the 1984 IEEE International Symposium on Circuits and Systems, Montreal, May 1984, pp. 856 - 859.

[10]. R. W. Newcomb and N. El-Leithy, "Chaos Generation Using Binary Hysteresis," Circuits Systems Signal Processing Journal, Vol. 5, No. 3, 1986, pp. 321 - 341.

[11]. B. Linares-Barranco, "Analog Neural Network VLSI Implementations," doctoral dissertation, Texas A&M University, December 1991.

[12]. PSpice and Design Center, Version 6.0, MicroSim Corporation, Irvine, CA, January 1994.

[13]. O. E. Rössler, "Continuous Chaos - Four Prototype Equations," Annals of the New York Academy of Science, Vol. 316, February 28, 1979, pp. 369 - 392.

[14]. T.-Y. Li and J. A. Yorke, "Period Three Implies Chaos," The American Mathematical Monthly, Vol., 82, No. 10, December 1975, pp. 985 - 992.

[15]. R. W. Newcomb, "Semistate Design Theory, Binary and Swept Hysteresis," Circuits, Systems, Signal Processing, Vol. 1, N. 2, 1982, pp. 203 - 216.

[16]. R. L. Geiger, P. E. Allen, and N. R. Strader, "VLSI Design Techniques for Analog and Digital Circuits," McGraw-Hill Publishing Co., New York, NY, 1990.

[17]. P. E. Allen and D. R. Holberg, "CMOS Analog Circuit Design," Holt, Rinehart and Winston, New York, 1987.

[18]. I. A. Khan and M. T. Ahmed, "OTA-Based Integrable Voltage/Current-Controlled Ideal C-Multiplier," Electronics Letters, Vol. 22, No. 7, March 27, 1986, ppl 365 - 366.

Appendix - SPICE Parameters for MOSIS Analog, BiCMOS, VLSI Fabrication

These are the MOSIS run parameters for run N21H SPICE LEVEL 2 parameters recorded 04/28/93 which were used for the simulations in order to design the circuit for VLSI fabrication.

A. The N Channel MOS Transistor Model

```
.MODEL MNMOSIS NMOS LEVEL=2 LD=0.250000U TOX=418.000008E-10
+ NSUB=9.236187E+14 VTO=0.858153 KP=5.048000E-05 GAMMA=0.198
+ PHI=0.6 UO=596.729 UEXP=7.029586E-02 UCRIT=10266.7
+ DELTA=2.7371 VMAX=65701.4 XJ=0.250000U LAMBDA=1.843384E-02
+ NFS=1.086360E+12 NEFF=1 NSS=1.000000E+10 TPG=1.000000
+ RSH=28.760000 CGDO=3.097916E-10 CGSO=3.097916E-10 CGBO=3.838441E-10
+ CJ=8.997900E-05 MJ=0.783638 CJSW=5.524800E-10 MJSW=0.285064 PB=0.800000
* Weff = Wdrawn - Delta_W
* The suggested Delta_W is -0.24 um
```

B. The P Channel MOS Transistor Model

```
.MODEL MPMOSIS PMOS LEVEL=2 LD=0.250000U TOX=418.000008E-10
+ NSUB=9.309300E+15 VTO=-0.889271 KP=1.908000E-05 GAMMA=0.6289
+ PHI=0.6 UO=216.28 UEXP=0.218144 UCRIT=62664
+ DELTA=0.164572 VMAX=100000 XJ=0.250000U LAMBDA=5.011626E-02
+ NFS=9.266623E+11 NEFF=1.001 NSS=1.000000E+10 TPG=-1.000000
+ RSH=66.820000 CGDO=3.097916E-10 CGSO=3.097916E-10 CGBO=3.727276E-10
+ CJ=2.981300E-04 MJ=0.556944 CJSW=3.002100E-10 MJSW=0.243045 PB=0.800000
* Weff = Wdrawn - Delta_W
* The suggested Delta_W is -0.17 um
```

C. The NPN Bipolar Transistor Model

```
.MODEL BN2X1 NPN
+ BF=82 IS=1.588E-16 NF=9.9563E-01 NE=1.3356 VAF=57.1
+ IKF=2.3067E-02 ISE=1.267E-16 RE=12.7 RC=420.00 RB=1.213E+03
+ RBM=7.53 ISC=3.363E-15 NC=1.0202
+ CJE=0.1952E-12 MJE=0.5050 VJE=0.85 CJC=0.18815E-12 MJC=0.4990 VJC=0.80
+ CJS=0.2326E-12 MJS=0.2033 VJS=0.70
* AREA OF TRANSISTOR
* AE=128 um^2  PE=64 um  AB=644 um^2 PB=102 um AC=2064 um^2 PC=182 um
```

EXPERIMENTING WITH CHAOS IN ELECTRONIC CIRCUITS

N. F. RULKOV
Institute for Nonlinear Science, UCSD,
9500 Gilman Drive, La Jolla, CA 92093-0402, USA

and

A. R. VOLKOVSKII
Radio Physics Department, University of Nizhny Novgorod,
23 Gagarin Av., Nizhny Novgorod 603600, Russia

Abstract

The results of investigations of nonlinear behaviors in electronic circuits with chaotic dynamics are presented. Bifurcations associated with transitions from regular to chaotic oscillations and the phenomena of chaos synchronization are analyzed via experiments on circuits. Chaotic circuits that generate smooth oscillations and chaotic pulse oscillators are considered. Applications of these nonlinear phenomena to communication and control of chaos are discussed.

1. Introduction

The behavior of nonlinear electronic circuit is one of the most attractive subject for applications in the theory of nonlinear oscillations.[1-5] Electronic circuits can generate periodic, quasiperiodic and even noise like (chaotic) oscillations. Complicated regular and chaotic oscillations of nonlinear circuits were encounted by physicists and engineers many years before a theory explaining the chaotic behavior of dynamical systems was available. Note, that complicated periodic oscillations and "irregular noise" generated by periodically forced oscillator with a neon glow lamp were observed in 1927 by B. van der Pol and J. van der Mark.[6] There are many nonlinear circuits which are known to generate chaotic oscillations. For example, chaotic behavior has been observed in a pairs of coupled periodic relaxation oscillators,[7, 8] in self-oscillators with delay in feedback,[9] in van der Pol oscillators with an additional tunnel diode,[10] in periodically driven nonlinear oscillators,[11, 12] in forced negative-resistance oscillator,[13] in a modified Van der Pol oscillator[14] and many other nonlinear electronic circuits. Chaotic oscillations may be generated in different ranges of frequencies. Examples of chaotic generators operating in low-frequency ranges have been considered in many papers.[15-19] There are also known oscillators which generate chaotic microwaves (see for example Reference 20). We believe that an electronic circuit generating chaos may be obtained by making minor changes to any circuit exhibiting periodic oscillations. The ability of electronic circuits to produce chaotic oscillations may be widely employed for different engineering applications. Nonlinear circuits with complex dynamics may also be convenient systems for modelling and studies of nonlinear phenomena.

In order to understand the nonlinear dynamics of an electronic circuit analysis of its behavior as a function of parameters of the circuit should be performed in *a phase space.* Phase space is a space of physical variables of the circuit. The dimension of the phase space may be determined as the minimum number of the variables required to define the time evolution of any portion of the circuit. To describe the nonlinear behavior of electronic circuits we will need to introduce a few definitions[a]. The phase space and the law of motion in this phase space define *a dynamical system.* A trace of the time evolution in the phase space with a well defined direction of motion is called *a trajectory* of the system. The velocity of the motion along the trajectories measured at any point of the phase space defines *the vector field of the phase velocity* of the system. A qualitative transformation of the vector field associated with a parameter variation is called *bifurcation* of the system. The phase space of a dynamical system may have a subspace of lower dimension containing a set of trajectories which never leave the subspace. This set of trajectories forms *an integral manifold.*

The post transient time image of the behavior of the system in the phase space is *an attractor.* If the system produces damped oscillations to a stationary state, than the attractor is a stable fixed point. Note, that the phase space may simultaneously contain several attractors. In this case the system exists in a multistable regime, where the final state of the system depends on its initial conditions. The region of initial conditions whose trajectories go to the same attractor is called *basin of the attractor.* Periodic oscillations of a dynamical system correspond to a stable *periodic orbit.* The image of quasiperiodic motion is a stable *torus* formed by an unclosed trajectory. We will use the term *chaotic attractor* to describe the image of chaotic oscillations measured from an experiment. For dynamical systems without noise, chaotic attractors contain an infinite number of unstable periodic orbits and nontrivial sets of unstable unclosed trajectories. The birth of these complicated images in the phase space is connected with the appearance of *homoclinic structure.* Homoclinic structures may arise as the result of intersections stable and unstable manifolds of an unstable periodic orbit.[b] Chaotic attractors may also contain stable periodic orbits if these orbits have very long periods and very narrow basins of attractions. In this case any small noise in the physical system will kick the trajectory of the system out of the basins of attraction of these stable orbits and the system will again produce chaotic behavior. When varying the parameters of a circuit any transition from regular behavior to chaotic behavior is the result of a bifurcation.

---

[a]Some of these definitions are not rigorous from a mathematical point of view, but may be sufficient for the purposes of our paper where these definitions are used to describe experimental observations.

[b]The stable manifold is formed by trajectories which approach the unstable periodic orbit with time. The unstable manifold is formed by trajectories originating from the periodic orbit and diverging from it due to its instability.

One of the goals of an experimental study of the nonlinear dynamics of a system with chaotic behavior is to find bifurcations which can explain transitions between different types of motion. In order to achieve this goal, one needs to employ experimental tools which allow one to study these bifurcations. The experimental tools can vary considerably depending upon the type of bifurcation and time characteristics of the circuits oscillations. In many cases, when the dimension of the chaotic attractor is not high, the experimental analysis may be based upon the study of one-dimensional maps as a function of the parameters of the system.

In this paper we will consider examples where experimental and theoretical analyses of bifurcations enable one to study nonlinear phenomena produced by electronic circuits with chaotic behavior. In Section 2 we present the results of an investigation of an electronic circuit which contains a nonlinear amplifier and linear feedback. This circuit is a convenient example for studding chaotic dynamics in experiments because it is capable of generating chaotic oscillations over a broad range of parameters. Using this circuit we will demonstrate different types of bifurcations which are responsible for the appearance of chaos in the circuit when it is operating in both in autonomous and non-autonomous regimes. In Section 3 we discuss the results of experimental analysis of the bifurcations which may provide the transitions to synchronous chaotic behaviors in two coupled circuits.

Experiments using an electronic circuits which generate pulses with chaotic intervals of time are presented in Section 4. We demonstrate a method for control of chaos in chaotic pulse oscillators with external periodic pulses and discuss the dynamical features of this control. Forced synchronization with chaotic driving pulses and mutual synchronization of pulse-coupled circuit are also discussed.

2. Dynamics of a circuit with chaotic behavior

In this Section we consider nonlinear phenomena demonstrated by electronic circuits generating smooth chaotic waveforms. The block diagram of the circuit is shown in Fig.1a. The chaotic circuit consists of a nonlinear converter[c] which transforms input voltage $x(t)$ into the output $\alpha f(x(t))$.

The output is then applied to the input of the nonlinear converter N through the low-pass filter RC' and the resonant circuit rLC shown in Fig.1a. The dynamics of the circuit is described by the following system of differential equations[16, 19]

$$\begin{aligned} \dot{x} &= y \\ \dot{y} &= -x - \delta y + z \\ \dot{z} &= \gamma[\alpha f(x) - z] - \sigma y. \end{aligned} \tag{1}$$

$x(t)$ is the voltage across the capacitor C, $y(t) = \sqrt{\frac{L}{C}}i(t)$ with $i(t)$ the current through the inductor L. $z(t)$ is the voltage across the capacitor C'. Time has been scaled by

[c]Details of this circuit are found in the Reference 21.

$\frac{1}{\sqrt{LC}}$. The parameters of this system have the following dependence on the physical values of the circuit elements $\gamma = \frac{\sqrt{LC}}{RC'}$, $\delta = r\sqrt{\frac{C}{L}}$, $\sigma = \frac{C}{C'}$. The shape of the

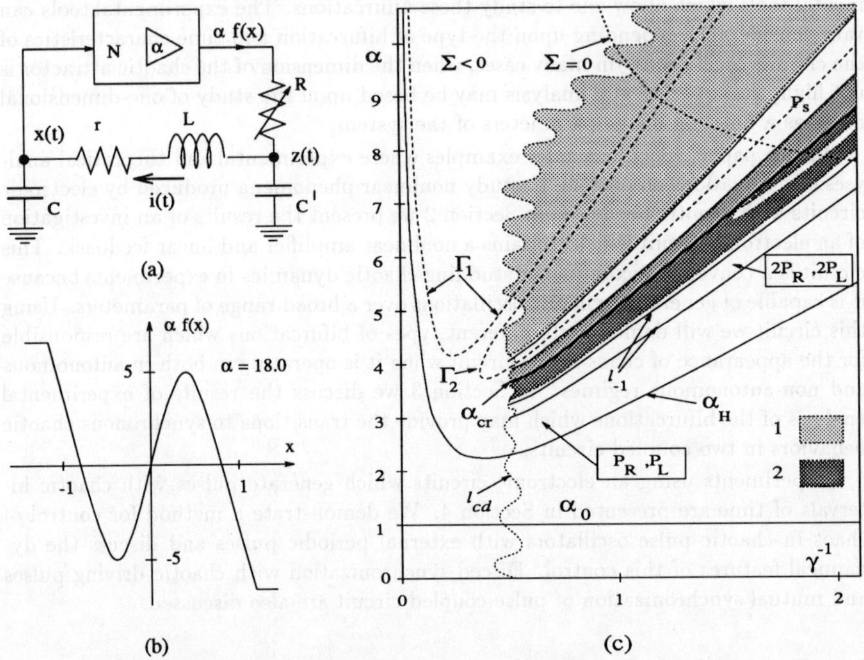

Figure 1: (a) Diagram of electronic circuit. (b) Function of transformation produced by nonlinear convertor N. (c) Bifurcation curves in the plane (γ^{-1}, α) obtained from experiment with the electronic circuit.

nonlinearity produced by the nonlinear converter, N, is shown in Fig.1b. In our theoretical analysis and numerical simulations we approximate $f(x)$ by the following function

$$f(x) \equiv \begin{cases} 0.528 & \text{if } x \leq -1.2, \\ x(1-x^2) & \text{if } -1.2 < x < 1.2, \\ -0.528 & \text{if } x \geq 1.2. \end{cases} \tag{2}$$

The control parameter α characterizes the gain of the nonlinear amplifier around $x = 0$.

2.1 Individual dynamics of the circuit

If α is less than 1, all trajectories in the phase space of the system given by Eq.1 approach a stable fixed point at the origin $O_0(0,0,0)$. When α becomes larger than unity, the fixed point O_0 is no longer stable, and two additional fixed points $O_L(x_L, 0, z_L)$ and $O_R(x_R, 0, z_R)$ appear in the phase space of the circuit.[d] The fixed points O_L and O_R are globally stable in the parameter region $1 < \alpha < \alpha_H \equiv 1 + \frac{(\delta+\gamma)(1+\sigma+\delta\gamma)}{2\gamma}$. The transition through the critical value α_H is accompanied by a supercritical Andronov-Hopf bifurcation from the fixed points O_L and O_R. The bifurcation gives rise to stable periodic orbits P_L and P_R. Note that due to the symmetry of the system (see Eq.1 and Eq.2) any asymmetric attractor coexists with another one which is topologically similar.[e]

For larger values of α the circuit may produce very complicated dynamics connected with bifurcations of periodic orbits and transitions to chaos. Note that Eq.1 can be rewritten in the form[22]

$$\dddot{x} + A_3\ddot{x} + A_2\dot{x} + A_1 x + A_0 f(x) = 0. \tag{3}$$

The chaotic dynamics of Eq.3 has been studied elsewhere.[23] The dynamics of the electronic circuit has been explored both experimentally[19] and in numerical simulations.[21] To describe the dynamics we will use the results of the experimental analysis. Here we consider only one typical cut of multi-dimensional parameter space. The cut corresponds to the following values for the coefficients in Eq.1: $\sigma = 0.72$ and $\delta = 0.43$. In the experiment this cut has been investigated by variation of the parameters α and R while the other parameters of the circuit were fixed.

Experimentally measured bifurcation curves separating regions with qualitatively different types of behavior are presented in the parameter plane (γ^{-1}, α), see Fig.1c. For small enough values of γ ($\gamma^{-1} > 0.6$), when α increases and intersects curve l_{-1}, the periodic orbits P_L and P_R lose stability, and two stable periodic orbits $2P_L$ and $2P_R$ appear as the result of a period doubling bifurcation. In the experiment we could clearly see up to 4 sequential period doubling bifurcations before the oscillations became chaotic. In order to study chaotic dynamics of the circuit we employed a number of auxiliary electronic circuits [f] which enabled us to analyze different projections of attractors, the structure of the attractors on Poincaré cross sections, one-dimensional (1-D) maps generating by the attractors and bifurcations associated with homoclinic orbits.

Figure 2 shows the projections of chaotic attractors CA_L and CA_R generated by the circuit when the control parameters α and γ are taken to be inside shaded region 2 in Fig.1c. The lower row of the photos in Fig.2 represent return maps $(z(t_n), z(t_{n+1}))$ plotted from a Poincaré cross section of CA_R which is shown in the upper row. The moments of time, t_n, are conditioned by the intersections of trajectories of the attractors with the Poincaré cross section. These intersections is marked by bright dots on the projections of CA_R. Figure 2a shows two-band chaotic attractors which

[d] The coordinates of these fixed points are given by $x_R = z_R = -x_L = -z_L \equiv \sqrt{\frac{\alpha-1}{\alpha}}$.

[e] These can be realized through the transformation $x \to -x, y \to -y, z \to -z$.

[f] The diagrams of the auxiliary circuits have been presented elsewhere.[24]

appear immediately after transition to chaos. The corresponding return map displays intervals of the return function covered by the trajectories of the attractor CA_R.

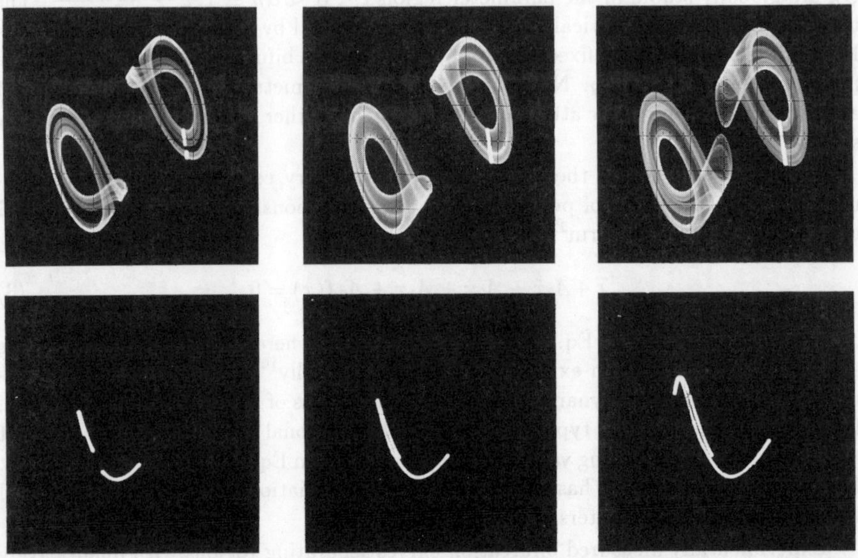

(a) (b) (c)

Figure 2: Projections of the attractors onto variable plane (x, z) with Poincaré cross sections marked by bright dots (above), and return functions $(z(t_n), z(t_{n+1}))$ generated by the trajectories of the attractors (bellow). $\gamma = 0.1$. (a) $\alpha = \alpha_1$, (b) $\alpha = \alpha_2 > \alpha_1$, (c) $\alpha = \alpha_3 > \alpha_2$.

When the parameter α increases, the attractors are transformed to those shown in Fig. 2b. The return map generated by the attractors can be well approximated by a 1-D map which is very close to the logistic map.[25] Using this connection between the dynamics of the circuit and the logistic map we conclude that despite the influence of physical noise, which prohibits the analysis of the whole chain of bifurcations, the transition to chaos is in accordance with the Feigenbaum scenario.

Varying the parameter α further towards more developed chaos shows a qualitative transformation in the shape of the return function of the attractor, see Fig. 2c. The appearance of a second extremum, which can be seen in the 1-D map, indicates a change in the topological properties of the attractor. This change marks the entrance into a parameter region[g] where different types of attractors may co-exist simultaneously.

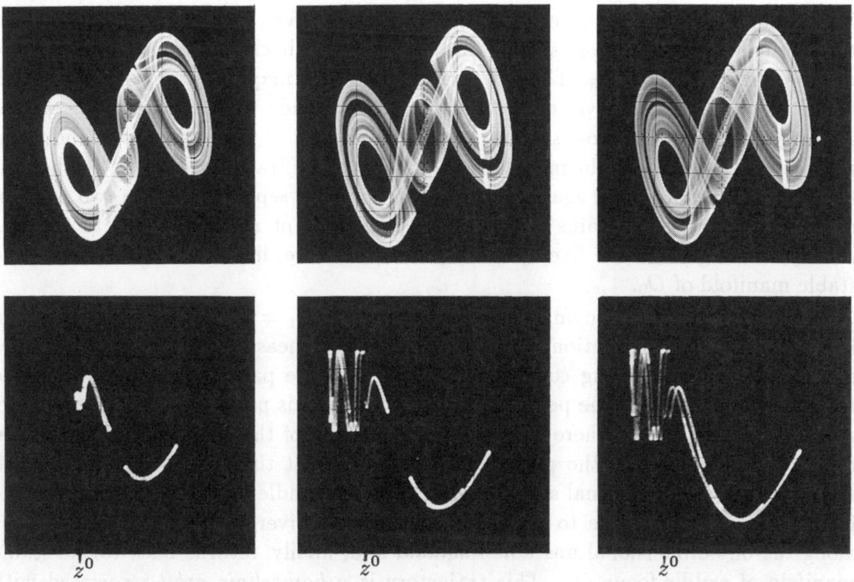

(a) (b) (c)

Figure 3: Projections of the attractors onto variable plane (x, z) with Poincaré cross sections marked by bright dots (above), and return functions $(z(t_n), z(t_{n+1}))$ generated by the attractors (bellow). $\gamma = 0.1$. (a) $\alpha = \alpha_4 > \alpha_{cr}$, (b) $\alpha = \alpha_5 > \alpha_4$, (c) $\alpha = \alpha_6 > \alpha_5$.

Below the α_{cr} curve in Fig 1c the basins of attractions of CA_R and CA_L are separated by the two-dimensional stable manifold of the fixed point, O_0. When parameters of the circuit cross the α_{cr} curve from shaded region 2 to shaded region 1 the twin chaotic attractors, CA_R and CA_L, merge together, thereby giving birth to the odd-symmetric chaotic attractor CA_S. This bifurcation is known as *a crisis of attractor.*[28, 29] A symmetry increasing bifurcation which is similar to the bifurcations in our case has been considered in references.[30-32] Projections of the odd-symmetric attractors measured for the parameters in region 1, and corresponding return maps,

[g]Experimental results of investigations of the main features of this sort of regions considered elsewhere.[24] The theoretical background for detailed analysis of bifurcations occurring in such regions of parameters can be found in References 26,27.

are presented in Fig.3. From the figures, one can see that the functions of the return maps contain two different parts. The right part of the function is qualitatively similar to the return function plotted from CA_R measured near the border α_{cr} (compare Fig.2c and Fig.3b). This interval of the function is given by the trajectories which starting from Poincaré cross section and then they go back similar to trajectories of the chaotic attractor CA_R. The left part of the return map has a more complicated, folded shape. This interval of the function corresponds to the trajectories which, starting from Poincaré cross section, go to the left part of the phase space and then, after several oscillations during which they behave like trajectories of CA_L, they intersect the Poincaré plain again. Usually, the point z^0, separating these two parts of the return function, contains a discontinuity. The point z^0 corresponds to the trajectory which goes to the fixed point O_0 and, therefore, lies in the two-dimensional stable manifold of O_0.

Qualitative parametric analysis of a 1-D map ($z_{n+1} = P^{(1)}(z_n, \alpha, \gamma)$), which can be used as an approximation of the experimentally measured return map, enables one to make the following conclusions: 1). There are parameter values when the trajectory started from the point z^0 returns back to this point after k iterations, i.e. $z^0 = P^{(k)}(z^0, \alpha_{\Gamma_k}, \gamma_{\Gamma_k})$, where z^0 is the left boundary of the right part of the return function. Therefore, in the phase space of the circuit there is a trajectory which starts from two-dimentional stable manifold of the saddle-focus O_0, approaches the fixed point O_0, then, due to the noise influence, it diverges from the saddle-focus along the one-dimensional unstable manifold and, finally, returns back to the stable manifold of saddle-focus O_0. This trajectory is *a homoclinic orbit* associated with the saddle-focus O_0. 2). The homoclinic orbits with different k can be found in parameter region 1 located above the bifurcation curve α_{cr}. 3). The return function $P^{(1)}(z, \alpha, \gamma)$ has many extrema. It is known that any periodic orbit which pass close enough to one of the extrema of 1-D mapping will be stable.[33, 34] Evolution of the return function with parameter variation indicates that many periodic orbits appear which pass close enough to one of the extrema. Therefore, shaded regions 1 and 2 have complicated detailed structure which contain an enormous number of very narrow domains corresponding to the existence of stable periodic orbits. [h]

The conclusion we have reached regarding the bifurcations associated with the appearence of homoclinic orbits can be confirmed by means of direct experimental observations. The homoclinic orbits associated with saddle-focus O_0 can be observed when we use periodic short-circuiting the input of nonlinear converter N. This short-circuiting makes the origin in the resulting system asymptotically stable. As the result the trajectory periodically starts from the close vicinity of O_0, and then moves along one-dimensional unstable manifold. Figure 4 shows intervals of the trajectories measured for different values of α. which are taken to be near the bifurcation values. Qualitative behavior of the trajectories clearly indicates that for these three cases the

[h] Figure 1c shows only two (relatively wide) domains of this type which correspond to stable orbits of period three $3P_R$, $3P_L$ (region 2) and to stable odd-symmetric periodic orbit P_S (region 1).

parameter crosses the bifurcation values of homoclinisity. In Figure 1c the bifurcation curves corresponding to homoclinic orbits are marked by Γ_1 and Γ_2.

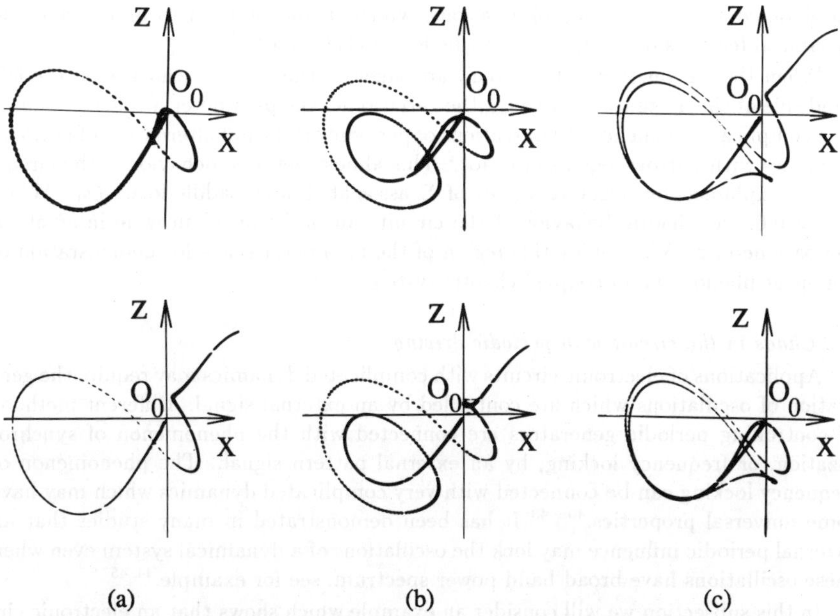

Figure 4: Intervals of the trajectories, started from the vicinity of the fixed point O_0, in the projections onto variable plane (x, z). $\gamma = 0.1$ (a) The parameter α close to intersection with bifurcation curve Γ_1: $\alpha < \alpha_{\Gamma_1}$ (above) and $\alpha > \alpha_{\Gamma_1}$ (bellow). (b) Close to intersection with lower branch of the curve Γ_2 $\alpha < \alpha_{\Gamma_2}$ (above) and $\alpha > \alpha_{\Gamma_2}$. (c) Close to intersection with upper branch of the curve Γ_2: $\alpha < \alpha_{\Gamma_2}$ (above) and $\alpha > \alpha_{\Gamma_2}$ (bellow).

Linearizing Eq.1 in vicinity of the fixed point O_0, one can find the values of the parameters where saddle-focus value $\Sigma = \lambda_1 + \mathrm{Re}\lambda_{2,3}$ becomes equal to zero. λ_i are eigenvalues of the fixed point O_0. In the (γ, α) parameters plane the values of the parameters where $\Sigma = 0$ are located on the curve given by the following equation: $\alpha_{\Sigma=0} = 1 + (\delta + \gamma)(1 + 2(\delta + \gamma)^2 + \sigma + \delta\gamma)/\gamma$. Analysis of asymptotic behavior of the curve $\alpha_{\Sigma=0}(\gamma)$ for $\gamma \to 0$ and for $\gamma \to \infty$ shows that it should intersect the bifurcation curves of homoclinicity Γ_k. In the region of the parameters $\alpha > \alpha_{\Sigma=0}$ the value of Σ is positive. Therefore, the chaotic behavior of the circuit is related to the well known Shilnikov's type of spiral attractor.[35] Since the divergence of the vector field $\mathbf{u} = (\dot{x}, \dot{y}, \dot{z})^T$ of the system Eq.1 is negative ($div\mathbf{u} = -\delta - \gamma$), the chaotic attractors generated by the circuit are quasiattractors,[36, 37] i.e. chaotic behavior of the system is characterized by generation of stable periodic orbits with very narrow basins of attraction, and alternation of stability and instability. This result confirms our third conclusion made from the analysis of the 1-D map. Theoretical studies of the detailed structure of the bifurcations, which should be used to understand the dynamical features of the system, can be found elsewhere[38, 39]

When the parameters of the circuit are taken in the domain which is on the left hand side of the dotted wavy line l_{cd}, the variation of the parameter α is accompanied by a complicated sequence of bifurcations of periodic orbits and alternation of periodic dynamics with narrow regions of chaos.[i] This almost periodic behavior of the circuit can be explained by negative values of Σ associated with saddle-focus O_0. When $\gamma^{-1} > 0.6$, the chaotic behavior of the circuit can be "robust" in wide intervals of the parameter α. We employ this region of the parameter space for demonsration of nonlinear phenomena in coupled chaotic systems.

2.2 Chaos in the circuit with periodic driving

Applications of electronic circuits with complicated dynamics may require the generation of oscillations which are controlled by an external signal. Different methods of controlling periodic generators are connected with the phenomenon of synchronization, or frequency locking, by an external pattern signal. The phenomenon of frequency locking can be connected with very complicated dynamics which may have some universal properties.[40 - 43] It has been demonstrated in many studies that an external periodic influence may lock the oscillations of a dynamical system even when these oscillations have broad band power spectrum, see for example.[44, 45]

In this subsection we will consider an example which shows that an electronic circuit may behave synchronously with an external periodic forcing. This synchronous behavior may include not only periodic oscillations, but also chaotic ones. In the experiment, the parameters were taken in the region where the circuit generates periodic orbits $2P_R$ and $2P_L$. Then, a periodic signal, $A\sin(2\pi ft)$, generated by an external oscillator was mixed with the voltage, $x(t)$, at the input of the nonlinear converter N, see Fig.1a. In the experiments we analyzed the projections of the attractors onto the (x,z) plane, cross sections of the attractors with the period of the external forcing and power spectra of the signals measured from the circuit. The bifurcations associated with these attractors have been considered in the (f,A) parameters plane. A typical bifurcation diagram, measured in vicinity of synchronization characterized by *winding number*[j] (1/1), is shown in Fig.5.

---

[i] In Fig.1c, the complicated structure of the bifurcations conditioning the dynamics in this domain is not presented.

[j] In the case of complicated dynamics of the non-autonomous system we define the winding number as the ratio between the main frequency of the locked oscillations and the frequency of the external driving. The main frequency of the signal corresponding to multiperiodic orbit may be defined by means of measurements of the period of the orbit which should then be divided by the number of oscillations of the signal within the period.

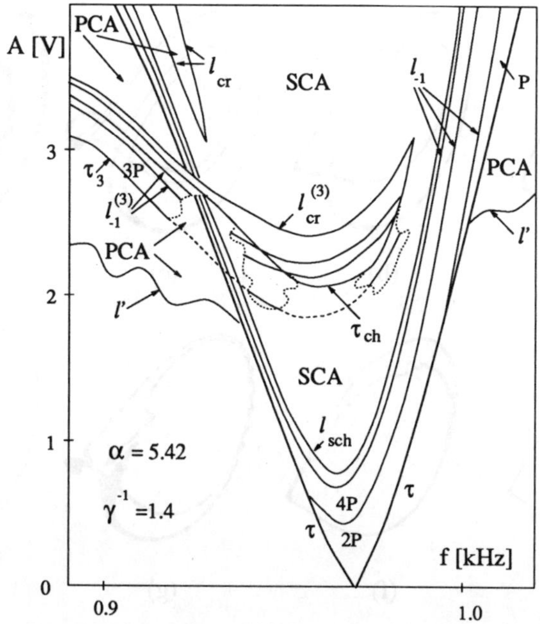

Figure 5: Diagram of bifurcations, associated with attractors, measured in the parameter plane (f, A).

For small amplitude driving, the synchronization zone is bordered by the bifurcation curves, τ, which form an Arnol'd tongue.[41] Synchronized periodic motion measured in the domain $2P$ is shown in Fig.6b. This periodic motion is the orbit $2P_L$ which is locked to the external driving. The locking is confirmed by the steady positions of dots shown against the background of the attractor. These dots mark the intersections between the periodic motion and the cross-sections S_T formed with the period of external driving. When A is small enough, the curves, τ, correspond to tangent bifurcation of the stable periodic motion $2P$. When the parameters leave the synchronization zone, $2P$ merges with the unstable periodic motion $2P^u$, both of them disappear, giving rise to quasiperiodic motions. These quasiperiodic motions are characterized by two incommensurate frequencies and ergodically cover a two-dimensional (2-D) torus. The appearance of these quasiperiodic oscillations indicates the lose of synchronization. It is known[40] that for small A a 2-D invariant torus exists inside the Arnol'd tongue. This invariant torus is formed by the unstable manifolds of the periodic $2P^u$ orbit. However, the torus is not an attractor because any trajectory on the manifold is attracted by the periodic orbit $2P$.

Tuning the parameters A and f inside the synchronization zone the circuit produce

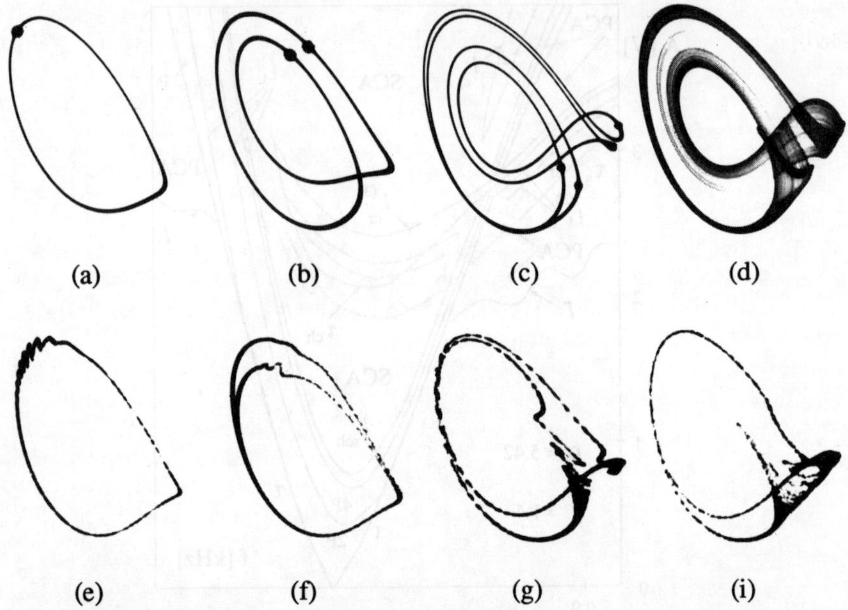

Figure 6: (a) - (d) projections of the synchronized attractors onto the variables plane (x, z). Black dots on the trajectories indicate the intersections with S_T. (e) - (i) maps of attractors measured via the period of external driving when the parameters of the driving are taken outside the synchronization zone.

a sequence of period doubling bifurcations. We can follow to these bifurcations both backward (down to the periodic orbit P, see Fig.6a) and forward (up to the transition to chaos at the border l_{sch}, see Fig.5). Note that all of these orbits are synchronized to the external driving, see Fig.6a-d. Synchronization for the chaotic orbit means that the difference between the "phases" of the chaotic oscillations and periodic driving does not grow. This conclusion follows from the steady structure of the map at the cross-section S_T which is located in only small portion of the two-dimensional projection of the attractor, see Fig.6d. Note that qualitatively the synchronized chaotic attractor (SCA) is the same as the chaotic attractor which can be generated by the circuit without the external influence.

Outside the synchronization zone chaos appears upon the base of quasi-periodic motion. Typical transition from regular quasi-periodic to non-synchronized chaotic motion can be seen from the maps presented in Figs.6f-i. For this kind of transition to chaos it is quite common that, before the transition, the 2-D torus is subjected to deformations connected with its folding followed by breaking of the torus.[46] In the diagram shown in Fig5 an approximate border of the transition to non-synchronized chaos is marked by wavy curves l'. When the parameters of the driving are close to the border, τ, the non-synchronized attractors look like synchronized ones but periodically pulsating in time. The closer the parameters to the border, the longer the period of these pulsations. In the region of chaotic dynamics we call these attractors *pulsating chaotic attractors* (PCA).

Experimental studies of different zones of synchronization show that their bifurcation structures can be quite complicated.[47] In many cases border of synchronization can not be described by a single bifurcation. For example, in the considered example of synchronization zone synchronized periodic orbit $3P$ that appeared inside the zone can keep its stability while the parameters of the driving leave the synchronization boundered by τ, see Fig.5. Mention should be made that periodic orbit, $3P$, appears as the result of typical evolution of the synchronized chaotic attractor, SCA, after the transition to chaos via a sequence of period doubling bifurcations.

The example considered in this subsection shows that circuit can be made to generate chaotic oscillations, both synchronized and non-synchronized, with external periodic driving. The regimes of synchronized regular and chaotic oscillations can be used to control the spectrum of the generated signals.

3. Synchronized chaos in two coupled chaotic circuits.

It is known that nonlinear coupled systems generating chaos may behave in a synchronous fashion.[48-50] The ability of chaotic circuits to synchronize to each other has many potential applications in engineering. Different ideas of employing synchronous chaotic behavior for communications, control and diagnostics have been already discussed in quite a few papers (see for example references[51-57]). In order to describe how synchronous chaotic behavior in two systems may be achieved, we will consider four different examples of coupled chaotic circuit.

3.1 Synchronized chaos in two mutually coupled identical chaotic circuit

As the first example we consider synchronization of chaos in two identical chaotic circuits. The synchronization is provided by mutual dissipative coupling between the circuits.[21] The diagram of the experiment is presented in Fig.7. The model describing the dynamics of two identical circuits with mutual coupling can be written in the form

$$\begin{aligned}
\dot{x}_1 &= y_1 + \tfrac{\epsilon}{2}(x_2 - x_1)\,, & \dot{x}_2 &= y_2 - \tfrac{\epsilon}{2}(x_2 - x_1)\,, \\
\dot{y}_1 &= -x_1 - \delta y_1 + z_1\,, & \dot{y}_2 &= -x_2 - \delta y_2 + z_2\,, \\
\dot{z}_1 &= \gamma[\alpha f(x_1) - z_1] - \sigma y_1\,, & \dot{z}_2 &= \gamma[\alpha f(x_2) - z_2] - \sigma y_2\,,
\end{aligned} \tag{4}$$

where subscripts 1 and 2 correspond to different circuits (systems). The nonlinearity, $f(x)$, and the dependence of the parameters in Eq.4 upon the physical values of the circuits elements were discussed in subsection 2.1. The coupling parameter, ϵ, is controlled by the resistor R_c:

$$\epsilon = \frac{2}{R_c}\sqrt{\frac{L}{C}}. \tag{5}$$

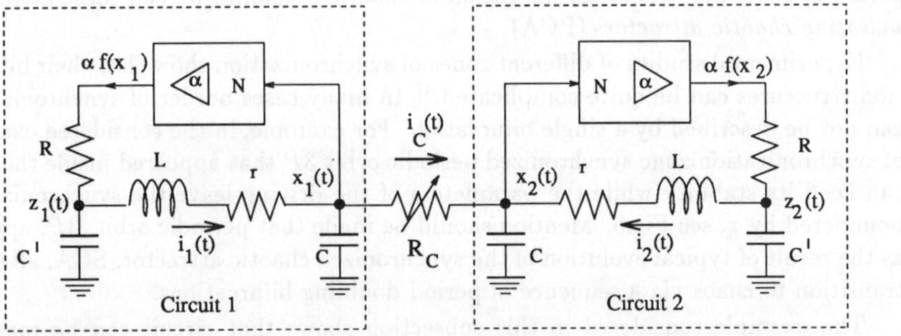

Figure 7: Diagram of the experiment with mutual dissipative coupling between the chaotic circuits.

It is easy to see that in the 6-dimensional phase space of Eq.4 there exist a 3-dimensional integral manifold

$$\begin{aligned} x_1 &= x_2, \\ y_1 &= y_2, \\ z_1 &= z_2. \end{aligned} \tag{6}$$

The trajectories of Eq.4 that are located on the manifold given by Eq.6 correspond to identical oscillations in both systems. One can easily see, from Eq.4, that when the systems behave identically the behavior of the coupled system is the same as that exhibited by the individual systems without coupling. Therefore, if the parameters of the circuits are chosen in a region associated with chaotic behavior and the manifold, Eq.6, is stable, then the circuits can demonstrate synchronized chaotic oscillations.[k]

In order to demonstrate synchronization in an experiment let us consider an example[21] where the parameters of the individual circuits are chosen in the region corresponding to a symmetric chaotic attractor, CA_s, shown in Fig.3c. Synchronized chaotic motion (SCM) associated with chaotic trajectories located on the manifold, Eq.6, are the only stable type of behavior found for coupling parameter, ϵ, higher than some critical value, ϵ^*. When projected onto the (x_1, x_2) plane SCM is along the diagonal, see Fig.8a. The regime of SCM remains stable as we decrease the value of the coupling down to ϵ_{SM} ($\epsilon_{SM} < \epsilon^*$) but other stable states coexist with SCM

[k]In experimental studies synchronization between identical chaotic circuits can be detected with the analysis of the projections of the generating attractor onto the variables planes $(x_1(t), x_2(t))$, $(y_1(t), y_2(t))$ and $(z_1(t), z_2(t))$.

in this region. For $\epsilon < \epsilon_{SM}$ the regime of SCM becomes unstable. Looking at the attractors outside the manifold, Eq.6, we found that, as ϵ decreases and crosses the boundary $\epsilon = \epsilon^*$, two stable fixed points, O^+ and O^-, appear outside the manifold. Afterwards they lose stability undergoing an Andronov-Hopf bifurcation, and two

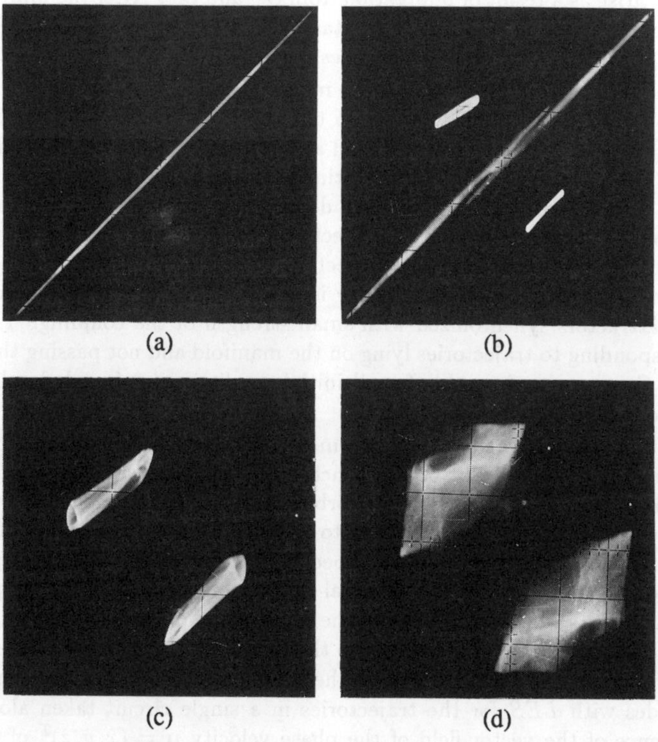

Figure 8: (x_1, x_2) projections of the attractors measured in the experiment with mutually coupled chaotic circuits. (a) Synchronized chaotic motions, $\epsilon > \epsilon^*$. (b) Multistable regime, $\epsilon_{SM} < \epsilon < \epsilon^*$. (c)-(d) Stable attractors when $\epsilon < \epsilon_{SM}$.

stable periodic orbits, P^+ and P^-, appear in the phase space. Projections of the attractors in the multistable regime are shown in Fig.8b. A further decrease of ϵ leads to a Hopf bifurcation of P^+ and P^-, and afterwards there is a transition to chaos via quasiperiodicity. The bifurcation scenario described above indicates the existence of hysteresis on the boundary of synchronization of chaotic oscillations. Thus, to estimate the bifurcation value, ϵ^*, which belongs to one of the boundaries of the synchronization regime, it suffices to investigate the stability and bifurcations associated with the fixed points O^+ and O^-.

The analytical study of Eq.4 shows that when decreasing the coupling parameter, a bifurcation from the unstable fixed point, O_0, located at the origin of the phase space takes place at $\epsilon = \epsilon_0 \equiv \frac{\gamma(\alpha-1)}{\sigma+\gamma\delta}$, and two other unstable fixed points, O^+ and O^-, appear outside the manifold, Eq.6. On the boundary $\epsilon^* = \frac{2\epsilon_0}{3}$, two additional pairs of unstable fixed points arise as a result of bifurcation from O^+ and O^-. After the bifurcation O^+ and O^- become stable. These points are stable down to $\epsilon = \epsilon_T \equiv \frac{\gamma(\alpha-1)-(\gamma+\delta)(1+\sigma+\delta\gamma)}{3(\sigma+\delta\gamma)}$, where the Andronov-Hopf bifurcation takes place.

These analytical results show that the manifold, Eq.6, is stable for $\epsilon > \epsilon_0$. In the region $\epsilon < \epsilon_0$, it becomes unstable due to the appearance of an additional unstable directions of the fixed point O_0 which lead away from the manifold. Therefore, there are narrow flows of trajectories which starting from the vicinity of the manifold, Eq.6, come close to the fixed point O_0 and then depart from the manifold along the unstable directions. The existence of these trajectories can provoke loss of synchronization for small intervals of time when the trajectory jumps from the manifold, Eq.6, and then returns back. This kind of behavior is observed in experiment for some types of chaotic attractors synchronized with small strength of the coupling. The oscillations corresponding to trajectories lying on the manifold and not passing through the vicinity of O_0 may remain stable for all initial conditions in the whole phase space until the boundary $\epsilon = \epsilon^*$ is reached.

The critical value of the coupling parameter, ϵ_{SM}, which defines the second point of hysteresis on the boundary of the synchronization zone can be estimated from the analysis of local stability of chaotic orbit lying on the manifold with respect to disturbances in the transversal direction to it.[21] To estimate the values of ϵ_{SM}, one need to analyze the Lyapunov exponent spectrum (LES) as a function the parameter ϵ. Owing to the assumption of the identical circuits, one can divide all components of LES into two parts which correspond to the evolution of perturbations along different directions. One part ($LES_{\|}$) is tangent to the manifold, Eq.6, and the other ($LES_{\perp}$) is transverse to it. For trajectories on the manifold, $LES_{\|}$ does not depend on ϵ and coincides with LES for the trajectories in a single circuit taken alone. Since the divergence of the vector field of the phase velocity $\mathbf{u} = (\dot{x}, \dot{y}, \dot{z})^T$ of Eq.1 does not depend upon the location in the phase space , $div\mathbf{u} = -(\gamma + \delta)$, the sum the Lyapunov exponents of $LES_{\|}$ equals the divergence, $-(\gamma+\delta)$. Therefore, the sum of the other three $LES_{\perp}$ is equal to $div\mathbf{u_c} - div\mathbf{u} = -(\gamma+\delta+\epsilon)$, where $\mathbf{u_c}$ is the vector field of the coupled systems, Eq4. Thus the introduction of the coupling, ϵ, between the systems breaks the equality between $LES_{\|}$ and $LES_{\perp}$.

In order to calculate the $LES_{\perp}$ one introduce new variables

$$\begin{aligned} \xi &= x_1 - x_2, \\ \eta &= y_1 - y_2, \\ \zeta &= z_1 - z_2. \end{aligned} \tag{7}$$

Then a linearized system for the behavior of the perturbations in vicinity of the

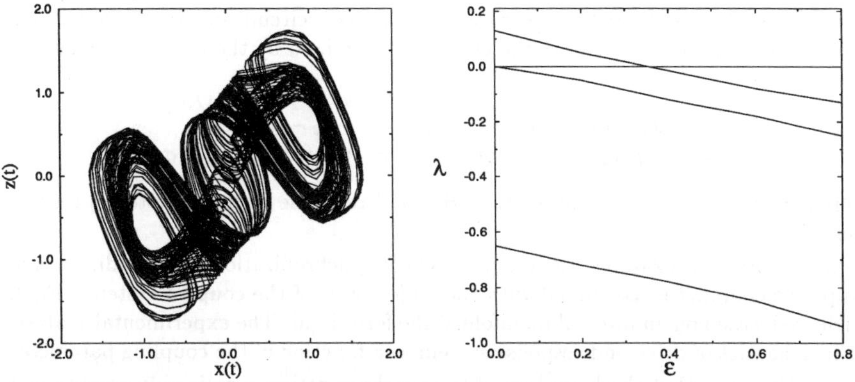

Figure 9: (x, z) projection of chaotic attractor of the system and dependence of $LES_{\perp}$ on ϵ calculated for this attractor. The parameters of the system are $\gamma = 0.1$, $\delta = 0.43$, $\sigma = 0.72$ and $\alpha = 22.3$.

manifold, Eq.6, has the form

$$\begin{aligned} \dot{\xi} &= \eta \\ \dot{\eta} &= -\xi - \delta\eta + \zeta \\ \dot{\zeta} &= \gamma[\alpha f'(x_1(t))\xi - \zeta] - \sigma\eta. \end{aligned} \tag{8}$$

where $x_1(t)$ is a solution corresponding to a chosen attractor of Eq.1, $f'(x) = \frac{df(x)}{dx}$. Calculation of Lyapunov exponent spectrum associated with Eq.8 gives us the $LES_{\perp}$. One can easily check that the Lyapunov exponent spectrum of Eq.8 calculated with $\epsilon = 0$ gives us the $LES_{\parallel}$. A dependence of the $LES_{\perp}$ components upon the coupling parameter which is calculated for chaotic trajectory of Eq.1 is shown in Fig.9. As it can be seen in Fig.9 when ϵ grows, all components of $LES_{\perp}$ monotonically decrease. For $\epsilon > \epsilon_{SM}$, all components of $LES_{\perp}$ are negative and, therefore, the chaotic trajectory on the manifold, Eq.6, is stable against the perturbations in the transversal directions to the manifold.

This example demonstrates how experimental analysis can help one to point out the bifurcations which condition transitions to the regime of synchronized chaotic oscillations. After the experimental observations these bifurcations can be directly studied by means of analytical or numerical methods. More details about analytical and numerical analyses of chaos synchronization with mutual coupling can be found elsewhere.[21]

3.2 Synchronized chaos in two identical chaotic circuit with unidirectional coupling

In the second example we consider the same coupled circuits as in Fig.7, but the voltage $x_1(t)$ is applied to the resistor of coupling R_c through a Unity- Gain Amplifier. This amplifier eliminate the influence of the coupling current $I_c(t)$ on the behavior of the circuit 1. As the result the circuit 1 is *a drive* circuit and the circuit 2 is *a response* circuit. The behavior of the circuits is described by the following equations

$$\begin{aligned} \dot{x}_1 &= y_1 , & \dot{x}_2 &= y_2 - \tfrac{\epsilon}{2}(x_2 - x_1) , \\ \dot{y}_1 &= -x_1 - \delta y_1 + z_1 , & \dot{y}_2 &= -x_2 - \delta y_2 + z_2 , \\ \dot{z}_1 &= \gamma[\alpha f(x_1) - z_1] - \sigma y_1 , & \dot{z}_2 &= \gamma[\alpha f(x_2) - z_2] - \sigma y_2 , \end{aligned} \tag{9}$$

where subscripts 1 and 2 stand for the drive and response circuits (systems) respectively.

Like in previous example, the regime of chaos synchronization with unidirectional dissipative coupling is associated with the trajectories of the coupled systems, Eq.9, which are located on an integral manifold of the form Eq.6. The experimental analysis of the dynamics of drive and response system as a function of the coupling parameter $\epsilon' = \frac{2}{R_c}\sqrt{\frac{L}{C}}$ shows that the boundary of the synchronization zone is only conditioned by the local stability of the regime of synchronized motion. The transition from the regime of $SCMs$ to non-synchronous behaviors of these circuits is associated with intermittency "synchronized chaos" - "non-synchronized chaos". The threshold of local stability of $SCMs$ against the perturbations in the transversal directions to the manifold is given by critical value ϵ'_{SM}, which can estimated with the linearized equations of the form Eq.8 where ϵ is replaced by $\frac{\epsilon'}{2}$. Therefore, the critical values of the local stability of $SCMs$ for unidirectional and mutual coupling are connected with the formula $\epsilon'_{SM} = 2\epsilon_{SM}$.

In both the first and the second examples, the synchronization of chaos appears due to the dissipative coupling of the chaotic circuits. Due to the instability of the trajectories of the chaotic attractors any, small difference between the states of the circuits tends to grow exponentially. We call this growth a transversal instability. However, the difference of the voltages $x_1(t) - x_2(t)$ gives rise to the current through the resistor R_c. If the resistance R_c is small enough, then the current results in the dissipation of the energy in R_c which suppresses the transversal instability and, therefore, tends to reduce the difference between the states of the circuits. Note, that the same mechanism for synchronization of chaos may have different implementations. For example, the analysis of the equations for the dynamical systems employed in the experiment by Newell et al[58] shows that the proportional feedback gives a dissipative term which can provide the suppression of the transversal instability in two chaotic nonautonomous circuits.

Unidirectional dissipative coupling may be employed to suppress chaos in the response circuit. The suppression of chaos may be achieved if one use waveform synthesizer instead of the driving circuit. It is important that chaotic attractors contain a countable set of unstable periodic orbits. When the synthesizer generates a periodic signal $x_1(t)$ which is a waveform corresponding to an unstable periodic orbit generated in the phase space of the response circuit, then the oscillations of the circuit may be synchronized to the signal $x_1(t)$. Note, that when the response system behaves in synchronism with the driving, $x_1(t)$, then $I_c(t) = x_2(t) - x_1(t) = 0$.

Therefore, the influence of the driving signal onto response circuit is exponentially small.[59] This mechanism of chaos suppression can also be implemented by means of continues proportional feedback.[60]

3.3 Synchronous chaotic behavior of a response system with a chaotic driving

A method of synchronizing a response system by a chaotic driving has been developed by Pecora and Carroll[50–].[62] We use a modification of this method to synchronize the response circuit in the third experiment. Let us consider the main idea of the modified method. We assume that an electronic circuit, which can generate chaotic oscillations, has terminals which are inputs of internal feedbacks.[l] This circuit is used as a drive circuit. The behavior of the circuit can be described by differential equations of the form

$$\dot{\mathbf{X}}_d = \mathbf{F}(\mathbf{X}_d, \mathbf{d}, \mathbf{X}_d), \tag{10}$$

where $\mathbf{X}_d$ is the vector of phase variables of the circuit, $\mathbf{d}$ is the vector of the parameters. The third argument of the nonlinear vector function $\mathbf{F}(\ ,\ ,\ )$ is introduced to specify the variables at certain inputs which are connected to the outputs of the internal feedbacks of the circuit.

To build a response circuit one use the circuit which has the same construction as the drive circuit. Then one takes the vector of output signals from the internal feedbacks (of the response circuit) which are specified above and mix it with the vector of driving signals derived from the same outputs of the drive circuit. We consider the form of the mixing which gives the following equations for the behavior of the response circuit

$$\dot{\mathbf{X}}_r = \mathbf{F}(\mathbf{X}_r, \mathbf{r}, \mathbf{X}_m), \tag{11}$$

where $\mathbf{X}_r$ is the vector of physical variables of the response circuit, $\mathbf{X}_m = \mathbf{X}_r - \epsilon(\mathbf{X}_r - \mathbf{X}_d)$ is the vector of mixed signals at the terminals and $\mathbf{r}$ is the vector of the parameters. The parameter ϵ is used to specify the strength of unidirectional coupling between the circuits.

When the parameters in drive and response circuits are identical, $\mathbf{d} = \mathbf{r}$, then an integral manifold of the form

$$\mathbf{X}_d = \mathbf{X}_r \tag{12}$$

exists in the phase space of the systems, Eq.10 and Eq.11. If the chaotic motion located on the manifold, Eq.12, is stable against the transversal perturbations, then the response circuit can behave in synchronism with the drive circuit. This type of transversal stability can be studied from the analysis of Lyapunov exponents spectrum generated by the linearized system

$$\dot{\boldsymbol{\xi}}_r = [\mathbf{DF}_1(\mathbf{X}_d, \mathbf{d}, \mathbf{X}_d) + (1-\epsilon)\mathbf{DF}_3(\mathbf{X}_d, \mathbf{d}, \mathbf{X}_d)]\boldsymbol{\xi}_r, \tag{13}$$

[l] For example, in our circuit we have the input of the nonlinear converter, N, see Fig.1a.

where $\mathbf{DF}_1$ is Jacobi matrix associated with the first argument and $\mathbf{DF}_3$ is Jacobi matrix associated with the third argument of the function $\mathbf{F}(\,,\,,\,)$.

In the third experiment we show how this method can be used for generating synchronous chaos in the electronic circuit. The diagram of the experiment is presented in Fig.10. The drive circuit, $G1$, generate chaotic signal, $x_1(t)$. This signal is applied to the response circuit, $G2$, where $x_1(t)$ is mixed with the output of the feedback, $x_2(t)$. The result of this mixing, $x_m(t) = x_2(t) - \epsilon(x_2 - x_1)$, is applied to the input of the nonlinear converter, N, of the response system. Assuming that certain parameters of the feedbacks in $G1$ and $G2$ satisfy the condition $L_1C_1 = L_2C_2$, the oscillations of coupled circuits are described by the following equations.[22]

$$\text{driving system} \quad : \quad \begin{cases} \dot{x}_1 &= y_1 \,, \\ \dot{y}_1 &= -x_1 - \delta_1 y_1 + z_1 \,, \\ \dot{z}_1 &= \gamma_1[\alpha_1 f(x_1) - z_1] - \sigma_1 y_1 \end{cases} \tag{14}$$

$$\text{response system} \quad : \quad \begin{cases} \dot{x}_2 &= y_2 \,, \\ \dot{y}_2 &= -x_2 - \delta_2 y_2 + z_2 \,, \\ \dot{z}_2 &= \gamma_2[\alpha_2 f(x_2 - \epsilon(x_2 - x_1)) - z_2] - \sigma_2 y_2 \end{cases} \tag{15}$$

where the indexes 1 and 2 are used to indicate the variables and the parameters in drive and response systems, respectively.

We will briefly consider two cases of synchronous chaotic behaviors. In the first case the parameters of the circuits are identical, and synchronization is studied as a function of the coupling parameter, ϵ. In the second case the circuits have different parameters values, and synchronization is studied with $\epsilon = 1$.

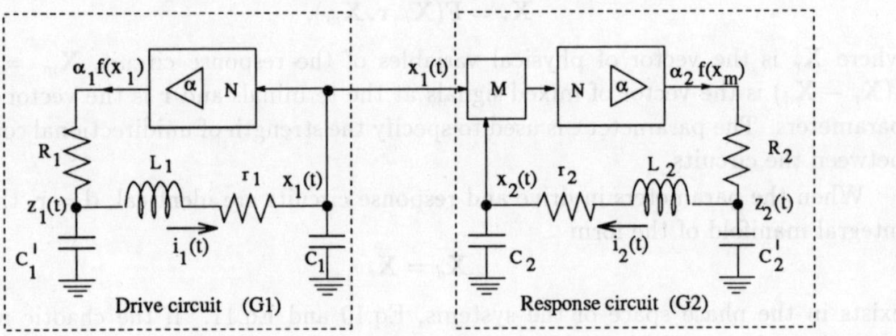

Figure 10: Diagram of the experiment with chaotic driving of the response circuit. M is a circuit which transforms inputs $x_1(t)$ and $x_2(t)$ into the output $x_m(t)$.

3.3.1. Let us assume that Eq.14 and Eq.15 have the same values of δ, σ, γ and α.[m] Despite the identity of the parameters the individual dynamics of the response

[m] Note, that Eq.14 and Eq.15 may have identical parameters even for a case when physical parameters of the circuit are different, but satisfy certain conditions.

system (with $x_1(t) = 0$) can be different from the drive system and depends upon the parameter ϵ. Using new variables $x_{new} = (1-\epsilon)x_2$, $y_{new} = (1-\epsilon)y_2$ and $z_{new} = (1-\epsilon)z_2$, one can reobtain the equations for the drive system, Eq.14, with normalized parameter $\alpha_n = \alpha_2(1-\epsilon)$. Without driving the response system behaves as the drive system would behave with different value of the parameter α_1.

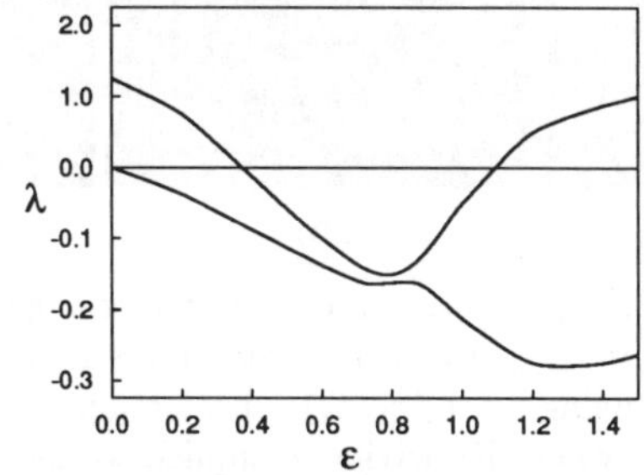

Figure 11: Dependence of two largest exponents of $LES_\perp$ upon the coupling parameter ϵ. The parameters of the circuits are the following $\alpha = 32$, $\delta = 0.43$, $\sigma = 0.72$ and $\gamma = 0.1$.

In spite of the difference between the individual behaviors of the circuits, these systems have an integral manifold of the form Eq.6, when they are coupled. Therefore, they have solutions in the form of identical chaotic oscillations, i.e. $SCMs$. The numerical analysis of Lyapunov exponent spectrum, $LES_\perp$, associated with the transversal perturbations of the trajectories on the manifold, Eq.6, shows that the regime of $SCMs$ may be stable within some interval of the values of the parameter ϵ, see Fig.11. In the experiment we have observed that there is a region of the parameter ϵ where $SCMs$ can be the only stable regime of oscillations generated by the circuits.

This experiment shows that identical synchronous oscillations can be observed even in the case when the response circuit without driving behaves differently from the synchronizing drive circuit. The synchronization of this type is illustrated in Fig. 12. Figures 12a,b show attractors generated by the drive circuit and the response circuit without driving. When the driving is applied the response circuit reproduces the behavior of the drive circuit, see Fig.12c.

3.3.2. Let us consider the case when $\epsilon = 1$ and the parameters in drive and response systems are different. One can see that when $\epsilon = 1$, then $x_m(t) = x_1(t)$. Now the linear portion $(C_2L_2C_2'R_2r_2)$ of the response circuit is not longer a feedback, see Fig.10. It is essential that this linear portion of the response circuit does not contain any active elements.

Taking into account that $f(x_m(t)) = f(x_1(t))$, it has been shown[22] that for the post transient behavior of the systems the physical variables of the drive and response

(a) (b) (c)

Figure 12: (a) (x_1, z_1)- projection of the attractor generated by drive circuit. (b) (x_2, z_2) - projections of the attractors generated by response circuit without driving. (c) (x_1, x_2) - projection of the synchronized motions measured when the driving is applied to the response circuit.

systems are connected by the following conditions

$$\begin{aligned} x_2(t) &= \hat{K}_{x2,x1}(\tfrac{d}{dt})x_1(t) , \\ y_2(t) &= \hat{K}_{y2,y1}(\tfrac{d}{dt})y_1(t) , \\ z_2(t) &= \hat{K}_{z2,z1}(\tfrac{d}{dt})z_1(t) , \end{aligned} \tag{16}$$

where $\hat{K}_{u2,u1}(\frac{d}{dt})$ are linear integrodifferential operators. These operators are rational functions of $\frac{d}{dt}$ which have coefficients depending upon the parameters of the circuits. It can be proved that $\hat{K}_{u2,u1}(p' + ip'')$ do not have poles with positive values of real part of its argument, p'. Therefore the connection between the values of physical variables in drive and response systems is stable. Note that in this case drive and response may behave not identically, but synchronously.[n]

For post transient behavior, it can be also shown that complex Fourier transforms of the signals $\{x_1, y_1, z_1\}$ and $\{x_2, y_2, z_2\}$ are proportional to each other[22]

$$\begin{aligned} \tilde{S}_{x2}(i\omega) &= \tilde{K}_{x2,x1}(i\omega)\tilde{S}_{x1}(i\omega) , \\ \tilde{S}_{y2}(i\omega) &= \tilde{K}_{y2,y1}(i\omega)\tilde{S}_{y1}(i\omega) , \\ \tilde{S}_{z2}(i\omega) &= \tilde{K}_{z2,z1}(i\omega)\tilde{S}_{z1}(i\omega) , \end{aligned} \tag{17}$$

where $\tilde{K}_{u2,u1}(i\omega)$ are complex functions of the frequency ω. These expressions show the existence of *a direct connection* between the spectra of chaotic oscillations in $G1$ and $G2$. Study of phase distributions of the continuous spectra of the chaotic realizations in $G1$ and $G2$ demonstrates that every harmonic of the response oscillations spectrum is "locked" to the harmonic with the same frequency contained in the driving signal. This property of coupled oscillators can also be employed as a definition

[n] This synchronous behavior is in accordance with mathematical definition of the synchronous chaotic oscillations which is given by Afraimovich *et al.*[49]

of synchronous behavior.[22] Thus if one knows the trajectory on the chaotic attractor in the phase space (x_1, y_1, z_1) of the driving system then projection of the chaotic trajectory located on the attractor in the phase subspace (x_2, y_2, z_2) of the response system is single-valued, and vice versa.

This example shows that synchronous chaotic behaviors of drive and response circuits can have a form of *nonidentical* oscillations. To demonstrate this we used a response circuit whose parameters values correspond to a particular case when the circuit behaves as a linear filter forced by chaotic driving. Therefore, synchronous oscillations in the circuits are caused by linear response to the chaotic driving, but not by locking of the oscillations in the response circuit. Next we will consider an example where nonidentical synchronous behaviors are observed in a response circuit which can generate chaos without driving.

3.4 Synchronized nonidentical chaotic oscillations.

As the fourth example of chaos synchronization we consider an experiment[63] where chaotic driving signal, $z_1(t)$, is applied to the response circuit and mixed with $x_2(t)$ at the input of nonlinear converter, N. In the experiment without coupling between the circuits the parameters of both circuits were chosen in the region where they generate chaotic attractors CA_R, see Figs. 1c and 2c.

The dynamics of the driving circuit is given by Eq.14. The dynamics of the response circuit is described by the following equations

$$\begin{aligned} \dot{x}_2 &= y_2 \, , \\ \dot{y}_2 &= -x_2 - \delta_2 y_2 + z_2 \, , \\ \dot{z}_2 &= \gamma_2[\alpha_2 f(x_2 + \epsilon z_1(t)) - z_2] - \sigma_2 y_2 \end{aligned} \tag{18}$$

where the $z_1(t)$ is the driving signal.

One can easily check that when the driving is applied (i.e. $\epsilon \neq 0$) the systems, Eq.14 and Eq.18, do not have any solution of the form Eq.6, even if the parameters in both system have the same values.

In order to observe chaos synchronization chaos in experiment we use the analysis of attractors in the cross sections conditioned by the Poincaré cross section of the attractor in the phase space of the drive circuit. Examining different projections of the attractors generated in 6-D phase space of coupled circuits, we mark the points on the trajectory at the moments of time, t_n, when the trajectory in 3-D phase space of the drive circuit intersects the plane of Poincaré cross section.

When $\epsilon = 0$, the chaotic oscillations in drive and response circuits are uncorrelated and marked points are uniformly distributed on the chaotic attractor generated in 3-D phase space of the response circuit. Figure 13 shows the projections of chaotic attractors onto the plane of variables $(z_1(t), z_2(t))$ measured in experiment with different values of the parameter ϵ. The cross section of the attractors $(z_1(t_n), z_2(t_n))$ are marked by bright dots against the background of trajectories. When the value of

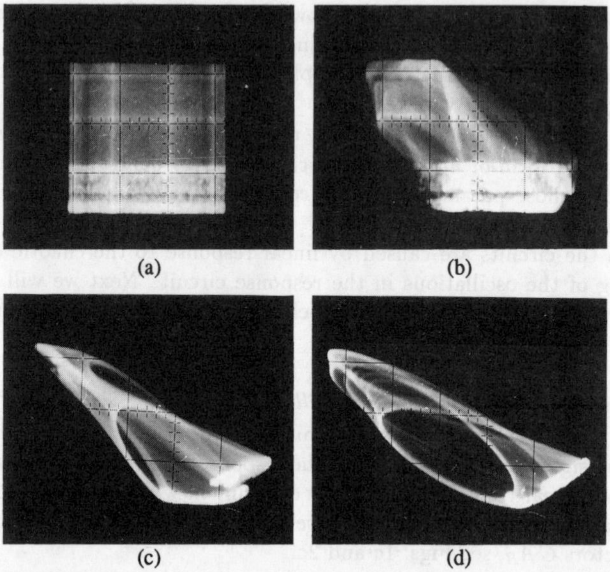

Figure 13: (z_1, z_2) projections of the attractors measured in the experiment with unidirectionally coupled chaotic circuits. Nonsynchronized chaotic motions: (a) $\epsilon = 0$; (b) $\epsilon = \epsilon_{u1} > 0$; (c) $\epsilon = \epsilon_{u2} > \epsilon_{u1}$. (d) Synchronized chaotic motions, $\epsilon > \epsilon_{cr}$.

ϵ increases within the interval $0 < \epsilon < \epsilon_{cr}$, the domain of the attractor covered by bright dots shrinks. When $\epsilon > \epsilon_{cr}$, the cross section of the attractor looks similar to that one measured in 3-D phase space of the drive system,[o] see Fig.13d and Fig.2b. Therefore, the transition via the critical value ϵ_{cr} corresponds to the onset of the synchronization between chaotic oscillations in the circuits.

This transition to the synchronization is characterized by "smooth" deformation of the attractor structure. It indicates that the synchronization is caused by internal bifurcations of chaotic attractor which is generated in 6-D phase space of the coupled systems.

In this Section we have considered four examples of synchronous chaotic behavior generated by coupled systems. These examples are quite typical and can be observed in the other dynamical systems which generate smooth chaotic waveforms. In the next part of the paper we briefly consider chaotic dynamics of coupled electronic circuits which generate chaotic pulses.

---

[o]Note that, moments of time when the intersections occur depend exclusively upon the behavior of the drive circuit.

4. Dynamics of pulse coupled chaotic relaxation oscillators

Dynamical properties of synchronized chaotic behavior depends on features of interacting systems. A particular case of synchronized chaos can be studied with relaxation oscillators. Relaxation oscillator is system which perform an alternation of slow motions and fast motions (pulsations). The transition from slow to fast motions occurs at the moments when the system reaches a certain threshold state. Immediately before the threshold the relaxation system is extremely sensitive to small external perturbations which may cause forced transition to fast motion. This property is widely used in electronics for synchronization of periodic pulse oscillators driven by periodic synchronization signals. This mechanism of synchronization is known as threshold synchronization.[64, 65] A similar type of synchronization phenomenon was considered for the problem of synchronization between pulse coupled biological periodic oscillators.[66]It was shown that the synchronization of the chaotic relaxation oscillators can also be achieved by means of pulse coupling between the oscillators.[55, 67] This type of synchronized chaos may be very useful for different engineering applications.

4.1 Dynamics of chaotic relaxation oscillator.

Different methods to achieve chaotic behavior generated by relaxation oscillator have been discussed in the paper by Bernhardt.[68] For the experimenting with coupled relaxation oscillators we designed an oscillator which can generate chaos in broad region of the parameter values. The chaotic dynamics of the relaxation oscillator can be modeled by the electronic circuit shown in Fig.14.

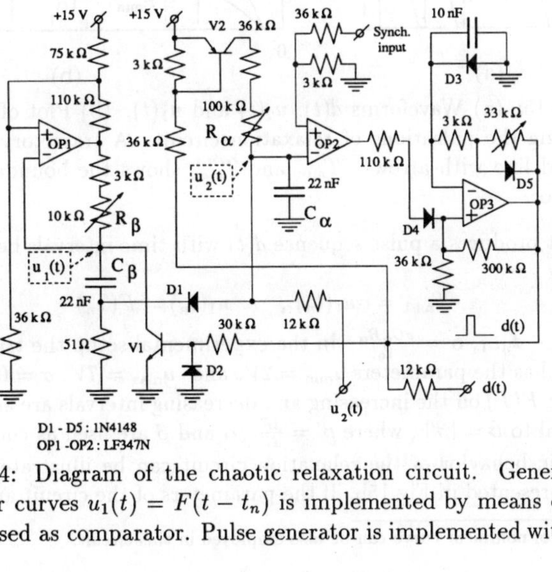

Figure 14: Diagram of the chaotic relaxation circuit. Generator of nonlinear curves $u_1(t) = F(t - t_n)$ is implemented by means of OP1. OP2 is used as comparator. Pulse generator is implemented with OP3.

Slow evolution of the system corresponds to increasing voltage u_2 on capacitor C_α as it is charged by current through the transistor $V2$ and the resistor R_α. When this voltage reaches the threshold level, $u_2(t) = 0$, the comparator (OP2) switches to another state and the pulse oscillator (OP3) generates pulse d_n at time t_n. This pulse switches off the transistor V2 and switches on the transistor V1. It provides recharging of the capacitor C_α to the voltage $u_2 = -\sigma u_1(t_n)$ and sets the nonlinear curve generator (OP1) to the initial state $u_1 = 0$. After the action of the pulse d_n, the capacitor C_α is charged from $V_0 = +15V$ and nonlinear curve of triangle shape[p], $u_1(t) = F(t - t_n)$, is generated by OP1 on the capacitor C_β. When the voltage u_2 reaches the threshold level again, the pulse oscillator (OP3) generates the next pulse d_{n+1}. The waveforms of the voltages $u_1(t)$, $u_2(t)$ and the output signal $d(t)$ are shown in Fig.15a.

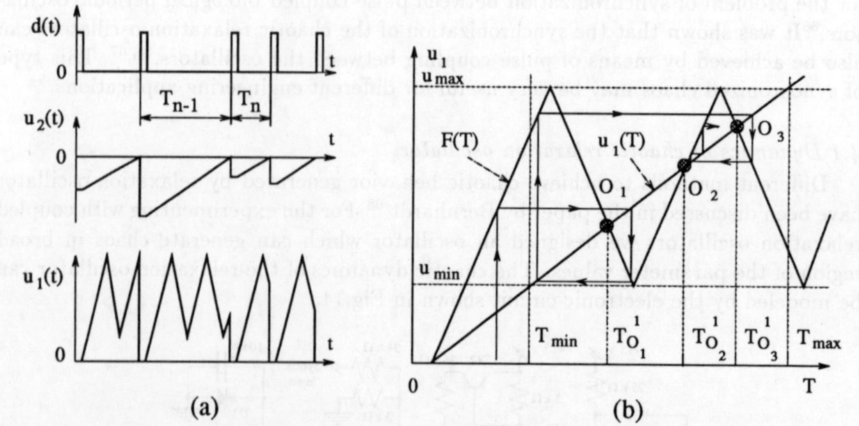

Figure 15: (a) Waveforms $d(t)$, $u_2(t)$ and $u_1(t)$. (b) Plot of 1-D map modeling the pulsations of relaxation circuit. A trajectory is shown by solid line with arrows. T_{min} and T_{max} shows the boundary of the attractor.

The circuit produces a pulse sequence $d(t)$ with time intervals between the pulses determined by

$$T_{n+1} = \alpha u_1(t_n) \; , \qquad u_1(t_n) = F(T_n) \tag{19}$$

where $T_n = t_n - t_{n-1}$, $\alpha = \frac{\sigma C_\alpha R_\alpha}{V_0}$. In the experimental setup the triangular function $F(T)$ (Fig.1c) has the parameters $u_{\rm min} = 2V$ and $u_{\rm max} = 7V$, $\sigma = 0.238$. The slopes of the function $F(T)$ on the increasing and decreasing intervals are assumed to be the same and equal to $\beta = |\beta'|$, where $\beta' = \frac{dF}{dT}$. α and β are used as control parameters.

The chaotic behavior of the relaxation circuit can be illustrated by the plot of 1-D map (1) presented in Fig.15b. If the parameters of the circuit are chosen so that

[p]Other types of nonlinear curve can be also employed in this circuit.

$\alpha\beta > 1$ then all trajectories in the phase space of the system (1) come to a strange attractor that is bounded by $T_{\min} = \alpha u_{\min}$ and $T_{\max} = \alpha u_{\max}$. The trajectory of the strange attractor have a positive Lyapunov exponent

$$\lambda = \lim_{n\to\infty} \frac{1}{n} \ln \prod_{m=1}^{n} \alpha \frac{dF(T)}{dT} |_{T=Tm} = \ln(\alpha\beta) > 0. \tag{20}$$

Using of a piece-wise linear function $F(T)$ enables us to design an oscillator which provides smooth variation of characteristics of chaotic output signals in a broad parameter region.

4.2 A mechanism of chaos suppression provided by external periodic pulses.

It is known[45, 69, 70] that periodic forcing applied to dynamical system with chaotic behavior can lead to suppression of chaos and transition to periodic behavior.

In order to study the chaos suppression in the experiment, external pulses d_{dr} with period T_{dr} are applied to the input "Synch", see Fig.14. During the pulse action the threshold level $u_{th} = 0$ changes down to the value $u_{th} = -E$, where E is a control parameter of the external influence. If at the moment of the pulse action the voltage u_2 is close enough to the unperturbed threshold level $u_{th} = 0$, i.e. the conditions $0 > u_2 > -E$ are satisfied, then the electronic circuit generates pulse d in step with d_{dr}. Otherwise the pulse d_{dr} has no influence on the operation of the circuit.

As it follows from dynamical model Eq.19, there exist many unstable periodic orbits in the phase space of the relaxation oscillator. Let T_P^N is the period of a periodic orbit $T_L = T_{L+N}$, where N and P correspond to the number of pulses generated by the oscillator during the period and the index of the orbit, respectively.[q]

Considering the case of small amplitude E ($E \ll \frac{\sigma T_{max}}{\alpha}$), it is easy to show that external pulses d_{dr} can force the circuit to oscillate with the period T_{dr} near the periodic orbit P, if the parameters of the external driving T_{dr} and E satisfy the following conditions

$$\alpha E > (T_{dr} - \sum_{m=1}^{N} T_m)(\sum_{k=1}^{N} \prod_{m=1}^{k} \alpha \frac{dF(T)}{dT} |_{T=Tm} - 1) > 0 \tag{21}$$

where $\sum_{m=1}^{N} T_m = T_P^N$ is the period of the unperturbed periodic orbit $P : \{T_{i+1}, T_{i+2}, ..., T_{i+N}\}$.

The results of measurements of the synchronization zones which provide chaos suppression near the fixed points O_i and corresponding synchronization zones obtained from Eq.21 are presented in the parameter plane (T_{dr}, E) in Fig.16. The diagrams show that the experimentally measured zones differ slightly from the theoretical ones. In particular, the experimental zones do not extend down to $E = 0$. This difference can be explained by the influence of small uncontrolled variations in the circuit parameters and the external signal, and by physical noise.

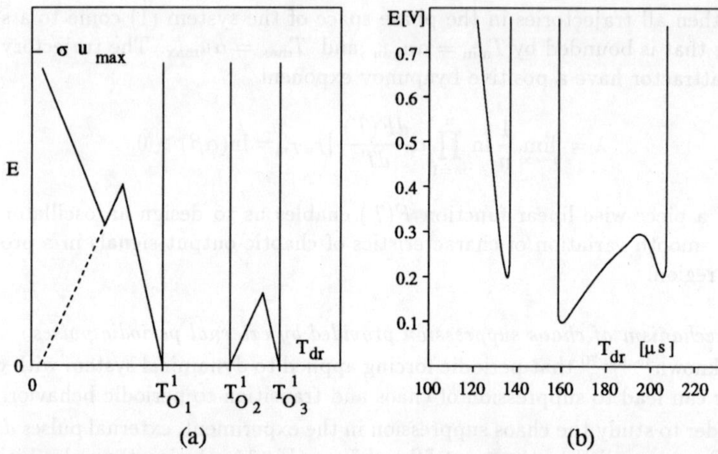

Figure 16: Synchronization zones $ST^{1,1}_{O_i}$ on the parameter plane (T_{dr}, E). (a) The theoretical zones with the parameters of the model Eq.19 taken to be the same as in Fig.15a. (b) The experimentally measured zones with $\sigma = 0.238$, $\alpha = 168\frac{\mu s}{V}$ and $\beta = 0.135\frac{V}{\mu s}$.

The measured waveforms of $d_{dr}(t)$ and $u_1(t)$ in the synchronization regimes are shown in Fig.17. In order to describe these regimes of the synchronization we use the indices $ST_P^{N,M}$, where N indicates the number of pulses generated by the circuit and M, the external oscillator of periodic pulses during the period of the synchronized oscillations related to the particular trajectory P,(see Fig.17).

The considered results show that chaos in relaxation circuit can be suppressed by *weak* periodic influence with the period close to the period of an unstable periodic orbit contained in the chaotic attractor of the relaxation oscillator. Since the chaotic attractors generated by the circuit do not contain any stable orbit this experiment presents an example where chaos suppression is not connected with enlargement of basins of attractions of stable periodic orbits.

4.3 Synchronization of chaotic pulsations in pulse coupled relaxation oscillators.

The threshold synchronization can provide locking of the oscillations of the relaxation system not only near an unstable periodic orbit but also near an unstable aperiodic trajectory. In the chaotic attractor of the model Eq.19 the aperiodic trajectories correspond to the chaotic pulsations of the circuit.

To investigate threshold synchronization with chaotic forcing, the chaotic sequence of pulses d_{dr} generated by drive circuit was applied to the input "Synch." (see. Fig.14) of an identical response circuit. The parameters $u_{\min}$ and $u_{\max}$ of the drive and

[q] For example, $T^1_{O_i}$ corresponds to the fixed points O_i of the map Eq.19, see Fig.15b.

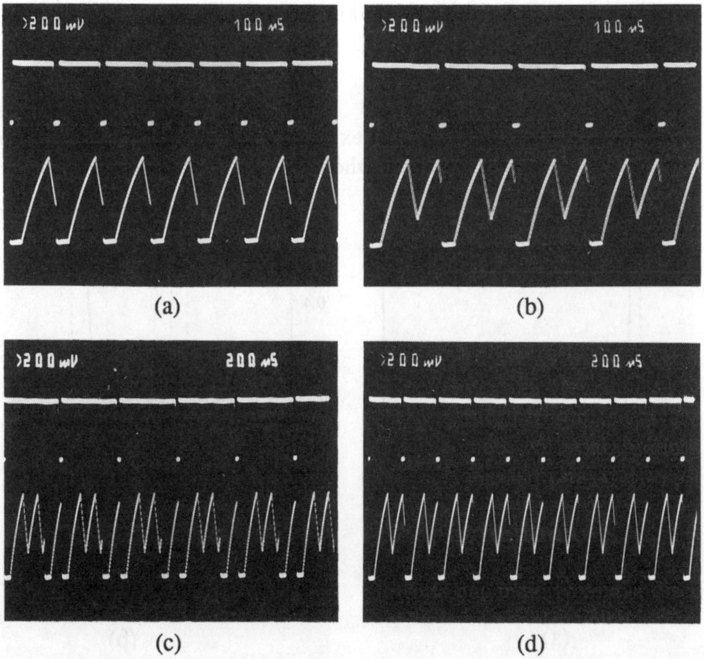

Figure 17: The waveforms of $d_{dr}(t)$ (above) and $u_1(t)$ (bellow) measured in the regimes of chaos suppression: (a) $ST^{1,1}_{O_1}$, (b) $ST^{1,1}_{O_3}$, (c) $ST^{2,1}_{P_2}$, (d) $ST^{3,4}_{P_3}$.

response oscillators were set to be close to each other and fixed. The parameters α_{dr}, β_{dr}, α_s, β_s and the normalized amplitude $\epsilon = \frac{E}{\sigma u_{\max}}$ of pulses d_{dr} were taken as control parameters.[r]

Figure 18 presents the synchronization zones measured in experiment (solid lines) and plotted on the parameter plane (α_s, ϵ) with fixed values of α_{dr} and $\beta_{dr} = \beta_s$. The diagrams in Fig.18a indicates that locking is possible if $\alpha_s > \alpha_{dr}$, which reflects the fact that, for threshold synchronization each synchronizing pulse d_{dr} should appear before the voltage $u_{2\ s}$ reaches the threshold level. The loss of synchronization with increasing of α_s is caused by the fact that the voltage $u_{2\ s}$ is so small that at the moment the pulse d_{dr} appears, the threshold level cannot be reached. It is quite easy to show that if the parameters $u_{\max}, u_{\min}, T_p$ and β are the same in both circuits, the

[r]Here and in the text below the subscripts "dr" and "s" denote the parameters of driving and response (synchronizing) oscillators, respectively.

synchronization zone is defined by the following conditions

$$\alpha_s > \alpha_{dr} \ , \qquad \epsilon > 1 - \frac{\alpha_{dr}}{\alpha_s} . \tag{22}$$

The spread in the parameters of the experimental circuits leads to narrowing of the synchronization zones which vanish when $\epsilon = \epsilon_{\min} > 0.$, see Fig.18a.

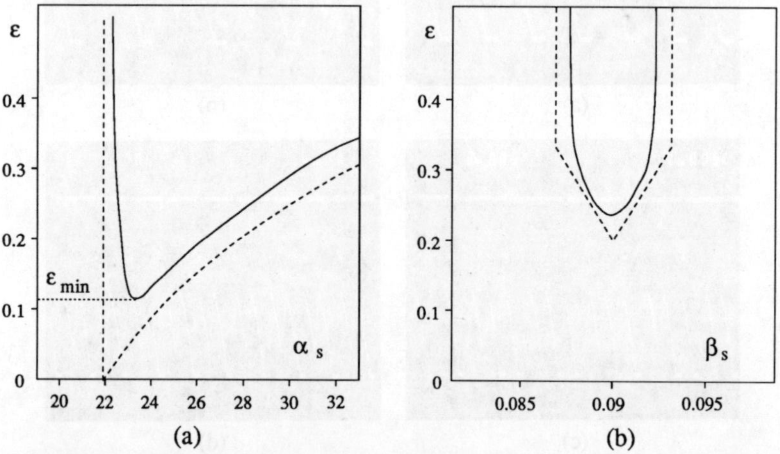

Figure 18: Parameter regions corresponding to regimes of synchronized chaotic oscillations. (a) Parameter plane (α_s, ϵ) with $\beta_{dr} = \beta_s = 0.09\frac{V}{\mu s}$ and $\alpha_{dr} = 21.9\frac{\mu s}{V}$. (b) Parameter plane (β_s, ϵ) with $\alpha_{dr} = 25.2\frac{\mu s}{V}$, $\alpha_s = 31.8\frac{\mu s}{V}$, and $\beta_{dr} = 0.09\frac{V}{\mu s}$. The theoretical boundary is marked by dashed line. Solid line shows the boundary found in experiment.

The synchronization of chaotic pulse oscillators has a property which differ it from threshold synchronization of periodic pulse oscillators. Unlike periodic oscillations for which only one parameter (period) is essential for the synchronization, chaotic behavior of relaxation oscillators is determined by many parameter of the circuit. Consequently the threshold synchronization of chaotic signals is possible only when *all* parameters of the oscillators are sufficiently close.[s] For example β can be considered as an additional control parameter. Analysis of the map Eq.19 shows that when the corresponding parameters $u_{\min}, u_{\max}$ and T_p are equal but $\beta_{dr} \neq \beta_s$, synchronization can be realized if

$$\epsilon > 1 - \frac{\alpha_{dr}}{\alpha_s} = \epsilon_{cr} \ , \ |\beta_s - \beta_{dr}| < \frac{(\epsilon - \epsilon_{cr})}{\alpha_s} \ , |\beta_s - \beta_{dr}| < \frac{u_{\min}\epsilon_{cr}}{u_{\max}\alpha_{dr}} \ . \tag{23}$$

[s]This property of the synchronization of the chaotic relaxation systems can be employed for the purposes of private communications.

The theoretical zones of the synchronized chaotic oscillations are shown in Fig.22a,b by dashed line.

The results of theoretical studies Eq.22 and Eq.23 can be easily applied for the case of mutual pulse-coupling between the relaxation oscillators. In this case, despite mutual coupling, one of them will play the role of drive oscillator.

These experiments show that relaxation chaotic oscillators with certain ratios of the parameters and sufficiently strong coupling can operate in the regime of synchronized chaotic pulsations. It is of fundamental importance that such a regime for identical oscillators can be achieved for arbitrarily weak coupling in spite of the fact that both oscillators may do not generate any stable regular behavior.

4.4 Application of the synchronized chaotic relaxation circuits for private communication.

Applications of chaotic relaxation oscillators modulated by information bearing signals for purposes of private communications were discussed in a number of recent papers, see for example references[55, 71]

Let us consider a method of communicating with the synchronized relaxation circuits shown in Fig.14. In order to modulate the chaotic pulsations of the relaxation circuit we can vary either the threshold level u_{th}^{dr} or the parameter α_{dr} in accordance with information bearing signal, [t] $v(t)$.

The chaotic pulses generated by the modulated relaxation circuit can be transmitted and then received by another relaxation circuit, which use these pulses for the synchronization. When the synchronized chaotic relaxation circuit oscillates in sync with the transmitter, the synchronizing pulses come before it ever reaches the threshold level u_{th}^{s}. To recover the information bearing signal we measure the voltages $v_r(t_n) = u_{th}^{s} - u_2^{s}$, where t_n is time when d_{dr}^{n} arrive. In the case of modulation of the threshold level $u_{th}^{dr} = v(t)$ with $\alpha_{dr} \approx \alpha_s$ and $u_{th}^{dr} > u_{th}^{s}$ we have $v_r(t_n) \cong v(t_n)$.

The same method can be used for the recovering the signal with the modulation of the parameter α_{dr}. The mean value of the voltage $u_{2\ s}$ at the moment a synchronizing pulse appears, $\overline{u}$ depends on the parameter difference (see Fig.19a). This dependence has a linear part where voltage $\overline{u}$ is proportional to α_{dr}. This property can be used for demodulation. The process of demodulation is shown in Fig.23 for the sinusoidally modulated parameter $\alpha_{dr}(t) = \alpha_0(1 + m\ \sin(\Omega t))$. It can be seen from the waveforms that the bending line of voltage $u_2^{s}(t_n)$ is a copy of the modulation signal $\alpha_{dr}(t)$.

The recovery of the information bearing signal is provided by the properties of the synchronization regime of chaotic relaxation circuits. Because the synchronization of chaotic pulsations is possible only in the case where *all* parameters of transmitting and receiving system are close enough, the information is not available for other receiving systems.

---

[t]We assume that the highest frequency of the modulation is much less then $\frac{1}{T_{max}}$.

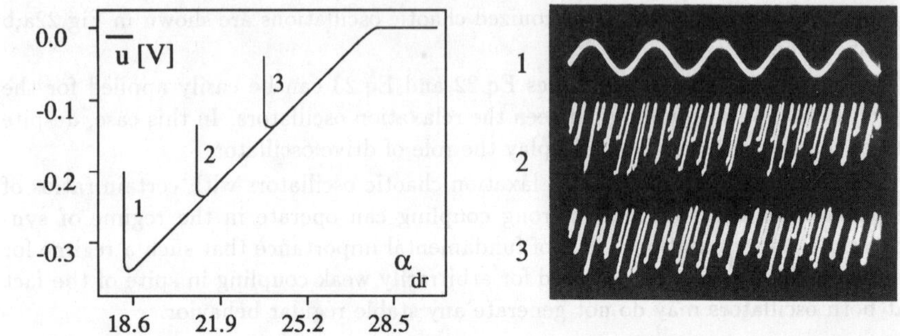

(a) (b)

Figure 19: a) Voltage $\overline{u}$ vs parameter α_{dr} with $\beta_{dr} = \beta_s = 0.09\frac{V}{\mu s}$ and $\alpha_s = 28.5\frac{\mu s}{V}$. Lines 1, 2 and 3 are obtained for E equal to $0.4V$, $0.3V$ and $0.15V$ respectively. (b) Waveforms in the regime of synchronized chaos with parameter modulation of the driving oscillator: 1) $\alpha_{dr}(t) = \alpha_0(1 + m\sin(\Omega t))$, 2) $u_{2\ dr}(t)$ and 3) $u_{2\ s}(t)$.

5. Conclusions

In this paper we have discussed different ideas which can be employed for investigations of chaotic behaviors generated by electronic circuits in physical experiments. All of these ideas are based upon well known results of qualitative theory of dynamical systems. We have tryed to apply these results for studies of the bifurcations which can explain the main properies of chaotic behavior observed in the experiments. To find these bifurcations we used experimental analyses based upon the examinations of different projections of the attractors, Poincaré cross sections of the attractors and return mappings generated by the trajectories of the attractors. When the bifurcations are defined, the results of the experiments enable us to derive the main goals of theoretical studies.

We employed electronic circuits to describe and experimentally demonstrate different methods for synchronizing chaotic behaviors in coupled chaotic oscillators. It has been shown that the methods of dissipative driving and threshold synchronization by external signals can be employed for the purposes of chaos suppresion.

6. Acknowledgements

The authors wish to express their gratitude to H. I. D. Abarbanel, I. S. Aranson, V. S. Afraimovich, M. I. Rabinovich, V. D. Shalfeev and L. S. Tsimring for stimulating discussions. Part of the results presented here have been obtained in collaboration with M. G. Velarde, A. Rodrigues-Losano and E. Del Rio. The authors are grateful to R. Brown, M. M. Sushchik, T. L. Carroll and L. M. Pecora for valuable comments.

This research has been partially supported by U.S.Department of Energy Contract No. DE-FG03-95ER14516, ARPA Contract No. 92-F141900-000, Office of Naval Research Contract N00014-D-0142 D0#15, DGICYT (Spain) Grant PB90-264, and by the European Union under Network Grant EXBCHRXCT 930107.

References

[1]A. A. Andronov, A. A. Vitt, and S. E. Khaikin *Theory of oscillators* (Oxford, New York, Pergamon Press,1966)

[2]M. I. Rabinovich, *Sov. Phys. Usp.* **21** (1978) 443.

[3]S. Vohra, M. Spano, M. Shlesinger, L. Pecora, and W. Ditto, *Proceeding of the 1st Experimental Chaos Conference*, (Wold Scientific, New Jersey, 1992).

[4]F. C. Moon, *Chaotic and fractal dynamics. An introduction for applied scientists and engineers* (A Wiley-Interscience Publication, NY, 1992).

[5]A. V. Gaponov-Grekhov and M. I. Rabinovich, *Nonlinearities in action: oscillations, chaos, order, fractals* (Springer-Verlag Berlin Heidelberg 1992) p.190.

[6]B. van der Pol and J. van der Mark, *Nature* **120**(1927) 363.

[7]J. P. Gollub, T. O. Bruener and B. G. Danly, *Science* **200**(1978) 46.

[8]J .P. Gollub, E. J. Romer and J. E. Soclar, *J. Stat. Phys.* **23**(1980) 321.

[9]V. Ya. Kislov, N. N. Zalogin and E. A. Myasin, *Radiotekh. Electr.* **24**(1979) 1118. [In Russian].

[10]S. V. Kiyashko, A. S. Pikovsky and M. I. Rabinovich, *Radio Engineering and Electronic Physics* **25** (1980) 74.

[11]P. S. Linsay, *Phys. Rev. Lett.* **47** (1981) 1349.

[12]J. Testa, J. Peres and C. Jeffries, *Phys. Rev. Lett.* **48** (1982) 714.

[13]Y. Ueda and N. Akamatsu, *IEEE Trans. CAS* **CAS-28**(1981) 217.

[14]M. Shinriki, M. Yamamoto and S. Mori, *Proc. IEEE* **69**(1981) 394.

[15]V. S. Anishchenko and V. V. Astakhov, *Radio Engineering and Electronic Physics* **28**(1983) 59.

[16]A. S. Dmitriev, V. Ya. Kislov and S. O. Starkov, *Sov. Phys.-Tech. Phys.***30**(1985)1439.

[17]T. Matsumoto, *IEEE Trans. CAS.* **CAS-31**(1984) 1055.

[18]G.-Q. Zhong and F. Ayrom, *Int. J. Circuit Theory Appl.* **13** (1985) 93.

[19]A. R. Volkovskii and N. F. Rulkov, *Sov. Tech. Phys. Lett.* **14**(1988) 656.

[20]I. S. Aranson and N. F. Rulkov, *Sov. Phys. Tech. Phys.* **33** (1988) 1003.

[21]N. F. Rulkov, A. R. Volkovskii, A. Rodrigues-Lozano, E. Del Rio and M. G. Velarde, *Int. J. of Bifurcation and Chaos* **2** (1992) 669.

[22]N. F. Rulkov, A. R. Volkovskii, A. Rodrigues-Lozano, E. Del Rio and M. G. Velarde, *Chaos, Solitons and Fractals* **4** (1994) 201.

[23]P. Coullet, C. Tresser and A. Arneodo, *Phys. Lett.* A**72** (1979) 268; A. Arneodo, P. Coullet and C. Tresser, *Commun. Math. Phys.* **79** (1981) 573; A. Arneodo, P. H. Coullet and E. A. Spiegel, *Phys. Lett.* A**92** (1982) 369.

[24]Ch. W. Wu and N. F. Rulkov, *IEEE Trans. Circuits Syst.-I: Fundamental Theory and Applications* **40** (1993) 707.

[25]M. J. Feigenbaum, *J. Stat. Phys.* **19** (1978) 25.

[26]P. Gaspard, R. Kapral and G. Nicolis, *J. Stat. Phys.* **35** (1984) 697.

[27]A. P. Kuznetsov, S. P. Kuznetsov, I. R. Sataev and L. O. Chua, *Int. J. of Bifurcation and Chaos* **3** (1993) 943.

[28]C. Grebogi, E. Ott and J. A. Yorke, *Physica* D**7** (1983) 181.

[29]V. S. Afraimovich, in *Nonlinear waves: Structures and Bifurcations*, ed. A. V. Gaponov-Grekhov, M. I. Rabinovich, (Nauka,Moscow,1987)

[30]P.Chossat and M. Golubitsky, *Physica* D**32** (1988) 423.

[31]G. P. King and S. T. Gaito, *Phys. Rev.* A**46** (1992) 3092.

[32]M. G. M. Gomes and G. P. King , *Phys. Rev.* A**46** (1992) 3100.

[33]H. G. Schuster*Deterministic Chaos, An Introduction* (VCH,New York,1988).

[34]A. J. Lichtenberg and M. A. Lieberman, *Regular and Stochastic Motion* (Springer-Verlag, New York,1983)

[35]L. P. Shilnikov, *Math. USSR Sbornik* **10** (1970) 91.

[36]V. S. Afraimovich and L. P. Shilnikov, in *Nonlinear Dynamics and Turbulence*, ed. G. I. Barenblatt, G. Ioos and D. D. Joseph (Pitman, Boston,1983)

[37]I. M. Ovsyannikov and L. P.Shilnikov, *Math. USSR Sbornik*, **73** (1992) 415.

[38]L. A. Belyakov, *Mathematical Notes* **36** (1984) 838.

[39]P. Glendinning and C. Sparrow *J. Stat. Phys.* **35** (1984) 645.

[40]V. I. Arnold, *Mathematical Methods of Classical Mechanics*, (Springer,Berlin,1974).

[41]J. A. Glazier and A. Lischaber, *IEEE Trans. Circuits Syst.* **35** (1988) 790.

[42]S. Ostlund, D. Rand, J. Sethna, and E. Siggia, *Physica* D**8** (1983) 303.

[43]D. G. Aronson, R. P. McGehee, I. G. Kevrekidis, and R. Aris, *Phys. Rev.* A**33** (1986) 2190.

[44]V. S. Anishchenko, *Dynamical Chaos in Physical Systems: experimental investigation of self-oscillating circuits*, (BSB B.G.Teubner Verlag, Leipzig, 1989).

[45]G. I. Dykman, P. S. Landa and Yu. I. Neymark, *Chaos, Solitons, and Fractals* **1** (1991) 339.

[46]V. S. Afraimovich and L. P. Shilnikov, *Amer. Math. Soc. Transl.* **149** (1991) 201.

[47]P. Bryant and C. Jeffries, *Physica* A**25** (1987) 196.

[48]H. Fujisaka and T. Yamada, *Prog. Theor. Phys.* **69** (1983) 32.

[49]V. S. Afraimovich, N. N. Verichev and M. I. Rabinovich, *Radiophysics and Quantum Electronics* **29** (1986) 747.

[50]L. M. Pecora and T. L. Carroll, *Phys. Rev. Lett.* **64** (1990) 821.

[51]T. L. Carroll and L. M. Pecora, *Physica* D**66** (1993) 126.

[52]K. M. Cuomo and A. V. Oppenheim, *Phys. Rev. Lett.* **71** (1993) 65.

[53]A. R. Volkovskii and N. F. Rulkov, *Tech. Phys. Lett.* **19** (1993) 97.

[54]K. S. Halle, Ch. W. Wu, M. Itoh and L. O. Chua, *Int. J. of Bifurcation and Chaos* **3** (1993) 469.

[55]N. F. Rulkov and A. R. Volkovskii, *Phys. Letters* A**179** (1993)332.

[56]R. Roy, K. S. Thornburg, Jr., *Phys. Rev. Lett.* **72** (1994) 2009.

[57]R. Brown, N. F. Rulkov and E. R. Tracy, *Phys. Rev.* E**49** (1994) 3784.

[58]T. C. Newell, P. M. Alsing, A. Gavrielides and V. Kovanis, *Phys. Rev.* E**49** (1994) 313.

[59]N. F. Rulkov, L. S. Tsimring and H. D. I. Abarbanel, *Phys. Rev.* E**50** (1994) 314.

[60]K. Pyragas, *Phys. Letters* A**170** (1992) 421.

[61]L. M. Pecora and T. L. Carroll, *Phys. Rev.* A**44** (1991) 2374.

[62]T. L. Carroll and L. M. Pecora, *IEEE Trans. CAS.* **38** (1991) 453.

[63]A. R. Volkovskii and N. F. Rulkov, *Sov. Tech. Phys. Lett.* **15** (1989) 249.

[64]Y. S. Tang, A. I. Mees and L. O. Chua, *IEEE Trans.Circ.Sys.* **30** (1983) 620.

[65]T. Allen, *IEEE Trans.Circ.Sys.* **30** (1983) 627.

[66]R. E. Mirollo and S. H. Strogatz, *SIAM J.Appl.Math.* **50** (1990) 1645.

[67]J. M. Perez, Z. Yu, J. M. Kovalskii, G. Albert, C. L. Littler and X. N. Song, in *Proceedings of the 1-st Experimental Chaos Conference* (World Scientific, 1991) 327.

[68]P. A. Bernhardt, *Physica* D**52** (1991) 489.

[69]V. S. Anishchenko and V. V. Astakhov, *Sov. Tech. Phys. Tech.* **28** (1983) 1326.

[70]R. Lima and M. Pettini, *Phys. Rev.* A**41** (1990) 726.

[71]P. A. Bernhardt, in *Chaos in Communications*, edited by Louis M. Pecora, SPIE Proc. **2038** (SPIE, Bellingham, WA, 1993) 162.

Part II: Applications of Circuits

ANALOG SIMULATIONS OF CHAOTIC AND STOCHASTIC SYSTEMS

Leone Fronzoni
Dipartimento di Fisica
Universita' di Pisa
Piazza Torricelli, 2
I-56100 Pisa, Italy

Frank Moss
Department of Physics and Astronomy
University of Missouri at St. Louis
St. Louis, MO 63121, USA

March 10, 1995

Abstract

Techniques for the study of chaotic and stochastic, or noise driven, systems by analog simulation are reviewed. Several example systems, including systems showing bifurcations to a limit cycle and to chaos, are employed in order to illustrate the general principles involved. Some symmetry properties of chaotic systems are illustrated using the Josephson junction model as an example. Chaotic systems which are also subject to high dimensional random disturbances are discussed. Simulations of a simple chaotic time delay system are put forth. Stochastic resonance is used as a generic example of the influence of purely random forces on certain weakly periodic nonlinear systems.

1 Introduction

The electronic circuits are a very useful devices for studying the behavior of dynamical systems. The first relevant application in this direction appeared just at the beginning of the science of electronics and before the arrival of digital computers[1]. These applications were devoted to the creation of an analogue simulation of systems that were not trivial to solve by analytical calculations. Stratonovich and Landauer in 1960 performed electronic simulations in order to study the effects of noise on nonlinear systems. Stratonovich[2] showed that the use of electronic circuits is a simple expedient for exploring the general problems of physical systems far from equilibrium. Landauer[3] studied for the first time a bistable system using a tunnel diode. This allowed him to study the activation process due to thermal fluctuations necessary to surmount the barrier between two states. In spite of the development of the digital computer, the analogue simulation remains a useful method of analysis in cases where long computing times are necessary to obtain the solutions, or when noise plays an important role in the systems. In the following sections we shall give some brief notions on how to use analogue devices, and then some relevant examples are shown.

Consider a simple ordinary differential equation (ODE) of the form

$$\dot{x} + a = 0 \tag{1}$$

The solutions of this equation can be related to the dynamics of an electric circuit (an integrator) consisting of a capacitance and resistance connected as show in Fig. 1.
The sum of the currents at the node provides an equation that formally is equivalent to eq. 1.

$$C\dot{V}_u + \frac{V_i}{R} = 0 \tag{2}$$

the voltage V_u is related to the input voltage V_i by an integration operation

$$V_u = \frac{1}{RC}\int V_i dt, \tag{3}$$

where $\tau = RC$ is the integration time constant of the circuit.

Connecting this integrator alone with other electronic elements results in changes of the values of the integration time which depend on the properties

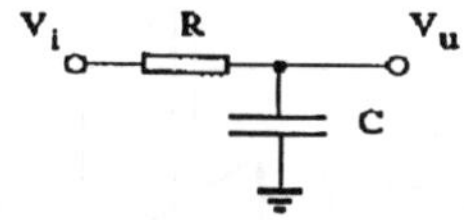

Figure 1: A simple RC integrator.

of the devices connected. In order to avoid this property, an operational amplifier is assembled as shown in Fig. 2. Considering the node A at the input of the amplifier, one obtains the same relation described by eq.1. A representation of more terms in this equation can be accomplished by the inclusion of additional input voltages by adding new resistances.

In many cases one is interested to simulate non-linear differential equations (NDE) where, for example , the variable of interest appears in a form such as x^2, x^3 etc. Multiplier devices in a chip configuration can be easily obtained at a modest price. A significant example of NDE is the Duffing equation which plays an important role in many models of bistable systems:

$$\ddot{x} - \alpha x + \beta x^3 + \gamma \dot{x} - A \cos \omega = 0, \tag{4}$$

The schematic diagram of the Duffing circuit simulator in a compact realization using a minimum number of electronic elements is shown in Fig. 3. With added noise (see below) this equation has been extensively studied[jung], and, in the limit of large damping, $\ddot{x} \rightarrow 0$ and $\gamma \rightarrow \infty$, this model has played a significant role in research on the phenomenon of Stochastic Resonance [17][18][19][20].

Two integrators are connected in a loop with two multipliers (ANALOG DEVICES Inc., model AD534). The multipliers are connect such as to result in the nonlinear term in the Duffing equation. The voltage, V, at the output

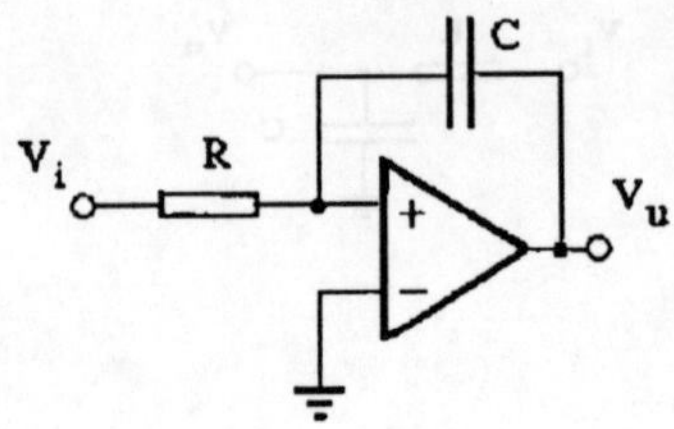

Figure 2: A Miller Integrator consisting of an operational amplifier with a feedback capacitor and input resistor.

of the second integrator, corresponds to the variable of interest, x. Sending this voltage to the two inputs of the first multiplier allows one to obtain the square quantity, while the second multiplier gives the voltage corresponding to the x^3 term in eq.4. The sum of the currents entering the node, A, results in the following equation:

$$\frac{V_\xi}{R_\xi} + \frac{V_\lambda}{R_\lambda} + \frac{V_\chi}{R_\chi} + \frac{V_u}{R} + \frac{\dot{V}}{R} = 0, \tag{5}$$

where

$$V_u = \frac{x^3}{100} - \frac{V_0}{10}x, \tag{6}$$

while at the node B we have

$$C_2\,\dot{x} + \frac{V}{R_2}V = 0. \tag{7}$$

Substituting eq.7 and its derivative in eq.5 one obtains an equation that corresponds to the Duffing equation 4, where

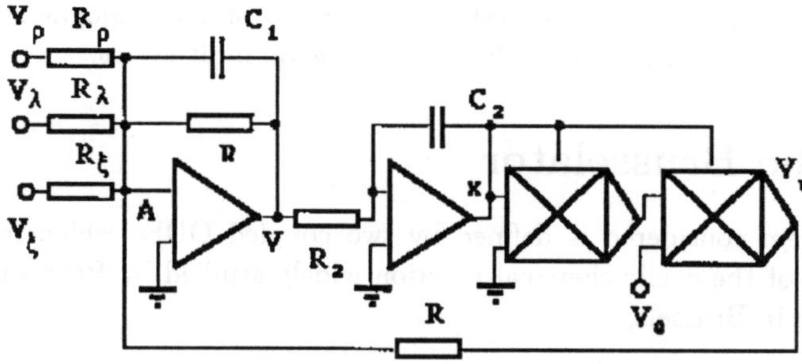

Figure 3: Schematic diagram of the Duffing-Holmes oscillator.

$$
\begin{aligned}
\alpha &= \frac{V_0}{10RR_1C_1C_2}, \\
\beta &= \frac{1}{100RR_1C_1C_2}, \\
\gamma &= \frac{1}{R_1C_1}, \\
\xi &= \frac{V_\xi}{R_\xi R_2 C_1 C_2}.
\end{aligned} \tag{8}
$$

These relations make a link between the values of the resistances and capacitances and the parameters present in the Duffing equation. A check of the reliability and accuracy of the simulator can be obtained from direct measurements of the resonance frequency ν_0, the stable points x_0, and the width of the resonance $\Delta\nu$.

$$
\begin{aligned}
\alpha &= \frac{1}{2(2\pi\nu_0)^2}, \\
\beta &= \frac{\alpha}{x_0^2}, \\
\gamma &= \pi\Delta\nu.
\end{aligned} \tag{9}
$$

In a subsequent section, we shall consider further the Duffing simulator in order to demonstrate how the addition of noise (a high dimensional random process) can influence the chaotic behavior near a crisis. First, however, we are going to show an interesting application of the analogue simulation technique for studies of the influence of noise on limit cycles.

2 The Brusselator

The system considered is defined by two coupled ODEs which model the behavior of the cyclic chemical reaction widely studied by Prigogine[4] and his group in Brussels:

$$\begin{aligned} \dot{x} &= A - (1+B)x + x^2 y, \\ \dot{y} &= Bx - x^2 y, \end{aligned} \tag{10}$$

where B is a control parameter, and A is a constant. The solution of this system of equations is characterized by a limit cycle that consists of a periodic solution of the two variables x and y. The dynamics in the phase space portraits, x-y, shows cyclic trajectories that arise when the parameter, B, exceeds a typical value, $B_c = 2$. We are interested to study the threshold B_{th} when noise is added to the control parameter

$$B_{th} = B_c + V_n, \tag{11}$$

where V_n is a Gaussian colored noise of intensity D, and correlation time τ, defined by

$$< V_{n(t)} V_{n(s)} >= \frac{D}{\tau} \exp \frac{-|t-s|}{\tau}. \tag{12}$$

In order to simulate the system of eqs.10 it is important to observe that in order to find the solutions $x(t)$ and $y(t)$, two integrations are necessary. The nonlinear terms present in the equation can be obtained by means of two multipliers. A single multiplier could assure a square term, x^2, and a second one serves to multiply this with the y variable . The schematic of this simulator is shown in fig.3.

2.1 Scheme of the Brusselator

Considering the currents at the inputs of the two integrators and using a suitable time rescaling $\tau_B = RC$, allows us to write two equations with the same form as eqs 10. The voltages at the output of the two integrators thus

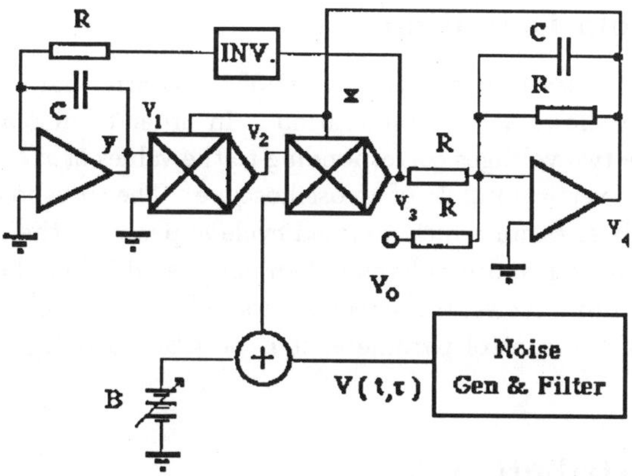

Figure 4: The Brusselator simulator. The variables x and y are given by the voltages V_4 and V_1.

give the x and y variables as shown in Fig.4. The values of the capacitances and resistances determine the limit cycle frequency, $\omega_B = \frac{1}{RC}$, while $V_0 = 1$ assures the value of the parameter $A = 1$. By changing the voltage B, one can vary the control parameter B, in eqs.10 .

The output voltage of a wide bandwidth noise generator is sent to a low-pass filter in order to obtain a colored noise with a well defined correlation time, τ_n . However, in eq. 10, the parameters are dimensionless, hence we must consider the dimensionless noise correlation time $\tau = \tau_n/RC$. In the experiment, we apply noise with different values of τ_n. One considers the noise to be colored when τ_n approaches the simulator time constant, $\tau_B \equiv RC$. The noise intensity, D, is obtained from the measured values of $< (V_n)^2 >= D\tau$.

2.2 Threshold Definition

Without noise it is easy to define the values of the control parameter B, for which the system shows a Hopf bifurcation . In order to display the limit cycle, we send the two voltages corresponding to the values of $x(t)$ and $y(t)$ to the horizontal and vertical inputs of an oscilloscope . The threshold is defined by the values of $B \equiv B_c$ for which a closed cycle appears on the oscilloscope screen . The situation is quite different when noise is added to the B voltage . Instead of clear limit cycles, the screen shows diffuse trajectories, and only for high values of the control parameter it is possible to recognize a closed cycle.

2.3 The Distribution

In the presence of noise it is useful to measure the distribution $P(x, y)$ which can be obtained by sending the voltages x and y to a data analysis system (for example, a PC with a suitable Analog-to-Digital Conversion board and software for acquiring and analyzing statistical samples of both variables). An example of the measured distribution is show in Fig. 5, where a typical, crater-like shape appears. By a sweep of the control parameter, increasing B_c, the distribution shows a transformation of its shape starting with a single peaked mountain structure followed by a volcano-like structure with a deepening crater . So we consider it natural to define the threshold in presence of noise as that value of B_{th} for which the cavity arising in the distribution becomes a crater able to "hold water". This is a bifurcation in the topology of the structure of the distribution from a cross-section parallel to the x, y plane which is singly connected (crater cannot hold water) to one which is doubly connected (crater holds water). Using this definition it is possible to study the behavior of this threshold for different values of the noise intensity, D, and correlation time τ . Figs. 6 and 7 show the results, respectively, for the threshold ratios B^*/B_c as function the noise intensity D^* and versus the noise correlation time, τ. Here B^* means the values of the control parameter at which the distribution assumes the shape previously defined and B_c is the threshold of the Hopf bifurcation without noise. As indicated by the results, a postponement of the threshold appears as a result of the influence of the noise. This result is in clear disagreement with Linear Theory (LT), which predicts a lowering of the Hopf bifurcation threshold when noise is applied

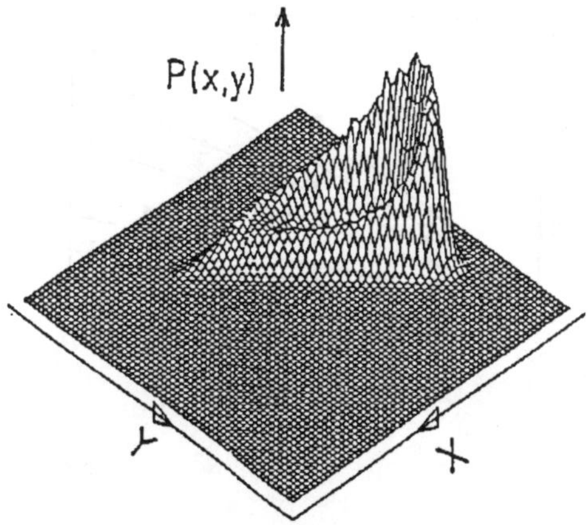

Figure 5: Two dimensional distribution $P(x,y)$ and its section in the x-y plane bor $B = 2.2, D = 0.008$ and $\tau = 0.10$.

to the system[5]. Noise induced postponements of bifurcation thresholds are now well known in a variety of nonlinear systems. We suppose that the linear prediction might be observable for very low values of noise intensity, where a linear approximation might be accurate. A test has been obtained using a description that considers the two-dimensional stationary density $P(\theta, u)$ with polar coordinates:

$$\begin{aligned} x &= 1 + \epsilon u^{\frac{1}{2}} \cos\theta, \\ y &= 2 + I\epsilon^2 + \epsilon u^{\frac{1}{2}} (\sin\theta - \cos\theta) \end{aligned} , \tag{13}$$

where

$$\epsilon = |< B > - B_c|^{\frac{1}{2}}. \tag{14}$$

The symbol, u, means the amplitude of the cycle . Integrating over θ, one obtains a one-dimensional probability $P(u)$. The LT gives a relation for the

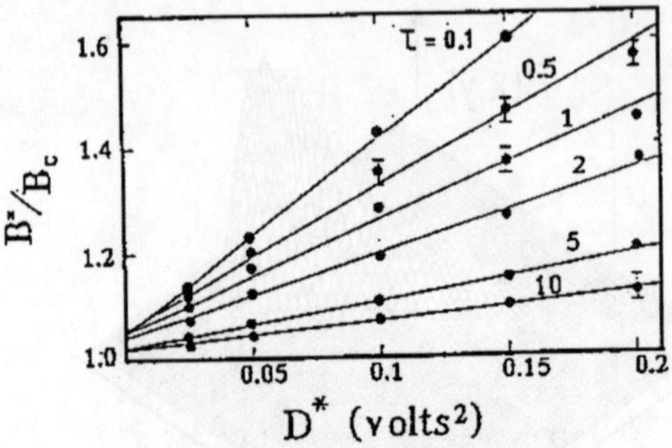

Figure 6: The control parameter ratio vs the noise intensity for various noise correlation times. B^* and B_c are respectively the threshold with noise and the threshold of the noise free Hopf bifurcation.

maximum amplitude as a function of the noise amplitude

$$u_m = \frac{4}{3} + \epsilon^2 \left(\frac{-352}{81} + \frac{47}{36} < V_n^{\,2} > \right). \tag{15}$$

The negative sign present inside the parentheses indicates that for very low noise one must observe a lowering of the threshold due to the fluctuations

By a computation of the u_m quantities from many experimental distributions we observed a qualitative agreement with this theoretical prediction . In Table I these results are summarized for very low amplitude noise , $\tau = 0.1$ and $\epsilon^2 = 0.1$. The experimental values show a non monotonous behavior as the noise amplitude is increasing, although the minimum does not correspond exactly to the amplitude predicted by the theory. For more information on applications involving the noisy Brusselator, the reader may consult reference[6] and reference[7].

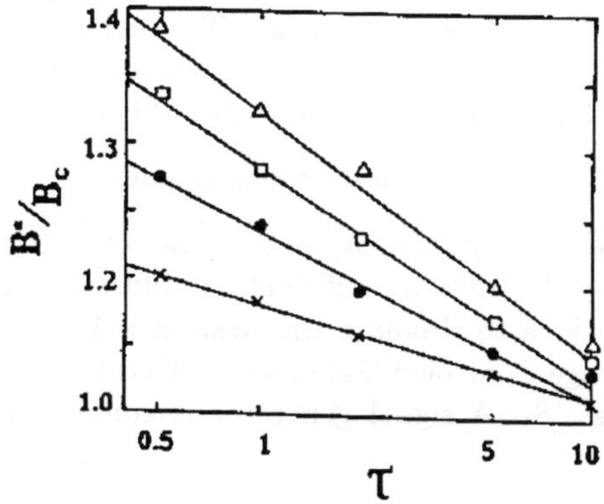

Figure 7: The control parameter ratio vs the noise correlation time for noise intensities: $D^* = 0.2$ (triangles); 0.15 (squares); 0.10 (circles); and 0.05 (crosses).

Table I. Theoretical and experimental values of u_m versus the mean square noise voltage.

.....$\langle V_n^2 \rangle$.....	.....u_m^{expt}.....	.....u_m^{theor}.....
0.02	0.69 ± 0.05	0.47
0.04	0.63	0.47
0.05	0.62	0.48
0.08	0.69	0.48
0.10	0.71 ± 0.08	0.49
0.15	0.75	0.50

3 The Josephson Junction Model

In this section we present another simple application of analogue simulation by means of electronic circuits . The system considered is the Josephson-

Junction model or the pendulum equation. This example is important, because it indicates the method to simulate a system with an infinitely extended periodic potential. The equation that describes this model is

$$\ddot{\phi} = -\sin\phi - \delta\dot{\phi} + \gamma\cos(\omega t). \tag{16}$$

The sine term indicates the periodicity in the space of the force, and δ , γ and $\omega/2\pi$ are respectively the friction coefficient , amplitude and frequency of the forcing . The main idea to simulate this system is based on an integrator controlled by a Voltage Controled Oscillator (VCO) followed by a multiplier M1 as shown in fig. 8. A signal generator is connected to the input of

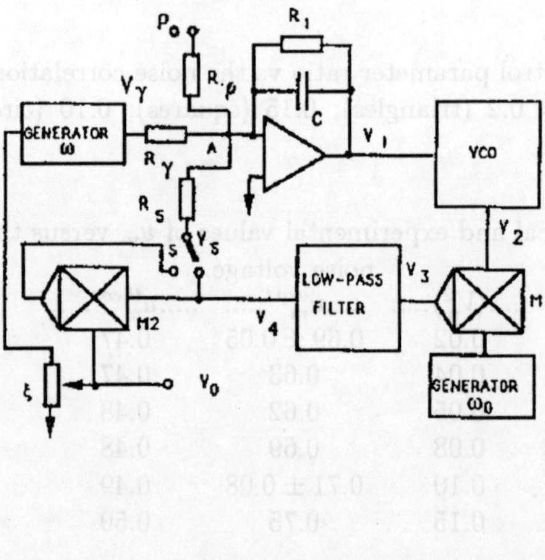

Figure 8: Schematic of the Josephson Junction simulator. To modulate, the voltage V_4 is applied to the second multiplier M_2 by placing the switch S in the left position.

the integrator and a second generator of frequency $\omega_0/2\pi$ is applied at the multiplier in order to get a multiplication with the VCO-output. By means of the switch S, it is possible to introduce a second multiplier M2 which serves to impose the modulation that will be discussed later.

Now , let us consider the currents in the A-node

$$\frac{V_5}{R_5} + \frac{V_1}{R_1} + \frac{V_\gamma}{R_\gamma} + C\dot{V}_1 = 0, \tag{17}$$

The phase ϕ of signal V_2 at the output of the VCO is driven by V_1 according to the relation

$$\dot{\phi} = kV_1, \tag{18}$$

where k is the frequency-modulation coefficient that characterizes the VCO. At the output of the multiplier M1, we get

$$V_3 = \sin(\omega_0 t + \phi)\cos\omega_0 t = \frac{1}{2}(\sin\phi + \sin(2\omega_0 t + \phi)). \tag{19}$$

A low-pass filter is applied at this voltage in such a manner as to exclude the component of frequency 2ω . Substituting this relation in the previous equation one obtains,

$$\frac{\sin\phi}{2R_5} + \frac{\dot{\phi}}{kR_1} + \frac{V_\gamma}{R_\gamma} + \frac{C}{k}\ddot{\phi} = 0, \tag{20}$$

which is the same as eq.16, which, however, is written in terms of dimensionless variables , and where V_γ represents the forcing term .

The study of the Josephson-Junction model involves many applications in physics and in particular it is interesting to analyze the chaotic properties of this system. Many studies are present in the literature regarding chaos in this system [8][9][10], but today a special interest is devoted to the influence of perturbations on the onset of chaos.

The circuit previously described is a good example in which to analyze the chaotic dynamics together with the influences of external perturbations. A general point of view arises when one considers the effect of a broken of symmetry due to the perturbation.

A simple method to break the symmetry in this system is to apply a small constant force. This means that an additive term ρ_0 must be included in eq.16

$$\ddot{\phi} = -\sin\phi - \delta\dot{\phi} + \gamma\cos\omega t + \rho_0. \tag{21}$$

Before studying the effects of the asymmetry on the onset of chaos, it is important to define such a threshold for the onset of chaos from the experimental point of view. The dynamics of the system can be followed by direct observation the phase-space portraits $(\sin\phi, \dot{\phi})$ on an oscilloscope screen. The Poincaré sections are obtained by means of a modulation of the z-axis of the oscilloscope and synchronized with the driving voltage . We consider the control parameter the amplitude γ of the voltage forcing and the threshold is defined by the value of γ where a strange attractor appears on the oscilloscope screen . The strange attractor is easily regcognized by its typical folded set look. We want to stress that measurements of the threshold of deterministic chaos involves some particular difficulties: a) Hysteresis and slowing-down effects make the determinations of the threshold depend on the direction and velocity in which the control parameter is varied. b) The presence of thin structures close to the boundary of the ordered and chaotic regions makes difficult the precise determination of the threshold. c) Because of the competition between both stochastic and deterministic noise, the statistics are not Gaussian, with the result that it is difficult to assign accurate amplitudes to the error bars. In any case one can determine the threshold approximately by very slowly increasing the ac driving voltage slowing even more as it approaches the threshold.

3.1 The Melnikov method to predict the onset of chaos

Cicogna and Fronzoni [11] have given a well developed method of calculation for predicting the threshold of chaos for this model . Here we report some relevant considerations expressed in this paper. This theoretical technique is based on the appearance of the typical Smale-horseshoe chaos that it is given by the vanishing condition of the following function

$$M(t_0) = \int_{-\infty}^{+\infty} dt \dot{\phi}_0(t) F(\phi_0(t), \dot{\phi}, t), t + t_0) = 0, \tag{22}$$

where $F(\phi_0(t), \dot{\phi}, t)$ includes the non Hamiltonian contributions as damping and forcing, $\phi_0(t)$ is the homoclinic orbit evaluated at the Hamiltonian limit. In the symmetric condition, $\rho_0 = 0$, there are two homoclinic orbits and the above Melnikov relation gives a well known result:

$$\frac{\gamma}{\delta} \geq \frac{4}{\pi} \cosh \frac{\omega\pi}{2} = R_0(\omega). \tag{23}$$

A comparison with the experimental data just reported in the ref[9] shows a disagreement with this relation . For instance, the experiments show a minimum of the threshold as a function of the driving frequency in contrast with eq.23. This disagreement is due to the fact that it is impossible experimentally, to have a very symmetric condition and the non-monotonous behavior arrises because of the presence of even a very small bias term ρ_0. Moreover, a computation of the Melnikov relation shows that the solution is very sensitive to the bias term, and this depends on the fact that ρ_0 drastically changes the global behavior of the homoclinic orbits. Note also that the Melnikov integral eq.22cannot be evaluated analytically when $\rho_0 \neq 0$, but precise numerical calculations are possible. In any case, an expression for the asymmetric potential that allows one to reach an analytical solution is considered . The choice is a potential of the form

$$V(\rho_0, \phi) = -\cos\phi - \pi\rho_0 \sin\frac{\phi}{2}, \tag{24}$$

For small values of ϕ this relation approaches the potential considered in the experiment. So the integral 22 can be explicitly evaluated, giving the following condition for chaos:

$$\gamma \geq \frac{4\delta}{\pi}(\cosh\frac{\pi\omega}{2v})\frac{\nu - \pi\rho_0\frac{\lambda}{4}}{|\sin\frac{\omega\lambda}{v}|}, \tag{25}$$

where

$$\nu = (1 - \pi\frac{\rho_0}{4})^{\frac{1}{2}}, \tag{26}$$

and

$$\lambda = ln(1+\nu) - \frac{1}{2}ln\pi\frac{\rho_0}{4}. \tag{27}$$

These relations are in complete agreement with both numerical and experimental results corresponding to the potential of the form eq.24. Figure 9 shows the threshold values of the driving amplitude for the appearance of chaos versus the dc bias term ρ_0 for fixed viscosity, $\delta = 0.25$ and frequency, $\omega = 0.75$.

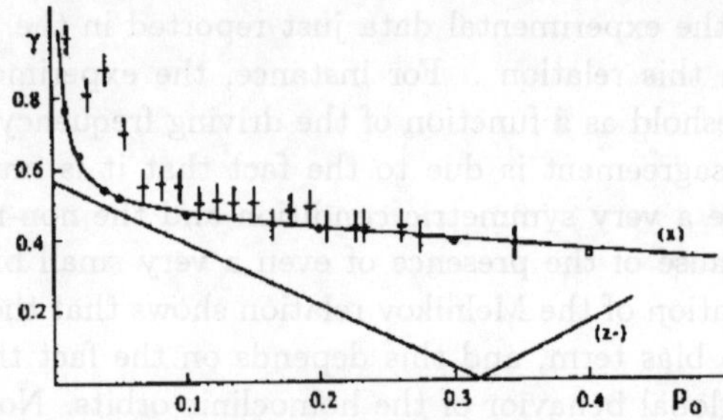

Figure 9: The threshold of chaos as a function of the bias term ρ_o for $\delta = 0.25$ and $\omega = 0.75$. Experimental data (crosses) are compared to the Melnikov condition eq.23 (circles) and theoretical results from the modified potential eq.24 (curve-x). The curve (z^-) is the lower threshold obtained by the approximate method described in Ref. [10].

3.2 Effect of the modulation

Now we consider a parametric modulation defined by the following equation

$$\ddot{\phi} = -[1 + \xi \cos(\Omega t + \theta)] \sin\phi + \rho_0 - \delta\dot{\phi} + \gamma \cos\omega t \tag{28}$$

The term in square brackets indicates a small periodic variation of the amplitude of the potential, where $\Omega/2\pi$ is the value of the frequency modulation and θ the phase relative to the forcing . In the case when $\rho_0 = 0$, the Melnikov function results in,

$$M(t_0) = -8\delta \pm 2\pi \,\text{sech}\, \frac{\omega\pi}{2} \cos\omega t_0 + 2\pi\xi\Omega^2 \,\text{csch}\, \frac{\Omega\pi}{2} \sin(\Omega t_0 + \theta) = 0, \tag{29}$$

which , assuming $\xi > 0$, gives the condition for which chaos does not appear

$$\gamma \,\mathrm{sech}\, \frac{\omega\pi}{2} + \xi\Omega^2 \,\mathrm{csch}\, \frac{\Omega\pi}{2} < \frac{4\delta}{\pi}. \tag{30}$$

It is interesting to note that to change the sign of ξ gives the same result as to change the sign in the eq.29. The latter corresponds to having homoclinic orbits with $\dot{\phi} > 0$ or $\dot{\phi} < 0$, which means that the presence of the modulation is equivalent to modifing the symmetry of the system . The resulting sign of the eq. 29 depends also on the sign of the differences of phase θ. For instance for $\theta = 0$ and $\xi > 0$, we can predict that the modulation introduces an asymmetry in the same way as the bias term ρ_0 does. Thus, the main result is a lowering of the actual threshold of chaos as is shown in fig. 10, where the experimental values of γ for the onset of chaos are reported as function the amplitude of modulation, ξ, at the resonance condition $\Omega = \omega$. The two lines mean the theoretical prediction deduced by the relation eq. 30 in the cases $\rho_0 = 0$ and $\rho_0 = 0.05$.

It is important to stress that the global effect of the modulation is to lower the threshold or to favor chaos depending on the relative phase of the modulation with the forcing. This is also in agreement with other results obtained with real Duffing-oscillators[12] and more recently in the experiments with lasers where the modulation, on the contrary, induces a suppression of chaos.

4 The effects of noise on a crisis in the Duffing-Holmes oscillator

An interesting application of simulation by means of electronic devices is the study of the effect of noise on deterministic chaos. In this section we show an experiment performed with the Duffing circuit discussed previously in section 1. We analyze the effect of noise on the system in the presence of crisis-induced intermittency[14][15]. A crisis appears when, by varying a control parameter, two attractors merge into a large attractor, and the trajectory of the system jumps ,very rarely, between two regions (repellors) corresponding to the two original attractors. The residence time distribution $G(t)$ obtained from the statistics of the time that the trajectory spends in

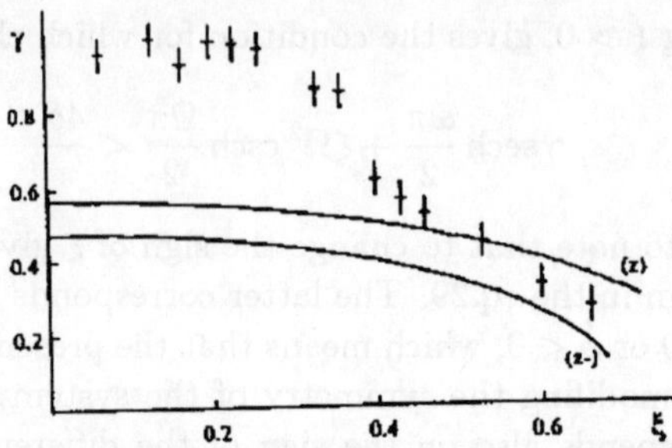

Figure 10: Experimental data and theoretical threshold for chaos versus the modulating amplitude ξ. In the experiment, $\rho_o, \theta = 0, \delta = 0.5$, and $\omega = 0.75$ were fixed. The curves (z) are the predictions and (z^-) is for the case $\rho_o = 0.05$ using the approximate method of Ref. 10.

one repellor has an exponential form

$$G(t) = G_0 \exp(-kt), \tag{31}$$

where k is the escape rate coefficient. In a recent paper Franaszek[16] studied the effect of noise on the repellors with computer experiments. The results of this work can be summarized by the following main points:

- 1- In presence of small noise the residence time distribution remains of the form expressed in eq.31 but with the escape coefficient k strongly dependent on the noise intensity σ.

- 2- The dependence $k(\sigma)$ shows a minimum corresponding to a critical value of noise amplitude σ^*.

- 3- The parameter k is also a function of a dimensionless parameter ρ, defined by

$$\rho = \frac{\sigma}{(\lambda - \lambda_0)}, \tag{32}$$

where λ is the control parameter, and λ_0 is the threshold of crisis-induced intermittence, without noise . $\rho = 2.5$ is the typical value which corresponds to the minimum of the escape rate k.

These results were obtained by means of a computer simulation for the case where a single repellor coexists with a stable attractor. In order to analyze this behavior we used the Duffing-Holmes electronic simulator previously described in section 1 .

The equation is given by the eq. 1.4 including a noise term $\xi(t)$ and a bias ρ.

$$\ddot{x} - \alpha x + \beta x^3 + \gamma \dot{x} - \lambda \cos(2\pi\nu t) + \rho + \xi(t) = 0, \tag{33}$$

where the control parameter is λ, which corresponds to the driving amplitude. The bias term ρ determines the asymmetry of the system, and the noise is a white Gaussian fluctuation of intensity D and variance $\Sigma = (<\xi^2>)^{\frac{1}{2}}$, in accord with the white noise definition,

$$<\xi(t)\xi(t+s)>= 2D\delta(s). \tag{34}$$

4.1 Experimental considerations

We determined the values of the parameters present in eq.33 by means of a direct measurement of the physical quantities, namely, the resonance frequency $\nu_0 = 300Hz$, the width of resonance $\Delta = 50Hz$, the equilibrium positions $x_{Left} = -3$ and $x_{rigth} = 3$. After a suitable time rescaling, one obtains the values of the parameters $\alpha = 1$, $\beta = 0.128$, and $\gamma = 0.23$. A choice of forcing amplitude $V_{pp} = 0.28V$ and forcing frequency $\nu_{ex} = 303.3Hz$ gives $\nu = 0.227$ and $\lambda = 1.11$. Special care was devoted to the bias term ρ , in particular, for a good control of the symmetry, we had to confine ρ to the range between 10^{-5} and 10^{-4}. Since the experimental noise has a high frequency limit, we consider that the dynamics mimics white noise when the time correlation of the noise, τ, is much less than the inverse of resonance frequency ν_0^{-1} (in this case $\tau = 7.510^{-3}/\nu_0$). The amplitude, D, of the noise is evaluated by means of a measure of the standard deviation $\sigma = < (V_n)^2 >^{\frac{1}{2}}$ at the output of the noise generator, according to $< (V_n)^2 > = D/\tau$. It is important to test the variance of the internal noise present within the simulator. We found $\sigma_{int} = 0.2mV$, much smaller than the external noise applied to the system.

For analyzing the effect of noise on the escape rate, k, it is necessary to measure the mean lifetime corresponding to the average time that the trajectory remains within a repellor. By means of a trigger, the output signal was converted to positive or negative square-pulses depending upon the sign of this voltage. Whence, the duration time of these pulses gives the residence time that the system spends in one repellor. We define the trajectory moving within the right repellor (R) for the positive pulse and within the left repellor (L) for the negative one. The distribution of lifetimes, $G(t)$, was computed by computer after the data were acquired by means of a digital oscilloscope (Data Precision model DATA 6000).

The rule of the asymmetry is determined by the dynamics of the system, and we defined the asymmetry ratio as $\phi = P_R/P_L$, where P_R and P_L are respectively the total area of the right half ($x > 0$) or of the left half ($x < 0$) of the probability distribution, $P(x)$, computed directly by the digital oscilloscope. When $\phi \rightarrow 1$, it means that the probability distribution is nearly symmetric in shape as showed in fig. 11a.

4.2 Experimental results

We observed that the noise does not destroy the symmetry in the case of $\rho = 0$, as is evidenced in figs(3.1 a,b), where the distribution $P(x)$ is shown for two values of noise intensity. Otherwise the noise influences strongly the symmetry of the system as is shown in fig (3.2 a,b,c,d). Note that a relatively weak noise transforms the asymmetric distribution (case a) into a symmetric one (case d).

Looking at the distribution $G(t)$, obtained in an asymmetric condition, fig.12 (L,R) shows that the distribution can be divided into two parts corresponding to short and long times. The escape rate, k, is defined as the slope of the second part of these distributions. It is important to observe that the right time distribution shows a nonmonotonic behavior as a function of the noise amplitude. In the asymmetric condition, the two repellors have two different thresholds λ_R and λ_L and so two different dimensionless parameters $\rho_R = \sigma/(\lambda - \lambda_R)$ and $\rho_L = \sigma/(\lambda - \lambda_R)$.

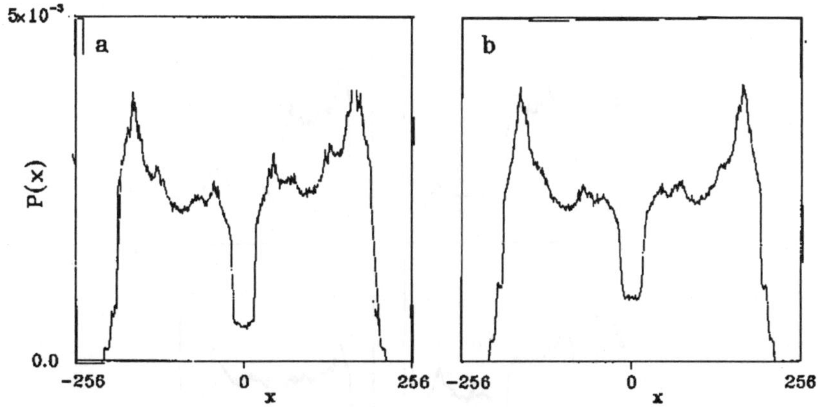

Figure 11: The statistical distribution $P(x)$ for (a) the symmetric case, $\varphi = 1.1$ and $\sigma = 0$; and (b) $\sigma = 6.25$ mV.

This gives different responses of the lifetime τ to the noise amplitude, see fig. 13 (L,R). For instance, the τ variation of the left repellor is much smaller than that of the right one. The non-monotonous behavior obtained for the right repellor generalizes the result obtained by Franaszek in ref [15]. In fact below a typical threshold of the noise intensity we get a stabilization of the transient chaos, and so it induces an increasing mean lifetime, otherwise, for noise intensities above this threshold, the tendency is reversed. It is particularly interesting to observe that the presence of a maximum in the mean lifetime versus the noise amplitude is quite similar to the stochastic resonance phenomenon [16-17-18], but with the substantial difference that in our case the noise amplitude is greater rather than less than the deterministic one. Figure 14 shows the normalized mean lifetime versus noise intensity.

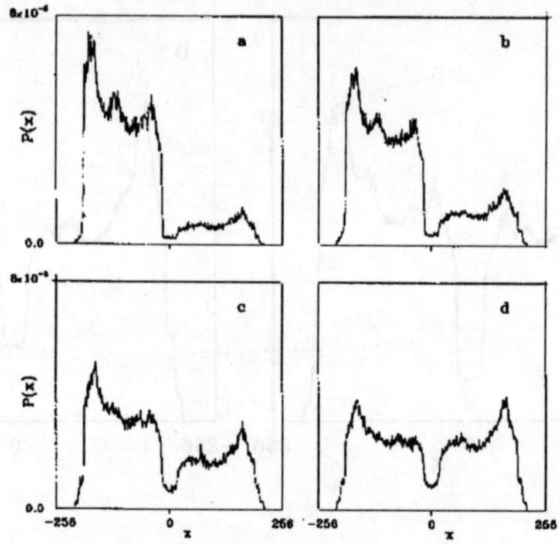

Figure 12: Statistical distribution $P(x)$ with bias $\rho = 2.5 \times 10^{-4}$ for the noise amplitudes: (a) $\sigma = 0$ mV, $\varphi = 0.16$; (b) $\sigma = 1.4$ mV, $\varphi = 0.30$; (c) $\sigma = 4.2$ mV, $\varphi = 0.56$; and (d) $\sigma = 7$ mV, $\varphi = 0.96$.

5 The Problem of Chaos in Biology

Chaotic behavior, as distinguished from random noise or high dimensional deterministic behavior, has not yet been rigorously and unambiguously demonstrated in any biological preparation, though it has been vigorously explored[21][22]. Recent work has, nevertheless, provided strong evidence for the existence of nonlinear low dimensional dynamical behavior in preparations involving simple neuronal circuits[23] and brain tissue[24]. These efforts have, however, made use of traditional time series analysis tools for the detection of low dimensional deterministic behavior[26][27][28][29][30][31]. These methods have two characteristics which render them difficult to use for analysis of biological data. They are notoriously sensitive to high dimensional noise contamination and they require large data files in order to yield satisfactory results. However, biological preparations nearly always yield data containing large

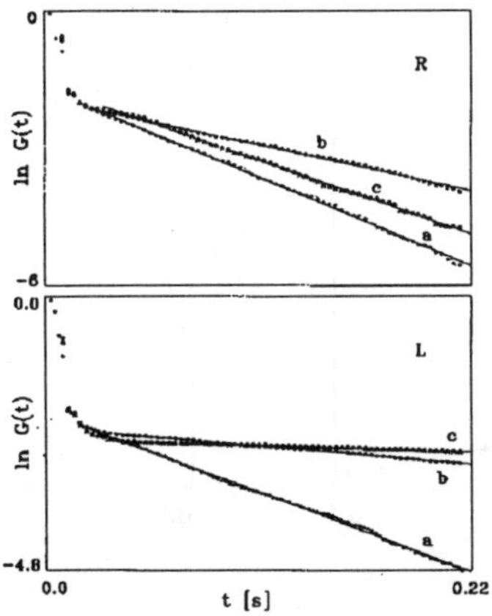

Figure 13: Distribution of lifetimes $G(t)$ in the left (L) and right (R) potential wells for the noise intensities: (a) $\sigma = 9.4$; (b) 2.4; and (c) 0 mV.

fractions of random noise. Moreover, they are not stationary for long periods of time. For example, an excised preparation, containing, say, neurons or excitable tissue, begins to die the moment it is extracted from some living structure. *In vivo* preparations hardly offer better behavior. The living structure does change in parameter space as a function of time. This means that biological data files are both noisy and short. The challenge is thus to detect chaos, or other deterministic behavior, and to distinguish it from the noise using such records. The difficulties associated with this enterprise has recently been highlighted[36].

Recently a new, and fundamentally different, approach has been put forth. Rather than analyze *all* the data points in a time series, as the traditional methods have all done, one searches the file for certain well defined rare events. Hopefully these events can be identified with sufficient precision to reliably ascertain that they are the signatures of some particular behavior being sought. In this way the majority of data points in the file - the noise

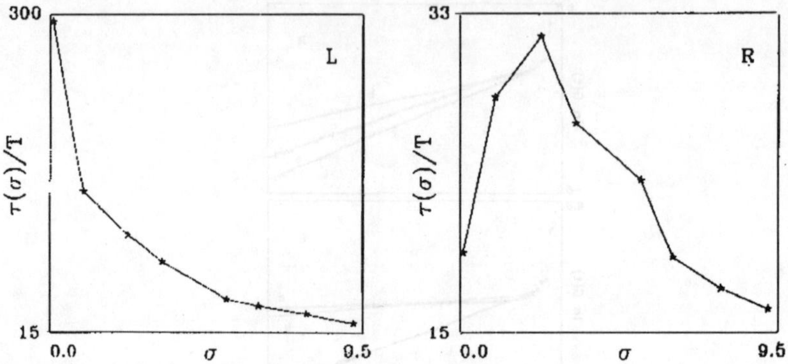

Figure 14: Mean lifetime versus noise intensity in left (L) and right (R) repellors.

- are ignored and hence do not contaminate the analysis. One such event, which has been used in the control of chaos in heart[32] and brain[33]tissue is the signature of a periodic unstable fixed point (PUFP). The theory of chaos control by manipulating trajectories near PUFPs is well established[34][35]. The statistics associated with the detection of rare events has recently been analyzed[37]. PUFPs occur in noisy files containing buried deterministic dynamical behavior, in their surrogates, and even in files made up from random numbers but with different probabilities. Moreover, the occurrence probability of some event depends strongly on the criterion used to define the event as well as on the background noise intensity. In order to reliably detect PUFPs in biological preparations, it is first necessary to study these probabilities in files obtained from systems with known and well defined behavior. Analog simulators are ideal for these studies, since their dynamics is well defined, and they can be made to mimic the behavior of biological preparations. Below we demonstrate how two example simulators - a driven

Van der Pol oscillator and a time delay Duffing system - can be used to measure the aforementioned probabilities.

6 Periodic Unstable Fixed Points (PUFPs)

The analysis herein presented stands upon a view of chaos as a structure built upon a skeleton of PUFPs[38][39]. Repeated encounters with PUFPs by trajectories lead to chaotic dynamics. As shown in the upper icon of Fig. 15, a PUFP is the intersection point of an unstable and a stable manifold. These manifolds lie on a saddle. Trajectories encounter the PUFP by approaches along or near the stable manifold and departures along or near the unstable manifold. Assuming that the dynamics can be embedded in some low dimension, say 2, then a variable $x(t)$ can be plotted against its prior value $x(t-1)$ as shown in the lower icon of Fig. 15. Linearizing the dynamics in the neighborhood of the PUFP, two eigenvalues determine the rates of approach along the directions of the stable and unstable manifolds.

6.1 The Signature of a PUFP

A discrete data file, $x(t_1), x(t_2)$...can be searched for sequences which display the correct eigenvectors which intersect at a specific PUFP[40]. A natural variable for biological data obtained, for example, from a time series of neural action potentials or heart beats, is the time interval between action potentials or neural spikes, T. Typically, one can plot the nth interval versus the $(n-1)$th on a scatter plot. Note that all periodic points, $T_n \equiv T_{n-1}$,lie on the line of identity, or 45° line shown dashed. A single encounter might appear as shown in Fig. 15 (b), where the approaching sequence is shown by the open circles labeled 1 - 4 and the departing sequence is depicted by the triangles labeled 5 - 8. The solid straight lines, which can be obtained by a fit to experimentally obtained points, show the directions of stable and unstable manifolds. The eigenvalues are given by the inverse slopes of these lines. Figure 15 (b) shows a flip saddle, for which sequential points fall on alternate sides of the 45° line, and for which the slopes, m, of the stable and unstable directions lie within the ranges $0 \succeq m_s \succ -1$ and $-1 \succeq m_{us} \succ -\infty$ respectively. The rates of departure, determined by the unstable eigenvalues, are thus always smaller than the rates of approach determined by the stable eigenvalues.

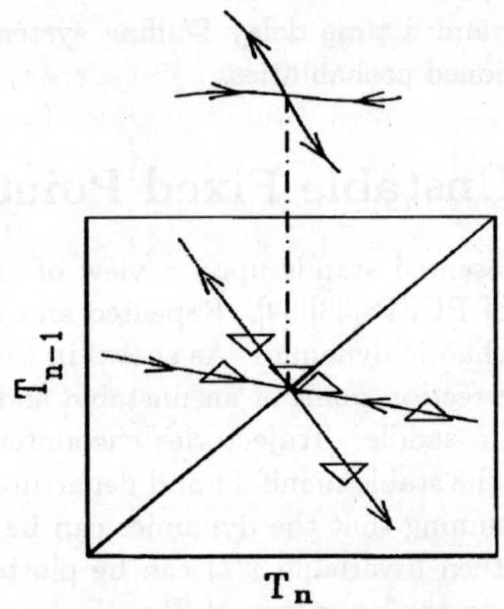

Figure 15: Representation of a saddle (upper) with stable and unstable directions shown by the arrows, and a signature (lower) shown by sequences of data points which approach (upright triangles) and depart (inverted triangles) the fixed point along the stable and unstable directions (arrows).

6.2 Noise Contamination and the Effects of Higher Dimension

Scatter plots of actual data files consist of many points and are usually more-or-less featureless. The vast majority of these points do not form sequences identifiable as PUFPs, or at least a single lower order PUFP. Indeed, chaotic attractors are build upon infinitely many PUFPs, and trajectories encountering these higher order PUFPs will project onto our two dimensional plane as a random appearing scatter of points. Fortunately, the PUFPs of lowest order are usually widely separated enough to be distinguishable. Moreover, even small noise contamination will blur the higher order PUFPs. Large noise will obscure even the lowest order PUFP. The problem in analyzing files from biological preparations is to find and identify the few encounters with the lowest order PUFP buried within this noise. In order to convincingly demonstrate chaos, it is also necessary to rigorously determine the statistics of the encounters not only in the original data files, but also in their surrogates and in random files. These statistics will be critically dependent upon the criteria used to define an encounter.

6.3 Encounter Criteria and the F Statistic

In the following demonstrations, we have used three levels of selection criteria varying from loose to more rigorous. In all cases, noisy data files, generated by the analog simulators are dragged in order to identify a number of encounters, N. Since the encounters are always defined by certain *sequences* of points, surrogate files can be produced by scrambling the sequence of all points in the original file. The scrambling should be random and, in the cases presented below, was accomplished with a Gaussian distribution. Upon each realization, the surrogate file is dragged for encounters using the same criteria as was used on the original file. The number of encounters so obtained from the surrogate file is N_s. The same original file can be scrambled many times (using different random seeds) and analyzed, so that the mean number of surrogate events $\langle N_s \rangle$ and its standard deviation σ are obtained. A statistical measure of the significance of N is thus given by

$$F = \frac{N - \langle N_s \rangle}{N + \langle N_s \rangle}. \tag{35}$$

Thus, detection of an event with perfect reliability might be $N = 1$ and $\langle N_s \rangle = 0$ so that $F = 1$. Moreover, for loose selection criteria, N will be large, but so also will be $\langle N_s \rangle$. For more rigorous criteria N will be smaller, but hopefully $\langle N_s \rangle$ will be moreso, so that F will be larger. This effect is demonstrated below.

Three selection criteria are used here. They are as follows:

- *Level 0:* Three points approach the 45° line at successively decreasing distances and are followed by three which depart at successively increasing distances.

- *Level 1:* The same as level 0, but now straight lines are fit to the two sets of points by linear regression and the slope conditions, $0 \geq m_s \succ -1$ and $-1 \geq m_{us} \succ -\infty$ are imposed.
- *Level 2:* The same but the lines must intersect within a circle of radius r_c centered on the PUFP.

7 Two Noisy, Chaotic Electronic Circuits

In order to demonstrate the techniques discussed above, we can consider two analog simulators. Both these electronic circuits output a continuous voltage, $x(t)$. However, in order to mimic typical neural firing events, or action potentials, we pass the outputs through threshold detectors. These output a single narrow pulse of standard shape whenever $x(t)$ crosses the threshold from below, but ignore the return threshold crossing from above. The circuits thus generate time sequences of narrow pulses in a way similar to that of an actual neuron which fires when its internal threshold is crossed from below by the membrane potential. The underlying process, $x(t)$, which mimics the membrane potential, can be either a limit cycle or a chaotic attractor with added noise. The time series of pulses are digitized and the time intervals between pulses are assembled into files. Data analysis is then carried out only on these time interval files.

The first analog simulator is that of a noisy, driven Van der Pol oscillator:

$$\ddot{x} + (\varepsilon - x^2)\,\dot{x} + x + A \sin \omega_d t + \xi(t) = 0, \tag{36}$$

where ε is the nonlinearity parameter ($\epsilon = 1.5$ in this work), $\xi(t)$ is the noise and ω_d is the driving frequency. Here the noise was quasi white, meaning that its dimensionless correlation time was small (0.1). Depending on the values of A and ω_d this system can exhibit a limit cycle or a chaotic attractor. Figure 16 shows phase space displays of chaotic and limit cycle attractors with and without noise. For this study we chose attractors which were close together in parameter space. We wanted the chaotic attractor to "look" similar to the limit cycle when even small amounts of noise were added. Figure 16 shows that the noisy attractors are indeed quite similar.

The second simulator was of a time delay differential equation:

$$\dot{x}\,(t) = x(t - T_d) - x^3(t - T_d) + \xi(t), \tag{37}$$

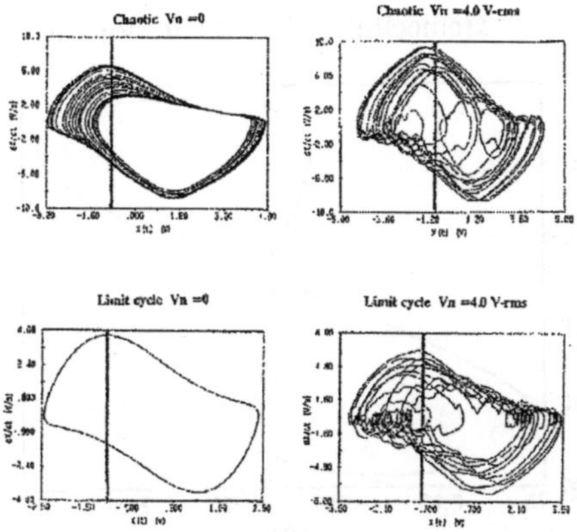

Figure 16: Phase portraits of the Van der Pol oscillator for a chaotic attractor (upper panels) and a limit cycle (lower panels). The noise free dynamics are shown on the left and noise contaminated dynamics are shown on the right.

where T_d is the time delay and the system's only bifurcation parameter. Note that the bistable forcing term on the right derives from the standard quartic potential. It is convenient to display the phase space of this system as a plot of the delayed coordinate, $x(t - T_d)$, versus the undelayed one, $x(t)$. Without delay, $T_d = 0$, the system relaxes to one of two stable fixed points of the potential on the 45° line. The single unstable fixed point of the potential bisects the 45° line. For $T_d \succ 0$, the system displays a complex set of bifurcations, which can be delineated for $\xi = 0$ by values of the dimensionless time delay. At $T_d = 1.179$ the fixed points lose stability giving birth to local limit cycles confined to the bottoms of the two potential wells. As the time delay increases, the limit cycles grow in amplitude. At $T_d = 2.110$, these orbits just touch the central unstable fixed point giving rise to a single homoclinic orbit as shown in Fig. 17. Beyond the homoclinic orbit lies a global limit cycle. We chose the one at $T_d = 2.392$ because it lies close to the

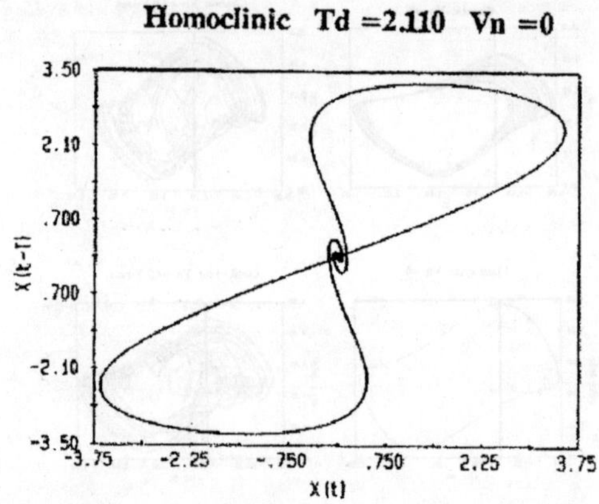

Figure 17: The homoclinic orbit of the delay Duffing system.

onset of a simple chaotic attractor at $T_d = 2.427$. We operated the experiment at the aforementioned limit cycle and just inside the chaotic region at $T_d = 2.472$. Phase space portraits of these attractors with and without noise are shown in Fig. 18. As before, it is quite difficult to distinguish "by eye" the two noisy attractors.

There were two reasons for choosing the time delay system, and operating it quite close to its homoclinic orbit. First, near the homoclinic orbit the time delay system is much more sensitive to the noise than was the Van der Pol oscillator which was quite robust to added noise comparable to the amplitude of the attractor itself. Compare the noise amplitudes in Figs. 16 and 18. Second, time delay systems are infinite dimensional systems for large time delay, and for small delay the dimension increases with the delay[41]. If increasing noise in any system represents increasing complexity, then increasing time delay in this system might play the same role as the noise. Moreover, we wanted to experiment with time delay differential equations as higher dimensional systems, because biological preparations are often de-

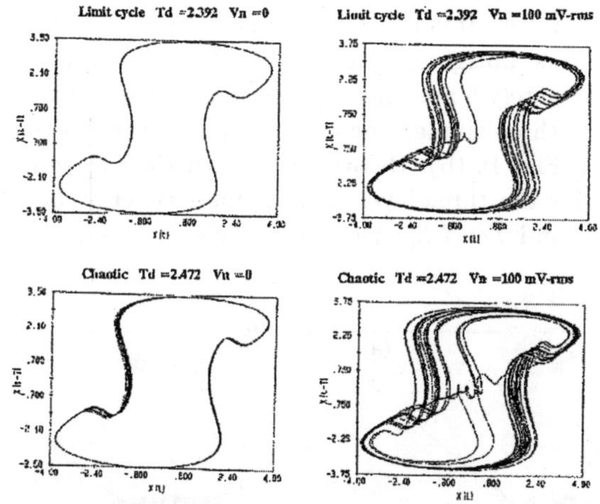

Figure 18: Phase portraits of the time delay Duffing system. Noise free (left panels) and noisy (right panels) dynamics are shown for the chaotic (upper) and limit cycle (lower) attractors.

scribed as high dimensional. Indeed, high dimensionality has often been offered as one of the reasons which have defeated attempts to detect chaos in such preparations.

8 Detecting Chaos and Limit Cycles within the Noise

Data from the two analog simulators were digitized and analyzed by computer. The basic measurement is the F statistic as a function of noise intensity. As described before, this statistic is formulated by counting encounters in both original and surrogate files with the encounters defined according to one of the three criteria.

In Fig. 19 (a) and we show three encounters identified from a file of 40,000 time interval measurements using the Van der Pol oscillator operated in the chaotic region. The upright triangles show the approaches to the PUFP located at approximately 5 ms, and the inverted triangles show the departures. For comparison, three similar encounters chosen from the surrogate of this file are shown in Fig. 19 (b). Below, the F statistic is shown as a function of noise intensity for the three levels of encounter criteria: *Level 0*, diamonds; *Level 1*, circles; and *Level 2*, inverted triangles. As expected, the statistic

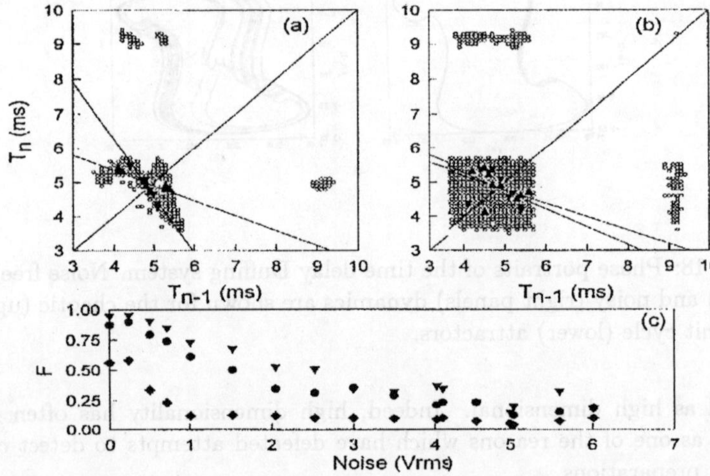

Figure 19: Encounters (upper panels) real (left) and surrogate (right). The F statistic for various selection levels (lower), see text.

shows a sharper ability to discriminate the chaotic attractor for the more stringent criteria. What is remarkable, is that even for the lowest level criterion, $F \succ 0$ even for quite large noise. This illustrates the power of the rare event approach to detecting chaos within large noise. Figure 20 illustrates the ability of the method to distinguish between chaotic and limit cycle attractors. The top panel (a) shows the statistic for the chaotic attractor (circles) and for the limit cycle (upright triangles). It is easy to distinguish out to about 50 % noise and possible even beyond that. The middle panel (b) shows the total numbers of identified encounters using the *Level 1* criterion in the 40,000 interval file (recall that each encounter constitutes 6 intervals, 3 approaching and 3 departing). The numbers of encounters identified in the surrogate file are shown in the bottom panel (c). Note that for the chaotic attractor the number of surrogate encounters is more-or-less independent of the noise intensity, as would be expected, and represent ≈1.5 % of the intervals in the file. By contrast the number of surrogate encounters for the limit cycle actually grows from about zero to somewhere around 1.5 % for large noise. Again, up to about 50 % noise, the method easily distinguishes between chaos and limit cycles. The open symbols shown in Fig. 20 are statistical tests. Each symbol is the average result for 6 files and the error bar represents the standard deviation of the 6 determinations. As an added point, tests have shown that for *Level 1* criterion about 0.7 % of intervals in *random* files are selected as encounters, and this fraction seems robust to the distribution from which the random files were made (Gaussian and gamma distributions with the same mean time intervals as the experimental files shown here were tested).

We turn now to the time delay system. Figure 21 (a) shows the statistic measured for the time delay system versus noise intensity, where again the chaotic attractor (circles) is compared to the limit cycle (upright triangles). The statistical properties of 6 independent measurements are again shown by the open symbols. Over most of the range shown, it is easy to distinguish the chaotic and limit cycle attractors. The panel below (b) shows F versus time delay (or dimension) for the noise free system. We can clearly see that in general, increasing time delay plays a role similar to noise. Specifically, it is difficult to distinguish low dimensional chaotic behavior in the high dimensional system using this method. It should be noted that the data shown in Fig. 21 (b) are repeatable. The system displays complex behavior including the shape of this graph. The break near $T_d = 2.530$ coincides with the onset of homoclinic crossings similar to those shown for high noise in Fig. 18. At larger time delays, more complex behavior is evident, including windows where $F \to 1$ again.

9 Conclusions

In this article we have shown several, widely diverse, example applications where electronic circuit analogs are useful for dynamical analysis. One of the main advantages is that, compared to numerical simulations, one can easily scan large regions of parameter space for "interesting" behavior. Analog simulators have also been useful when exploring the behavior of nonlinear systems with large added noise: a condition which has been notoriously difficult theoretically and time consuming numerically. However, perhaps the most useful aspect is that analog simulators mimic real experiments. For the experimentalist contemplating some dynamical measurements on a natural system, it is immeasurably helpful and instructive to first make those measurements on a model analog system. Moreover, analog experiments conducted in parallel with natural ones often result in new insights into the dynamics and thus are able to prompt further measurements, perhaps in new regions of parameter space, on the natural system.

10 Acknowledgments

We are grateful to the Office of Naval Research for providing support and encouragement for both analog and natural experiments with stochastic and chaotic nonlinear dynamical systems.

References

[1] W.Aspray, A. G. Bromley, M. Campbell-Kelly, P. E. Ceruzzi and M. R. Williams, Computing Before Computers, Iowa State University Press, Ames, 1990.

[2] R. L. Stratonovich, Topics in the Theory of Random noise , Gordon Breach, New York, vol 1, 1963, vol 2, 1967.

[3] R. Landauer, J. Appl. Phys. 33, 2209 (1962).

[4] G. Nicolis and I. Prigogine, Self-Organization in Nonequilibrium Systems (Wiley, New York, 1967).

[5] R. Lefever and J. Wm. Turner, Phys. Rev. Lett. 56, 1631 (1986).

[6] L. Fronzoni , R. Mannella, P.V.E. McClintock, and Frank Moss, Phys. Rev. A 36, 834 (1987)

[7] L. Fronzoni, Frank Moss and P.V.E. McClintock, Phys. Rev. A 36, 1492 (1987)

[8] A. Huberman, J. P. Crutchfield, and N. Packard, Appl. Phys. Lett. 37, 750 (1980).

[9] N. F. Petersen and A. Davidson, Appl. Phys. Lett. 39, 830 (1981).

[10] D. D'Humieres, M. R. Beasley, B. A. Huberman and A. Libchaber, Phys. Rev. A 26, 3483 (1982)

[11] G. Cicogna and L. Fronzoni, Phys. Rev. A 42, 1901 (1987)

[12] L. Fronzoni , M. Giocondo and M. Pettini, Phys. Rev. A 43, 6483 (1991).

[13] F. T. Arecchi, S. Coccaletti, G. B. Mindlin and C. Perez Garcia, Phys. Rev. Lett. 69, 3723 (1992)

[14] C. Grebogi, E. Otto, F. Romeiras and J. A. York, Phys. Rev. A36, 5365 (1987)

[15] F. Romeiras, C. Grebogi and J. A. York, Phys. Rev. A 38, 463 (1988)

[16] M. Franaszek, Phys. Rev. A 44, 4065(1991)

[17] R. Benzi, A. Sutera and A. Vulpiani, J. Phys. A 14, L453 (1981)

[18] R. Benzi, G. Parisi, A. Sutera and A.Vulpiani, Tellus 34, 10 (1982)

[19] F. Moss, Stochastic resonance: From the ice ages to the monkey's ear, in Some Problems in Statistical Physics, G.H.Weiss, ed (SIAM, Philadelphia, 1994)

[20] K. Wiesenfeld and F. Moss, Nature, 373, 33 (1995)

[21] A. L. Goldberger, D. R. Rigney and B. J. West, Sci. Am., 262, 42 (1990)

[22] L. Glass, P. Hunter and A. McCulloch, eds. Theory of Heart (Springer, Berlin, 1990)

[23] T. Chang, S. J. Schiff, K. Jerger, T. Sauer, J-P. Gossard and R. E. Burke, Biophys. J. 67, 671 (1994)

[24] S. J. Schiff, K. Jerger, T. Chang, T. Sauer and P. G. Altken, Biophys. J. 67, 684 (1994)

[25] S. J. Schiff, K. Jerger, D. H. Duong, T. Chang, M. L. Spano and W. L. Ditto, Nature 370, 615 (1994)

[26] G. Mayer-Kress, ed. Dimensions and Entropies in Chaotic Systems (Springer, Berlin, 1986)

[27] N. B. Abraham, A. M. Albano, A. Passamante and P. E. Rapp, eds. Measures of Complexity and Chaos (Plenum, New York, 1989)

[28] N. B. Abraham, A. M. Albano, A. Passamante, P. E. Rapp and R. Gilmore, eds. Proc. Second Workshop on Measures of Complexity and Chaos, Intern. J. Bifurcation and Chaos 3, 485 (1993)

[29] P. Grassberger and I. Procaccia, Physica 9 D, 189 (1983)

[30] A. M. Albano, J. Muench, C. Schwartz, A. I. Mees and P. E. Rapp, Phys. Rev. A 38, 3017 (1988)

[31] D. T. Kaplan and L. Glass, Phys. Rev. Lett. 68, 427 (1992)

[32] A. Garfinkel, M. L. Spano, W. L. Ditto and J. N. Weiss, Science 257, 1230 (1992)

[33] S. J. Schiff, K. Jerger, D. H. Duong, T. Chang, M. L. Spano and W. L. Ditto, Nature 370, 615 (1994)

[34] T. Shinbrot, C. Grebogi, E. Ott and J. A. Yorke, Nature 363, 411 (1993)

[35] E. Ott, C. Grebogi and J. A. Yorke, Phys. Rev. Lett. 64, 1196 (1990)

[36] D. Ruelle, Physics Today 47, 24 (1994)

[37] D. T. Kaplan, Physica D 73, 38 (1994)

[38] P. Cvitanovic, Phys. Rev. Lett. 61, 2729 (1988)

[39] R. Artuso, E. Aurell and P. Cvitanovic, Nonlinearity 3, 325 (1990)

[40] W. L. Ditto, M. L. Spano and J. Lindner, Physica D, in press

[41] D. Farmer, Physica 4 D, 366 (1982)

SYNCHRONIZING CHAOTIC CIRCUITS

THOMAS L. CARROLL and LOUIS M. PECORA
Code 6341, U. S. Naval Research Lab
Washington, DC 20375
USA

ABSTRACT

By choice of the proper coupling, it is possible to create chaotic systems that are synchronized. To create the synchronized circuits described in this work, the original chaotic circuit (the drive circuit) is divided into subsystems. When one or more of these subsystems is driven with a signal from the drive system, they will (under the right conditions) synchronize to it. Combining these subsystems in the right way creates a sort of "chaotic filter" that recognizes only the proper chaotic signal. These synchronized chaotic circuits may be used for broad band communications, and signal separation for broad-band carrier signals is demonstrated.

1. Introduction

1.1. Types of synchronization

The study of how to synchronize electronic systems is an old branch of engineering. Periodic systems must be synchronized to keep control systems in sync, to synchronize communications systems for the exchange of information, to synchronize pseudo random number generators for secure communications, and for other applications. The problem of synchronizing periodic systems is the problem of matching the phase of two periodic signal sources. A periodic dynamical system has a largest Lyapunov exponent of 0, in the direction along the trajectory. In order to synchronize a periodic system to some master periodic signal, the largest Lyapunov exponent of the periodic signal source in the receiver must be made less than 0, usually by finding the phase difference between the receiver signal and the master signal and using this difference to generate a feedback signal. Other information necessary for keeping the periodic signals synchronized, such as the frequency of oscillation, is already known in advance. The theory of synchronizing periodic systems is a linear theory.

Chaotic systems have been much less studied than periodic systems, owing mostly to the fact the nonlinearities necessary for generating chaos make chaotic systems harder to understand and engineer. A chaotic system has one or more Lyapunov exponents greater than 0, making the synchronization problem more difficult. As with periodic systems, to synchronize a chaotic receiver (by synchronization, we mean that all signals in the receiver converge to the corresponding signals in the transmitter), its Lyapunov exponents must all be made less than 0. Because a chaotic system has a positive Lyapunov exponent, the adjustment to the receiver must be continuous-otherwise, the receiver signal will drift away from the transmitter signal at an exponential rate.

There are several techniques for synchronizing a chaotic receiver to a transmitter. The oldest techniques involve a mutual coupling between the transmitter and receiver[1, 2]. The transmitter sends a signal to the receiver, and the receiver sends a signal to the transmitter. This type of coupling is also considered to be important in biological systems or neural networks.

In other methods for synchronizing chaotic systems, the receiver does not send a signal back to the transmitter. One of these methods resembles the technique commonly used for synchronizing periodic systems. The difference between the chaotic signal from the transmitter and the corresponding chaotic signal from the receiver is applied to the receiver to match the receiver signal to the transmitted signal[3]. Another method for synchronizing the chaotic receiver to the transmitter to the receiver involves forcing one of the signals in the receiver to be the same as one of the signals in the transmitter[4]. If the receiver is properly stable, it will then synchronize to the transmitter.

Our technique[5-12] for synchronizing chaotic circuits is different from these other techniques, although one can write all of these synchronization techniques as specific examples of a more general coupling technique. In our technique, a subsystem of the full chaotic system is copied and used as a receiver. This subsystem is chosen so that when driven with the proper signal from the full chaotic system, all of its Lyapunov exponents are less than 0. All of the signals in the subsystem will then synchronize with the corresponding signals in the drive system. No extra feedback signal is needed, as the choice of the subsystem makes the Lyapunov exponents all negative (although for some driving signals one or more may become positive). By cascading subsystems it is possible to reproduce the drive signal as the receiver output. In this case, the receiver output may be compared to the input to confirm synchronization.

In comparing different synchronization methods, it does not make any sense to say that any one technique is better than any other unless one has some specific application in mind for synchronization. To keep this chapter to a finite length, we will restrict further discussion mostly to the synchronization technique developed by us, as this is the one that we are most familiar with.

1.2 Why use circuits?

The principles of chaotic synchronization may be defined mathematically, and as such, may be applied to any physical system. Electronic circuits are a particularly easy physical system to work with. By using analog computer techniques, one may design a circuit with a wide range of different properties. All signals in an electronic circuit are accessible, and most or all variables may be manipulated. Electronic circuits are also easy to interface with computers, allowing computer control of experiments and rapid collection of data. It must be emphasized that electronic circuits don't just simulate physical systems; they are physical systems. We have heard the opinion expressed occasionally that electronic circuits are somehow not real physical systems. An electronic circuit is as real a physical system as a pendulum or a magnetic system or a fluid experiment. The ease with which circuits may be manipulated makes them very good for testing theoretical ideas and numerical experiments on a real, analog physical system.

2.0 Theory of Drive-Response Synchronization

2.1 Stability, dynamics and chaos, all you need to understand synchronization in chaotic systems

2.1.1 The behavior of chaos

Chaos is a label we put on a certain kind of behavior in systems that are evolving in time. Most people are familiar with the usual labels used to describe circuit behavior like steady state or equilibrium when the voltages settle down to a constant, unchanging value. A circuit can also be an oscillator (usually cyclic) when the voltages go through a cycle and repeat their values after some time called the period of oscillation. Most of us are used to thinking of these states as stable in the sense that slight changes in the voltage will still lead us back to the same place in the circuit operation.

The label of chaos is applied to systems in which the voltages' detailed future behavior depends sensitively on where the circuit starts off. Slight changes lead to big differences and there are two mechanisms at work which cause this. One is that there are local instabilities near the oscillation pattern. Small deviations from the pattern result in the system being pushed farther away from where it would have gone. This leads to the sensitivity of the circuit on the initial conditions: where it ends up depends strongly on where it starts. The second part of chaotic behavior is that the system does not continue to diverge forever, but because of nonlinearities in the behavior of the circuit the motion is re injected back into the general pattern it wants to execute, but at an entirely different place from where it would have gone in the same interval of time.

What we have in a chaotic system is a constant stretching and folding action caused by local instabilities and nonlinearities. Nearby points are pushed apart then later folded back into the final pattern. Figure 1 shows this pattern. The figure is a plot of the trajectory of a system's state points (e.g. the voltages that define a circuit) in the state space, which is also called a phase space. Final patterns are referred to as *attractors* in nonlinear dynamics. In a chaotic system it is impossible to tell where the circuit will end up on the pattern after a long time.

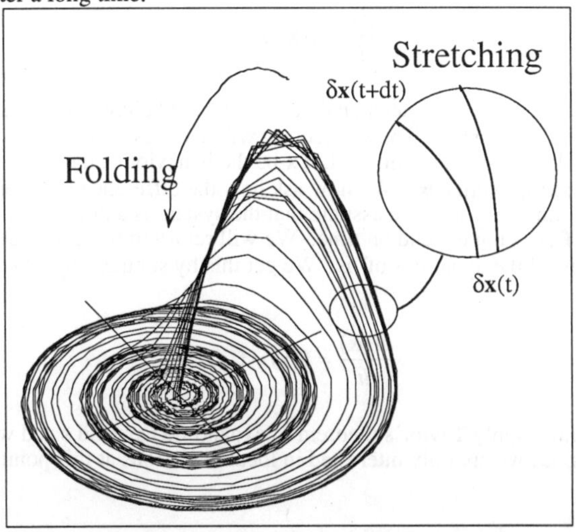

Figure 1 The Rössler strange attractor showing evidence of stretching and folding.

Figure 1 is referred to as a *strange* attractor. The adjective strange refers to the "multilayered" structure of the attractor. This pattern is fractal in nature, which means it has structure on an infinitesimal level. Compare Figure 2 (a) and (b) to see the difference between (a) a cyclic or periodic attractor and (b) a strange attractor.

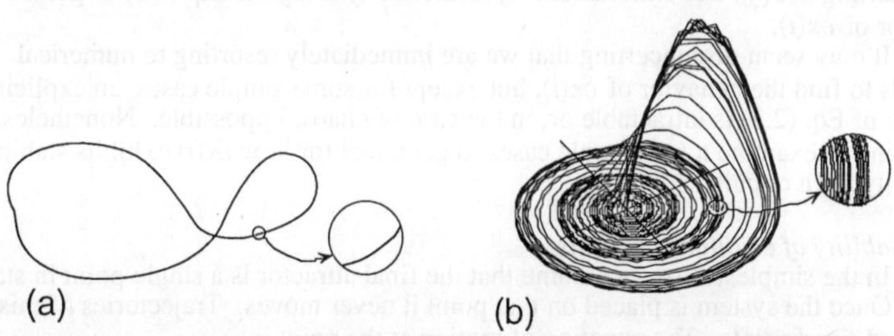

Figure 2 Views that zoom in on (a) a periodic attractor and (b) a strange or chaotic attractor showing that the periodic attractor will eventually look like a simple line at high enough magnification and the chaotic attractor will always have structure--it is geometrically fractal.

We can quantify the initial stretching action of the attractor and this quantification leads to a test for chaotic behavior that is sometimes useful in determining whether a system is chaotic, and also in the design of systems to accomplish other tasks. We will see this below.

We formulate our stretching problem in state space by collecting all the dynamical variables, those quantities that define the system's behavior, into a single vector $\mathbf{x}(t)$. If we have d dynamical variables, then we have a d-dimensional vector $\mathbf{x}$. The behavior of the system can be visualized as a curve or trajectory in the state space, like Figs. 1 and 2. We will assume for now that the behavior of the system is governed by ordinary differential equations (ODE's). Other systems of motion, like iterated systems, will follow the same general argument. So we should have a law of motion or equations of motion like

$$\frac{d\mathbf{x}}{dt} = \mathbf{f}(\mathbf{x},t) \quad . \tag{2.1}$$

Now we let the d-dimensional vector **s** be the behavior of the system when started from the initial point $\mathbf{s}(t_0)$. The **s** trajectory will serve as a fiducial trajectory against which we will measure deviations. Let **x** be the behavior of the same circuit started at a nearby point $\mathbf{x}(t_0)=\mathbf{s}(t_0)+\delta\mathbf{x}$. We now ask how the difference between **s** and **x** will evolve with time. For now we assume that the system is autonomous which means the vector field **f** does not depend on time. We will return to this point later.

We need the dynamics of $\delta\mathbf{x}$. We get this by subtracting the equations of motion for **s** and **x**:

$$\frac{d(\delta\mathbf{x})}{dt} = \mathbf{f}(\mathbf{x}) - \mathbf{f}(\mathbf{s}) \quad . \tag{2.2}$$

We now apply Taylor's theorem to expand $\mathbf{f}(\mathbf{x})$ around **s** and we assume that $\delta\mathbf{x}$ is very small since we are only interested in local deviations at this point. Then Eq. (2.2) becomes

$$\frac{d(\delta\mathbf{x})}{dt} = D\mathbf{f}(\mathbf{s}) \cdot \delta\mathbf{x}, \tag{2.3}$$

where $D\mathbf{f}(\mathbf{s})$ is the Jacobian of **f** evaluated on the trajectory $\mathbf{s}(t)$. The Jacobian is the matrix of partial derivatives $(D\mathbf{f}(\mathbf{s}))_{ij} = (\partial f_i/\partial x_j)|_{\mathbf{s}(t)}$. This makes Eq. (2.3) a linear, time-dependent equation of motion. One way to solve Eq. (2.3) is to numerically solve Eq. 1 starting at $\mathbf{s}(t_0)$ and simultaneously numerically integrate Eq. (2.3) to get the behavior of $\delta\mathbf{x}(t)$.

It may seem disconcerting that we are immediately resorting to numerical methods to find the behavior of $\delta\mathbf{x}(t)$, but except for some simple cases, an explicit solution of Eq. (2.3) is intractable or, in the case of chaos, impossible. Nonetheless, it is interesting to examine a few simple cases to get a feel for how $\delta\mathbf{x}(t)$ exhibits stability or instability for a circuit.

2.1.2 Stability of equilibrium points

In the simplest case we assume that the final attractor is a single point in state space. Once the system is placed on that point it never moves. Trajectories of this nature are called *fixed* points. The equation of motion at the point is

$$\frac{d\mathbf{x}}{dt} = 0 \quad . \tag{2.4}$$

So the fiducial trajectory **s** is a constant and $\mathbf{f}(\mathbf{s})=0$. Therefore the Jacobian in Eq. (2.3) is a constant and we can explicitly solve for $\delta\mathbf{x}(t)$,

$$\delta\mathbf{x}(t) = Ce^{t \cdot D\mathbf{f}}\,\delta\mathbf{x}(0), \tag{2.5}$$

where C is a constant of integration.

If the fixed point is stable, deviations like $\delta\mathbf{x}(t)$ should go to zero as time progresses. We see from Eq. (2.5) that $\delta\mathbf{x}(t) \to 0$ if the matrix $D\mathbf{f}$ has eigenvalues that are all negative. Then $\delta\mathbf{x}(t)$ will decrease like $e^{t\cdot\lambda_{\max}}\delta\mathbf{x}(0)$, where $\lambda_{\max}$ is the largest eigenvalue. If $\lambda_{\max}$ is positive, $\delta\mathbf{x}(t)$ will increase and the fixed point will be unstable. In the latter case in a real system any attempts to keep the system on the fixed point will surely fail since noise or other perturbations will knock it off and the system will move away from that point, exponentially fast.

2.1.3 Stability of periodic orbits

If we are dealing with a periodic trajectory (see Fig. 2 (a)), then we can sometimes get an estimate of the stability of that trajectory by calculating Floquet multipliers [13, 14]. This subject is quite well developed and we will give only a general idea of what is going on here.

Since Eq. (2.3) is linear it has, in principle, a solution that is a linear operator which is usually written as $\mathrm{X}(t) = \mathrm{Z}(t)e^{t\mathrm{R}}$, where X, Z, and R are $d{\times}d$ matrices and Z(t) is periodic in t with a period T, the same as the period of the trajectory [13-15]. It is standard to choose X(0)=Id, the identity matrix. A particular solution given a starting vector $\delta\mathbf{x}(0)$ is given by the matrix multiplication $\mathrm{X}(t)\,\delta\mathbf{x}(0)=\delta\mathbf{x}(t)$. Obviously, whether $\delta\mathbf{x}(t)$ grows or shrinks determines whether the periodic trajectory is stable or unstable. This in turn depends on the behavior of X(t).

Given the properties of X(0) and Z(t) we see that after one period of motion we will have $\delta\mathbf{x}(T) = \mathrm{Z}(T)e^{T\mathrm{R}}\delta\mathbf{x}(0)$, after two periods of motion we will have $\delta\mathbf{x}(2T) = \mathrm{Z}(T)e^{2T\mathrm{R}}\delta\mathbf{x}(0)$ and after n periods $\delta\mathbf{x}(nT) = \mathrm{Z}(T)e^{nT\mathrm{R}}\delta\mathbf{x}(0)$. Obviously, the factor $e^{nT\mathrm{R}}$ controls the growth or shrinkage of the deviation $\delta\mathbf{x}$. This factor is often referred to as the Floquet muitiplier, and its eigenvalues tell the story of the stability of the orbit. If the factor has eigenvalues all less than 1, the motion is stable and $\delta\mathbf{x}(t) \to 0$ as $t \to \infty$. If the factor has an eigenvalue greater than 1, the trajectory is unstable; deviations will grow exponentially fast as the system moves away from periodic orbit. If the multiplier is exactly 1, then the periodic orbit is called neutrally stable. In the latter case, on average, the deviations remain the same size forever.

The connection to the exponents that we get for the fixed point situation come from considering a Poincaré section of the motion. Simply put, we set up a surface that is pierced by the trajectory and note that on that surface from one piercing to the next we have, in principle, a map that tells us where the next piercing will take place. See Fig. 3 for an example of this.

We can denote this map as an iterated system, which is another kind of dynamical system. The next point is $\mathbf{x}_{n+1}$ and it is generated from a prior point $\mathbf{x}_n$by a function $\mathbf{F}$: $\mathbf{x}_{n+1} = \mathbf{F}(\mathbf{x}_n)$, where $\mathbf{F}$ is directly related to X(t) above. Then, since a periodic orbit will pierce the surface in exactly the same place each time, it will be represented by a fixed point of the map $\mathbf{F}$. We can analyze the stability of that fixed point as we did before, only this time we will look for the eigenvalues of $D\mathbf{F}$ at the equilibrium point. These eigenvalues are the Floquet multipliers and their logarithms are the exponents analogous to the stability exponents in the ODE fixed point case.

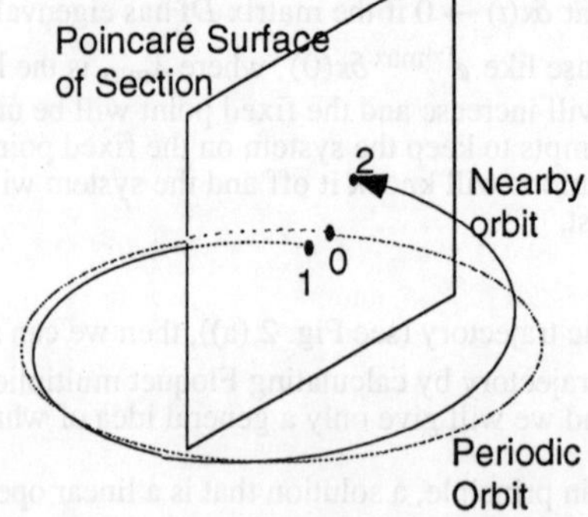

Figure. 3. A surface defining a Poincaré section of the flow near a periodic orbit. The map on the surface maps point 1 to point 2 in one time period. The nearby orbit is unstable in this figure--the map points are moving away from the fixed map point 0 defined by the periodic orbit

2.1.4 Stability of chaotic trajectories

The case of a chaotic trajectory is more difficult. Here we have no equilibria or even any periodicity. Chaos is aperiodic motion. However, we can take a similar approach to stability as before. This approach is based on the seminal work of A.M. Lyapunov [16] done many years ago and put on a firm foundation for general dynamical systems by Oseledec [17]. We start with a variational equation, namely Eq. (2.3) and integrate it along with the equation of motion Eq. (2.1). We then calculate the limit

$$\lim_{t\to\infty}\frac{1}{t}\ln\frac{\delta x(t)}{\delta x(0)} \tag{2.6}$$

which gives the largest Lyapunov exponent. We abbreviate Lyapunov exponent as LE from here on since it is used so often.

When the LE is positive we have an unstable trajectory. Since in infinite time starting at a typical point we will cover almost all the attractor (except for a set of measure zero) we have a good measure of the stability of the entire attractor. If there is a long time instability (at least one positive Lyapunov exponent), the motion is chaotic. All other motion, periodic, equilibria, quasiperiodic (more than one frequency of periodicity) all have their largest Lyapunov exponent ≤ 0.

Pictorially, Fig. 4 shows this situation. Two nearby state space points on a chaotic attractor move away from each other, exponentially fast at first. The points remain on the attractor, but any correlation between them rapidly goes to zero. This illustrates why chaotic systems have sensitivity to initial conditions – they are continually unstable.

We now have a way to detect chaos in some cases, especially for those systems for which we have a model. Sometimes it is also possible to determine the largest exponent from time-series data [18-26], although we don't go into that here. There are better ways to calculate the exponents than Eq. (2.6). Namely, we can employ a matrix approach similar to the Floquet method above. We just define a matrix $X(t)$ which starts as the identity matrix and evolves according to

$$\frac{dX}{dt} = Df(s)\cdot X \tag{2.7}$$

which is similar to Eq. (2.3). After a long time we see what the eigenvalues of X are. This will give the full set of eigenvalues and hence Lyapunov exponents. Note that there is *no* decomposition of X into Z and e^{Rt} components as there is for periodic orbits. We are also guaranteed by Oseledec's theorem [17] that we will almost always get the same result no matter where the motion starts on the attractor. There are plenty of numerical problems that have to be dealt with since, on a chaotic attractor we will have positive and negative eigenvalues, which means that certain components of X will be growing exponentially in time and others will be shrinking exponentially in time. A good reference on which an algorithm can be built is Eckmann and Ruelle [26, 27].

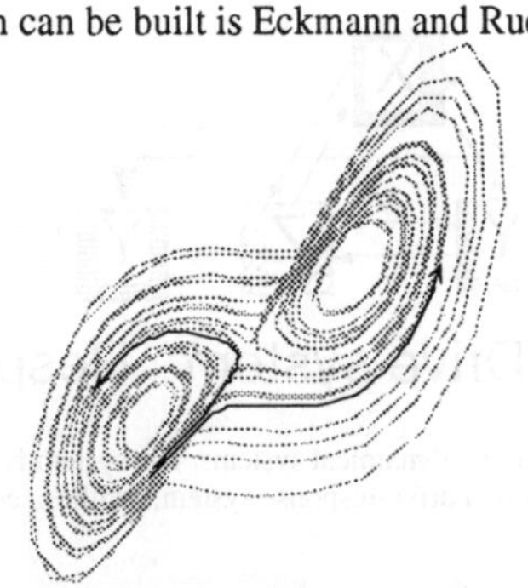

Figure. 4 Motion of nearby chaotic trajectories, exponentially diverging, but staying on the attractor.

2.2 The need to synchronize to transmit information ; transmitting one signal

We are now in a position to understand synchronization in chaotic systems. Imagine two isolated, identical chaotic systems. In any real systems we can never start them at exactly the same point in state space and we can never make them exactly identical. This means that we have a situation as illustrated in Fig. 4. Each system is represented by a point on the chaotic attractor and the points are never exactly the same.

The result? The systems soon become unsynchronized and uncorrelated. Where one is on the attractor has no relation to where the other one is. The central question we faced is can we do anything about this? The answer is yes.

The first thing we realize is that we cannot keep the systems isolated. Chaos will defeat us every time. We must send information of some kind to cause the systems to synchronize. We took the simplest approach we could think of, namely sending a signal that was actually one of the dynamical variables. Thus, we pick a component of one system's state vector **x** and transmit it to the other system. The question is what to do with it at the receiving end? We again chose the simplest approach. We had that transmitted signal completely replace that part of the receiving system that would normally generate the same component.

Let's look at a simple situation and draw some block diagrams. Suppose we have a chaotic system that has a three-dimensional state vector (x,y,z). Each component is associated with some part of the system and it will have it own laws of motion which will depend simultaneously on the other two components. In other words the components will all interact. Fig. 5 (a) shows this in block diagram form. We see that the signal that comes from each block is sent to the other two blocks. Each signal is a component of the state vector.

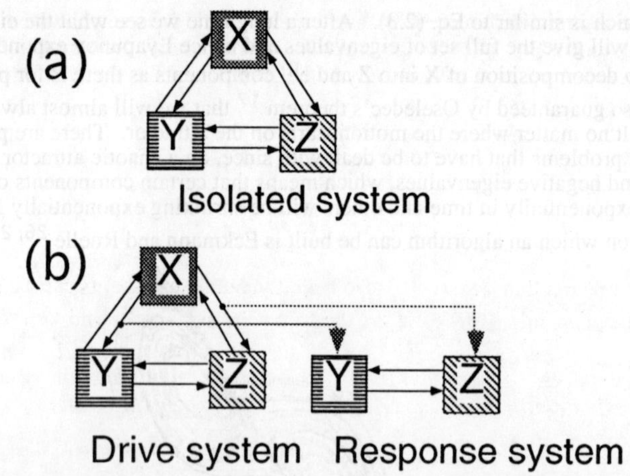

Figure 5. (a) Diagram of a dynamical system. The system is made up from the three coupled, interacting parts. (b) Diagram of a drive-response system, synthesized from the original system.

The prescription for setting up two systems for synchronization by transmitting a single dynamical signal is shown in Fig. 5 (b). It consists of a transmitter, which we call a *drive*, and a receiver, which we call a *response*. The drive is simply the original system, operating autonomously. The response is a duplicate of the drive, except that the block corresponding to the signal that is transmitted is removed.

We can express this in ODE form as follows. We start with the equations for a three-dimensional system,

$$\frac{dx}{dt} = f(x,y,z)$$
$$\frac{dy}{dt} = g(x,y,z) \tag{2.8}$$
$$\frac{dz}{dt} = h(x,y,z)$$

and we create a new system by using two of the three equations in Eq. (2.8),

$$\begin{array}{ll} \dfrac{dx}{dt} = f(x,y,z) & \\ \dfrac{dy}{dt} = g(x,y,z) & \dfrac{dy'}{dt} = g(x,y',z') \\ \dfrac{dz}{dt} = h(x,y,z) & \dfrac{dz'}{dt} = h(x,y',z') \\ \text{drive} & \text{response} \end{array} \tag{2.9}$$

Note that the response equations also depend on x from the drive. The variable x is the transmitted variable. If we have done things correctly, (y',z') will synchronize with (y,z). To test this we will now use the stability theory we introduced above.

We want the differences $\Delta y = y' - y$ and $\Delta z = y' - z$ to go to zero. If we subtract the y and z equations in Eqs. (2.9) we get a variational or stability problem like before only with a twist. We are now looking only at the Jacobian of the y-z subsystem:

$$\begin{pmatrix} \frac{d\delta y}{dt} \\ \frac{d\delta z}{dt} \end{pmatrix} = \begin{pmatrix} \frac{\partial g}{\partial y} & \frac{\partial g}{\partial z} \\ \frac{\partial h}{\partial y} & \frac{\partial h}{\partial z} \end{pmatrix} \begin{pmatrix} \delta y \\ \delta z \end{pmatrix} . \tag{2.10}$$

It turns out that Eqs. (2.10) also have Lyapunov exponents associated with them [11], only they depend on $x(t)$. We say they are conditional on x and we refer to them as *conditional Lyapunov exponents* [12, 28]. We use the abbreviation CLE for conditional Lyapunov exponent. The meaning is still the same as before regarding stability. Namely, (y',z') will synchronize with (y,z) providing the two exponents (CLEs) associated with Eqs. (2.10) are negative.

The CLEs are *not* the same as the LEs for the full system. In general, they are not even a subset of the LEs. In fact, given the LEs, it is not generally possible to predict the CLEs.

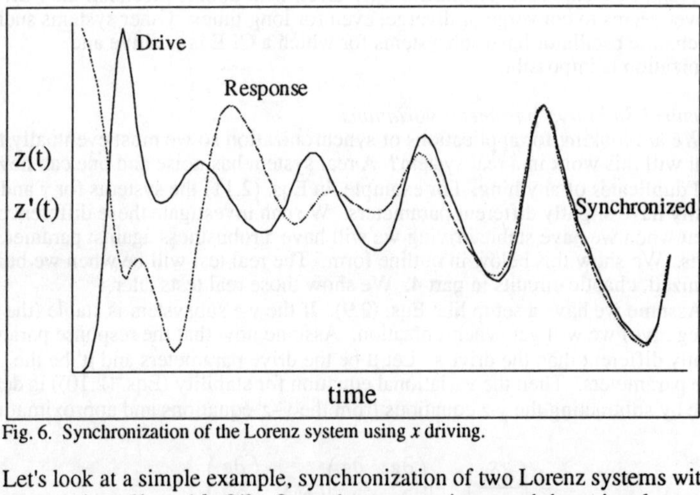

Fig. 6. Synchronization of the Lorenz system using x driving.

Let's look at a simple example, synchronization of two Lorenz systems with parameters σ=10, r=60, and b=8/3. One scheme to try is to send the x signal:

$$\begin{aligned} \frac{dx}{dt} &= -\sigma x + \sigma y & & \\ \frac{dy}{dt} &= -xz + rx - y & \frac{dy'}{dt} &= -xz' + rx - y \\ \frac{dz}{dt} &= xy - bz & \frac{dz'}{dt} &= xy' - bz' \\ &\text{drive} & &\text{response} \end{aligned} . \tag{2.11}$$

Fig. 6 shows the results of starting the drive and response at different (y, z) points. Obviously they synchronize. A calculation of the largest CLE of the response gives −1.805 which confirms that the synchronization is stable. Note that the CLE is of the same order of magnitude as the LEs which range from about -22 to +1.2. This is important since we want the CLEs to be numerically meaningful and not due to roundoff or other numerical problems.

On the other hand if we use z driving we have the setup

$$\begin{array}{ll} \frac{dx}{dt} = -\sigma x + \sigma y & \frac{dx}{dt} = -\sigma x' + \sigma y' \\ \frac{dy}{dt} = -xz + rx - y & \frac{dy'}{dt} = -x' z + rx' - y' \\ \frac{dz}{dt} = xy - bz & \\ \text{drive} & \text{response} \end{array} \tag{2.12}$$

which has a maximum CLE of 0.00481. This shows that the stability is nearly neutral and if the systems synchronize, it will be a slow process and probably not very robust. Numerical calculations confirm this. Even with double precision the z drive setup never seems to converge or diverge, even for long times. Other systems such as the Rössler chaotic oscillator have subsystems for which a CLE is positive and synchronization is impossible.

2.3 Structural Stability; parameter variations

We are looking for applications of synchronization so we must eventually face the question: will this work in a real system? A real system has noise and one can never get identical duplicates of anything. For example, in Eqs. (2.11) the systems for y and y' will necessarily have slightly different parameters. We can investigate these differences and argue that when we have stable driving we will have a robustness against parameter variations. We show this below in outline form. The real test will be when we build synchronized, chaotic circuits in part 4. We show those real tests later.

Assume we have a setup like Eqs. (2.9). If the y-z subsystem is stable (the CLEs are all negative) we will get synchronization. Assume now that the response parameters are slightly different than the drive's. Let μ be the drive parameters and μ' be the response parameters. Then the variational equation for stability (Eqs. (2.10)) is derived as before by subtracting the y-z equations from the y'-z' equations and approximating $\mu' = \mu + \Delta\mu$,

$$\begin{pmatrix} \frac{d\delta y}{dt} \\ \frac{d\delta z}{dt} \end{pmatrix} = \begin{pmatrix} \frac{\partial g}{\partial y} & \frac{\partial g}{\partial z} \\ \frac{\partial h}{\partial y} & \frac{\partial h}{\partial z} \end{pmatrix} \begin{pmatrix} \delta y \\ \delta z \end{pmatrix} + \begin{pmatrix} \frac{\partial g}{\partial \mu} \\ \frac{\partial h}{\partial \mu} \end{pmatrix} \cdot \Delta\mu \tag{2.13}$$

where $\partial/\partial\mu$ accounts for the variation in parameters. Eq. (2.13) has the form

$$\frac{d\mathbf{x}}{dt} = \mathrm{A}(t)\mathbf{x} + \mathrm{B}(t) \tag{2.14}$$

which is a linear, time-dependent equation. The formal solution for Eq. (2.14) is

$$\mathbf{x}(t) = \int_0^{\infty} \mathrm{X}(t-\tau)\mathrm{B}(\tau)\,d\tau \tag{2.15}$$

where X(t) is the principal matrix solution introduced before. The important point is that for long times X(t) is bounded by e^{-ct} and if B(t) is bounded then so is $\mathbf{x}(t)$ and thus ($\delta y,\delta z$) [12]. So for long times we expect that the differences will not be large as long as the parameter differences $\Delta\mu$ are small.

We note that recently interesting and unexpected short-time behavior can change this picture. Long-time Lyapunov exponent analysis is not the whole story. We postpone presenting this until later. For many circuit implementations these are not issues, but any circuit builder or emulator should be aware of them.

2.4 General setup, sending multiple signals

The previous approach can be generalized [12]. We show an outline of that here. Suppose we have an n dimensional dynamical system

$$\frac{d\mathbf{x}}{dt} = \mathbf{F}(\mathbf{x}) \tag{2.16}$$

and we divide it into two subsystems, $\mathbf{u}$ of dimension k and $\mathbf{w}$ of dimension m:

$$\frac{d\mathbf{u}}{dt} = \mathbf{f}(\mathbf{u},\mathbf{w}), \qquad \frac{d\mathbf{w}}{dt} = \mathbf{g}(\mathbf{u},\mathbf{w}) \tag{2.17}$$

If we duplicate the $\mathbf{w}$ subsystem we get a $\mathbf{w}'$ system driven by the multiple signals $\mathbf{u}$. We have actually created a new $n+m$ dimensional system. The stability conditions are analogous to before, namely the CLEs of the variational equations

$$\frac{d\delta\mathbf{w}}{dt} = D_{\mathbf{w}}\mathbf{g}(\mathbf{u},\mathbf{w})\cdot\delta\mathbf{w} \tag{2.18}$$

should be negative, if not $\delta\mathbf{w}$ will diverge and synchronization will not be possible.

We can even go further and attach some other system to the ($\mathbf{u},\mathbf{w}$) system that is not a duplicate of the $\mathbf{w}$ subsystem. This might look like

$$\frac{d\mathbf{u}}{dt} = \mathbf{f}(\mathbf{u},\mathbf{w}), \qquad \frac{d\mathbf{w}}{dt} = \mathbf{g}(\mathbf{u},\mathbf{w}), \qquad \frac{d\mathbf{v}}{dt} = \mathbf{h}(\mathbf{u},\mathbf{v}) \tag{2.19}$$

and is $n+l$ dimensional, where l is the dimension of the $\mathbf{v}$ subsystem. We call systems that can be decomposed into separate, driven subsystems *drive decomposable*. Our drive-response approach is a special case of Eqs. (2.19) where $\mathbf{g}$ and $\mathbf{h}$ are identical.

In the more general case when $\mathbf{g}\neq\mathbf{h}$ we can ask the question when does the $\mathbf{v}$ system fall into the same pattern, given the same drive, $\mathbf{u}$? This is also a stability question. We won't go any further with this general approach here, but simply point out that obtaining the same output signal given an input signal may have a use, for example, in signal identification, especially for complex signals like chaos. Furthermore, this shows that one can take a constructive approach to creating new chaotic systems with interesting properties.

2.5 Use of synchronization in new scenarios

Once one has grasped the constructive approach of synchronization, new scenarios can quickly be created. One that may have use in communications and signal processing is that of a cascaded synchronized chaotic drive-response system [6, 7]. Fig. 7 shows a block diagram for this setup. For a three dimensional system the equations will look like

$$
\begin{array}{lll}
\frac{dx}{dt} = f(x,y,z) & & \frac{dx''}{dt} = f(x'',y',z'') \\
\frac{dy}{dt} = g(x,y,z) & \frac{dy'}{dt} = g(x,y',z') & \\
\frac{dz}{dt} = h(x,y,z) & \frac{dz'}{dt} = h(x,y',z') & \frac{dz''}{dt} = h(x'',y',z'') \\
\text{drive} & \text{R1} & \text{R2}
\end{array}
\qquad (2.20)
$$

where the x signal drives the (y',z') subsystem (response R1) and the y' signal in turn drives the (x'',z'') subsystem (response R2). If the R1 and R2 subsystems are stable their components will synchronize with the system driving each one, i.e. $y' \to y$, $z' \to z$, $z'' \to z'$, and $x'' \to x$.

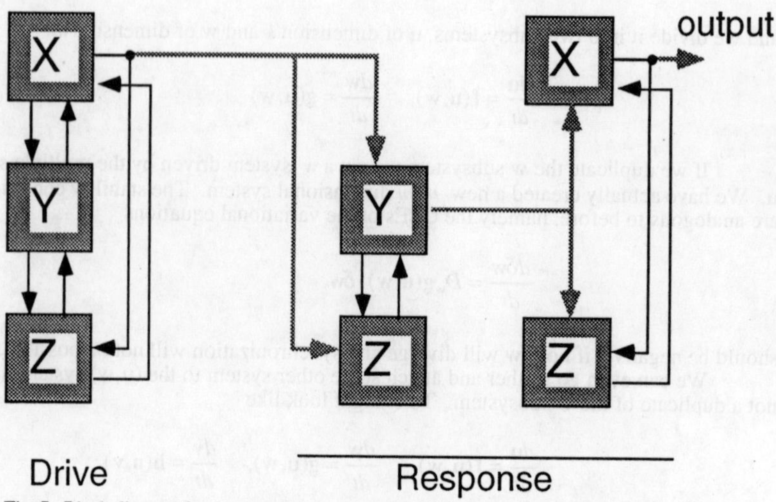

Fig. 7. Block diagram for the cascaded, synchronized drive-response system. The input signal, x is recreated at the output only if the parameters in the response match those in the drive.

To form such a cascaded system we need to start with a chaotic system that has two stable subsystems. For example, the Rössler system with the usual parameters (a=b=0.2, and c ∈ [4.7, 20] has only one stable subsystem, the (x, z) one. The Lorenz has two, the (x, z) and the (y, z).

The relation $x'' \to x$ shows that we can reproduce the incoming signal. This will happen if all the parameters are the same in the subsystems. If the parameters in, say, R1 change then $\Delta x = x'' - x$ will never go to zero. However, this has possible uses for sending signals in chaos. A system based on cascading of autonomous chaotic systems is shown in section *4.2*.

Cuomo and Oppenheim [29] have employed a different version of cascaded synchronization to achieve signal masking in communications. Their results are presented later in this volume.

We can broaden the synchronization approach above to nonautonomous systems by adding a time dependence to the vector fields. We would write, for example, $f(x,y,z,t)$. Now synchronization will take place provided the drive and response both have the same time dependence. If the time dependence comes from being driven by another signal (e.g. sinusoidal forcing), then we will not have synchronization unless the external forcing for each subsystem is the same. If it's sinusoidal, all forcing must be in phase, for example. This deviations from synchronization can be used to synchronize the periodic forcing parts of chaotic systems which can be cascaded. In section 7.1 we introduce the nonautonomous synchronized circuits. In section 7.2 we show that the drive signal can even be filtered and then reconstructed at the response in a stable way to maintain synchronized behavior.

3.0 Unidirectional and Bi-directional Coupling Synchronization

The drive-response synchronization scenario is just an extreme version of other coupling scenarios. Although drive-response setup is often easy to implement the other approaches have advantages and their study connects us to the general problem of synchronization in systems of many nearly identical chaotic oscillators. We start with the simplest generalization of drive-response, one-way or unidirectional coupling.

3.1 Unidirectional coupling

The drive-response scenario can be changed from the complete replacement of a response variable with a drive variable to a form in which the response variable that corresponds to the drive variable being sent is slaved to that incoming signal. This is considered unidirectional coupling and is a form of feedback control often called dissipative coupling. In general, we can write a unidirectional coupling system as

$$\frac{d\mathbf{x}}{dt} = \mathbf{F}(\mathbf{x}) \qquad \frac{d\mathbf{x'}}{dt} = \mathbf{F}(\mathbf{x'}) + \mathbf{C}(\mathbf{x} - \mathbf{x'}) \tag{3.1}$$

where C is a matrix. A simple example for one component unidirectional coupling in a three-dimensional system is

$$C = \begin{pmatrix} c & 0 & 0 \\ 0 & 0 & 0 \\ 0 & 0 & 0 \end{pmatrix}$$

$$\begin{aligned} \frac{dx}{dt} &= f(x,y,z) & \frac{dx'}{dt} &= f(x',y',z') + c(x - x') \\ \frac{dy}{dt} &= g(x,y,z) & \frac{dy'}{dt} &= g(x',y',z') \\ \frac{dz}{dt} &= h(x,y,z) & \frac{dz'}{dt} &= h(x',y',z') \end{aligned} \tag{3.2}$$

The relation of this unidirectional coupling to the drive-response scenario is seen from Eqs. (2.22): drive-response is the limit as $c \to \infty$ of the unidirectional coupling [3, 30, 31].

Several researchers have used the unidirectional approach with circuits [4, 32-36]. Recently, the relation between unidirectional driving, drive-response setups and chaos control was made explicit in the paper by Wu and Chua [37].

It is also possible to synchronize models of dynamical systems to time-series that drive them using such coupling [23, 38]. Such models if close to the physical system's true dynamics should provide an indirect measure of a system's parameters.

3.2 Bi-directional coupling

The move from a unidirectional coupling to a mutual coupling is now relatively straightforward. We replace Eqs. (2.21) with

$$\frac{d\mathbf{x}}{dt} = \mathbf{F}(\mathbf{x}) + \mathbf{C}(\mathbf{x}' - \mathbf{x}) \qquad \frac{d\mathbf{x}'}{dt} = \mathbf{F}(\mathbf{x}') + \mathbf{C}(\mathbf{x} - \mathbf{x}') \quad . \tag{3.3}$$

Now both systems "drive" each other. The three-dimensional example becomes

$$\begin{aligned} \frac{dx}{dt} &= f(x,y,z) + c(x'-x) \qquad & \frac{dx'}{dt} &= f(x',y',z') + c(x-x') \\ \frac{dy}{dt} &= g(x,y,z) & \frac{dy'}{dt} &= g(x',y',z') \\ \frac{dz}{dt} &= h(x,y,z) & \frac{dz'}{dt} &= h(x',y',z') \end{aligned} \quad . \tag{3.4}$$

Such systems have been studied in the context of a few coupled oscillators, usually two [35, 39-46].

We can continue the generalization to a large array of n chaotic oscillators, one oscillator at each node of the array. The equations of motion become

$$\frac{d\mathbf{x}^j}{dt} = \mathbf{F}(\mathbf{x}^j) + \mathbf{G}^j(\mathbf{x}) \tag{3.5}$$

where $\mathbf{x}^j$ is the vector of dynamical variables at the jth node, $\mathbf{G}^j$ is the coupling function at the jth node and $\mathbf{x}=(\mathbf{x}^1, \mathbf{x}^2, ..., \mathbf{x}^n)$ represents an aggregate vector made from dynamical variables of each node. Typical $\mathbf{G}$'s studied are

$$\mathbf{G}^j(\mathbf{x}) = \mathbf{C}(\mathbf{x}^{j+1} + \mathbf{x}^{j-1} - 2\mathbf{x}^j) \tag{3.6}$$

and

$$\mathbf{G}^j(\mathbf{x}) = \frac{\mathbf{C}}{n}\sum_{i=1}^{n} \mathbf{x}^i \quad . \tag{3.7}$$

The first is diffusive coupling and the second is global or all-to-all coupling.

A few earlier works studied synchronized chaos in these contexts [2, 39, 46-51]. Some of these papers show that the stability problem can be simplified (the Jacobian can be put into block form) by special transformations for coupling $\mathbf{G}$'s with a certain form (as in Eqs. (2.26) and (2.27). Heagy *et al* [52] showed that the simplification of the Jacobian stability equations can always take place when the coupling has a shift symmetry in the synchronized state. That is, globally shifting the indices of the nodes will leave the equations of motion invariant. The transformation which will in general always block diagonalize the Jacobian is the discrete Fourier transform applied in such a way as to treat each node as a whole. The details of this approach and its consequences are beyond the scope of this paper and can be found in reference [52].

One consequence of studying chaotic synchronization in systems of coupled oscillators has been the discovery that the Lyapunov stability is not the whole story. In real systems (with noise and parameter differences) where the coupled oscillators never fully synchronize we always have the possibility of local instabilities (locally positive Lyapunov exponents) causing the nodes to burst apart. This was first seen experimentally by Ashwin *et al* [41] and they referred to the phenomenon as bubbling of the attractors near the synchronized state. Later Heagy *et al* [53] showed directly that riddled basins of attraction could exist in these systems.

Later in this article we show the results of experiments operating near the Lyapunov threshold of chaotic synchronization. We show how to map out the riddled basin that ensues.

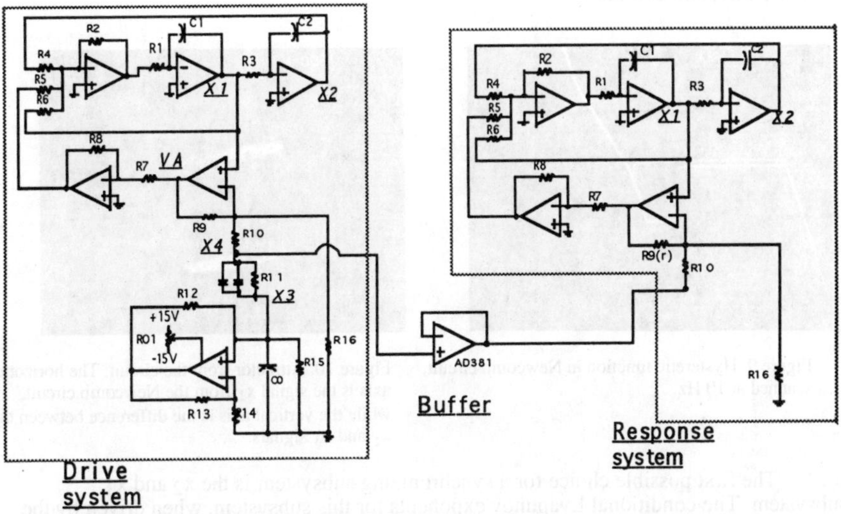

Figure 8. Newcomb circuits used to demonstrate chaotic synchronization. R1 = R3 = R7 = R8 = R16 = 100 kΩ; R2 = 330 kΩ; R4 = 220 kΩ; R5 = R6 = 150 kΩ; R9 = R15 = 20 kΩ; R10 = 3 kΩ; R11 = 50 kΩ; R12 = R13 = R14 = 10 kΩ; R01 = 10 kΩ potentiometer; C0 = C2 = 0.001 μF; C1 = 0.01 μF. The diodes are type 1N4739A. All op amps were type 741 unless otherwise noted. Points labeled X1, X2, X3, and X4 are where the corresponding voltages were measured.

4. Circuit Examples

4.1 The Newcomb circuit

A schematic for the first circuit in which we demonstrated chaotic synchronization is shown in Fig. 8. This circuit[5] is based on a design by Newcomb[54, 55]. The heart of this circuit is an unstable 2nd order oscillator, producing variables x1 and x2. The rest of the circuit is a hysteretic switching element that flips sign when x1 becomes too large. Figure 9 shows the response curve for the hysteretic part of the circuit. The unstable oscillator causes the variables x1 and x2 to spiral outward, while the hysteretic part adds a constant to this spiral. When x1 exceeds a certain threshold, the hysteretic element switches sign, changing the additive constant. Newcomb shows how the proper arrangement of the additive constant and expansion rate in the unstable oscillator may lead the value of x1 after the switching to be smaller than the value before the switching, and shows how this circuit can act chaotically. The equations describing this circuit are:

$$\frac{dx_1}{dt} = x_2 + \gamma x_1 + c x_3 \quad (4.1.1)$$

$$\frac{dx_2}{dt} = \omega x_1 - \delta_2 x_2 \quad (4.1.2)$$

$$\varepsilon \frac{dx_3}{dt} = \left(1 - x_3^2\right)\left(S x_1 - D + x_3\right) - \delta_3 x_3 \quad (4.1.3)$$

where $\gamma = R_2/R_6$, $c = R_2/R_5$, $\omega = R_1C_1/(R_2C_3)$, and δ_1 and δ_2 are damping parameters that model losses in the system. Equation (4.1.3) is based on a function used by Rossler to model hysteresis. For this circuit, $\gamma = 0.2$, $c = 2.2$, $\delta_2 = 0.001$, $\delta_3 = 0.001$, $\varepsilon = 0.3$, $\omega = 10$, $S = 1.667$, and $D = 0.0$.

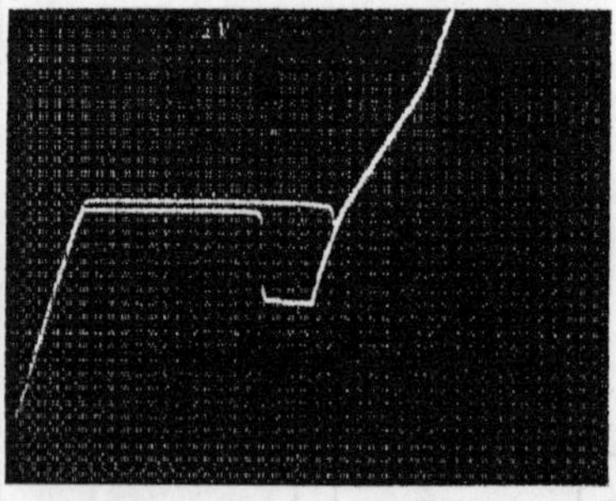

Figure 9. Hysteretic function in Newcomb circuit, scanned at 10 Hz.

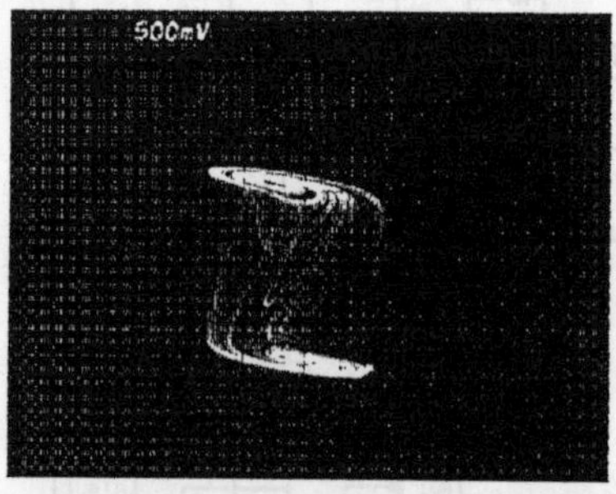

Figure 10. Attractor from the circuit. The horizontal axis is the signal x_1 from the Newcomb circuit, while the vertical axis is the difference between the x_1 and x_3 signals.

The first possible choice for a synchronizing subsystem is the x_2 and x_3 subsystem. The conditional Lyapunov exponents for this subsystem, when driven by the x_1 signal, are -9.98×10^{-4} and -5.01×10^{-3} s^{-1}. One may synchronize this system numerically, but synchronizing this subsystem in an actual circuit is difficult. The hysteretic part of the circuit depends on the nonlinear properties of two diodes and an op amp. Most electronic devices are engineered for their linear behavior, so matching the nonlinear characteristics of these devices is usually not easy, and this circuit is not an exception. It would be much easier to synchronize the response circuit to the drive circuit if the response circuit did not contain difficult to match nonlinearities.

One may find a circuit that is easier to synchronize by removing the part of the circuit containing the hysteretic nonlinearity. The response circuit that we actually used is shown in Fig. 8. To describe this circuit, a new variable x_4 is introduced:

$$x_4 = \frac{1}{\alpha}\left(x_3 + \beta x_1\right) \quad (4.1.4)$$

where $\alpha = R_9/R_{10}$ and $\beta = 1 + R_9/R_{02} + R_9/R_{10}$. The equations of motion for this circuit are then:

$$\frac{dx_1}{dt} = x_2 + \gamma x_1 + c\left(\alpha x_4 - \beta x_1\right) \tag{4.1.5}$$

$$\frac{dx_2}{dt} = -\omega x_1 + \delta_2 x_2 \tag{4.1.6}$$

$$\frac{dx_4}{dt} = \frac{1}{\alpha}\Big[\left(1 - (\alpha x_4 - \beta x_1)^2\right)\left(Sx_1 - D + \alpha x_4 - \beta x_1\right)/\varepsilon - \delta_3(\alpha x_4 - \beta x_1)/\varepsilon - \beta x_2 - \beta\gamma x_1 - c\beta(\alpha x_4 - \beta x_1)\Big] \tag{4.1.7}$$

where $\alpha = 6.67$ and $\beta = 7.87$. The conditional Lyapunov exponents for the x_1, x_2 subsystem are -16.587×10^{-3} and -0.603×10^{-3} s^{-1}.

The attractor for this circuit is shown in Fig. 10, where x1 is plotted vs. x1 - x3. Figure 11(a) is a plot of x_2' from the response circuit vs. x_2 from the drive circuit. The two voltages match to within about 1% of each other. The effect of mismatch on synchronization may be seen by replacing R_9 in the response circuit by a different fixed resistor. Figure 11(b) shows x_2 from the response circuit vs. x_2 from the drive when R_9 = 30 kΩ, so the mismatch between drive and response is 50%.

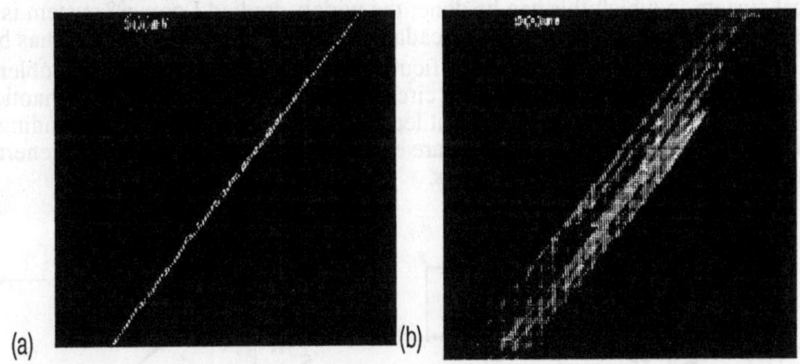

Figure 11. (a) Oscilloscope photo of the voltage x_2' from the response circuit vs. the corresponding voltage x_2 in the drive circuit when the circuits are well matched. (b) The corresponding signals when the circuits are mismatched by 50%.

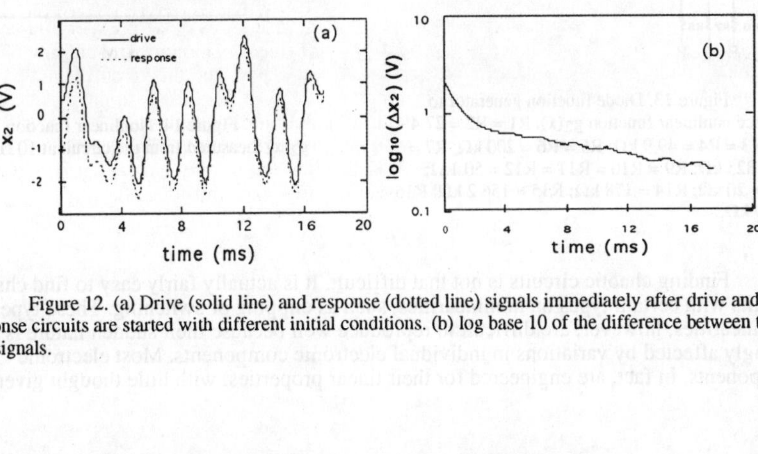

Figure 12. (a) Drive (solid line) and response (dotted line) signals immediately after drive and response circuits are started with different initial conditions. (b) log base 10 of the difference between these two signals.

As another comparison between experiment and theory, it is possible to estimate the conditional Lyapunov exponents for the response circuit and compare to the values calculated from the circuit equations (eqs. (4.1.1-4.1.3)). Figure 12(a) shows the voltage x_2 in the drive and response circuits as a function of time, when the two circuits have been started with different initial conditions. It can be seen that the x_2' voltage in the response converges to the x_2 voltage in the drive. Figure 12(b) is the log of the difference between x_2 voltages. The log curve shows two different straight line regions. Initially, the response circuit approaches the drive circuit at an exponential rate with a decay constant of $-0.667 \times 10^{-3}\ s^{-1}$, while at later times the decay constant is $-0.0609 \times 10^{-3}\ s^{-1}$. The faster decay time should correspond to the more negative of the conditional Lyapunov exponents, while the slower decay rate is seen after the difference between the drive and response in the direction corresponding to the most negative conditional Lyapunov exponent has decayed to a small value.

4.2 Cascaded synchronization.

The synchronization of the Newcomb circuit demonstrates the principles of synchronizing chaotic circuits, but the actual usefulness of this circuit is limited. To use this circuit it is necessary to send two signals; a drive signal to synchronize the response circuit, and a comparison signal to confirm that the two circuits are synchronized. It was shown above that it is possible to cascade stable subsystems of a circuit to create a response circuit in which the original drive signal is reproduced. It is easy to find a numerical system in which this can be done; the widely studied Lorenz[56] system is one example. It is not as easy to construct cascadable circuits. The Lorenz system has been built as a cascadable circuit[29], but it is difficult to build and work with. The problems involved in building a cascadable chaotic circuit are finding circuits that are chaotic, finding circuits that may be divided into at least two stable subsystems, and finding nonlinear elements for these circuits that are easy to reproduce. There are no general technique for achieving any of these goals.

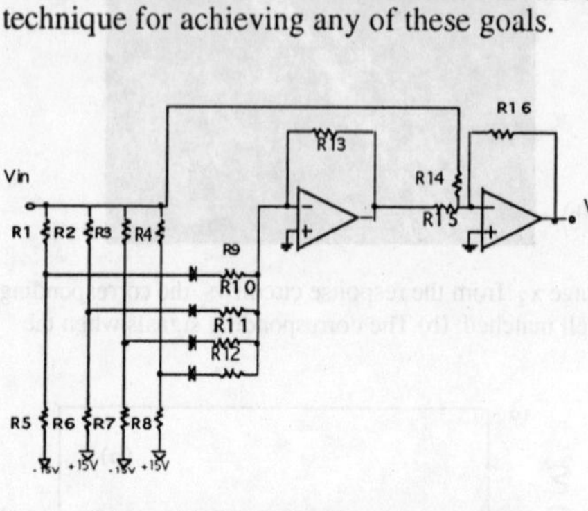

Figure 13. Diode function generator to produce nonlinear function $g_2(x)$. R1 = R2 = 27.4 kΩ; R3 = R4 = 49.9 kΩ; R5 = R6 = 200 kΩ; R7 = R8 = 825 kΩ; R9 = R10 = R11 = R12 = 50.1 Ω; R13 = 20 kΩ; R14 = 178 kΩ; R15 = 156.2 kΩ; R16 = 100 kΩ..

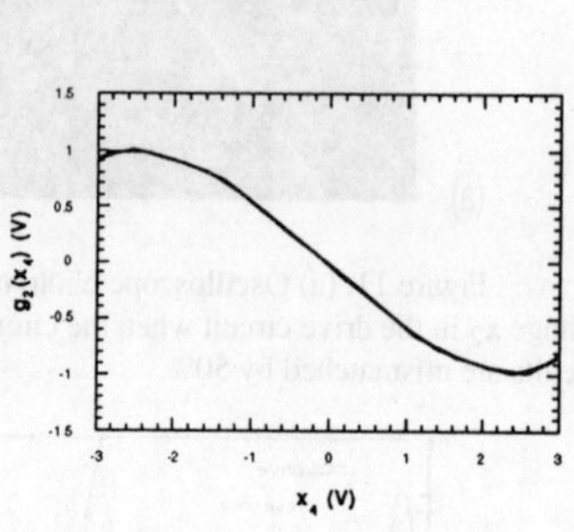

Figure 14. Nonlinear function $g_2(x)$ measured from the circuit at 10 Hz.

Finding chaotic circuits is not that difficult. It is actually fairly easy to find chaotic circuits with certain types on nonlinearities, such as clipping or switching. These types on nonlinearities, however, are difficult to reproduce well because their sudden nature is strongly affected by variations in individual electronic components. Most electronic components, in fact, are engineered for their linear properties, with little thought given to their nonlinear behavior, since most engineers avoid the nonlinear regime. From studying old analog computer techniques we have found one type on nonlinear element that is easy to build and reproduce is the piecewise linear diode function generator[57-59].

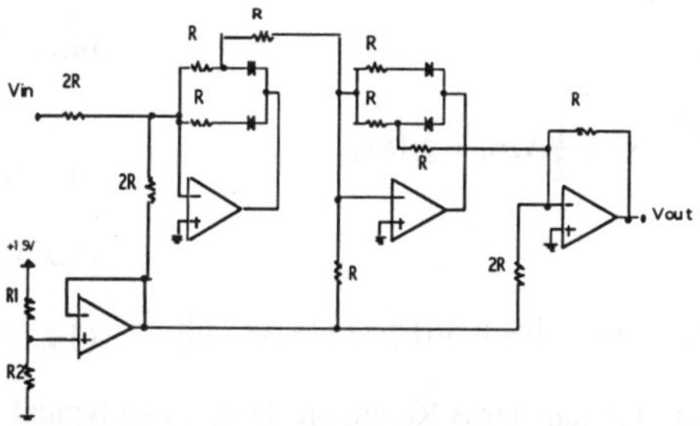

Figure 15. Diode function generator to produce function $g_1(x)$. R = 10 kΩ; R1 = 100 kΩ; R2 = 20 kΩ.

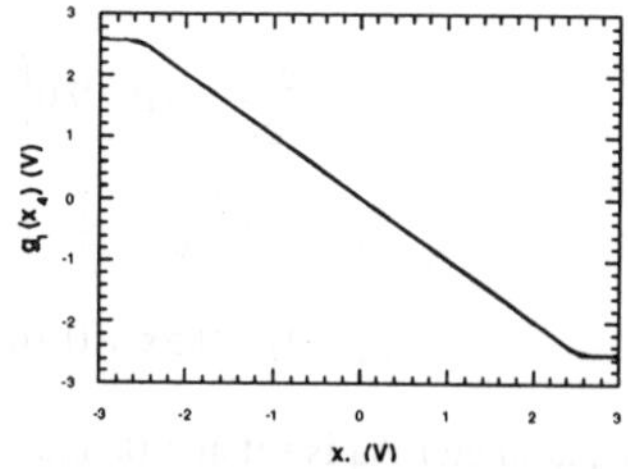

Figure 16. Nonlinear function $g_1(x)$ measured from the circuit at 10 Hz.

Figure 13 is a schematic of a diode function generator used to produce the piecewise linear function $g_2(x)$ shown in Fig. 14. The incoming voltage V_{in} is biased upward or downward by a pair of resistors, such as R_1 and R_5. The biased signal will be amplified with a gain determined by the feedback resistor R_{13} when it is large or small enough to turn on the diode. The only nonlinearity used is the diode turn-on, and the voltage at which this occurs may be compensated for by using potentiometers for resistors R_5 to R_8. Figure 15 is a schematic of a diode function generator that produces a clipping function[59], and Fig. 16 shows the output of this circuit.

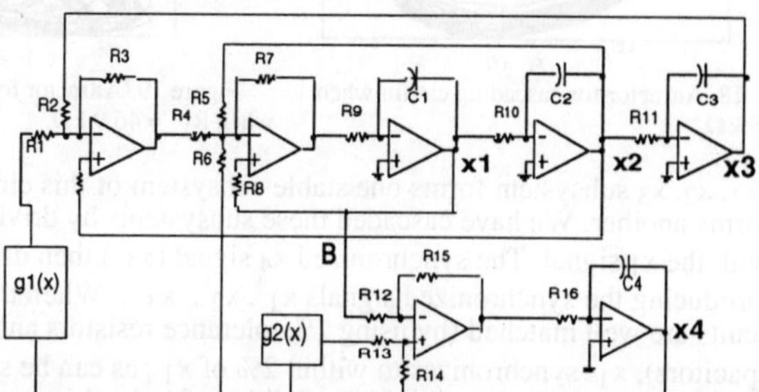

Figure 17. Chaotic circuit used to study cascaded synchronization. The boxes $g_1(x)$ and $g_2(x)$ are nonlinear functions. A and B are the points at which the circuit may be broken to produce subsystems. R1 = R2 = R3 = R4 = R6 = R7 = R10 = R13 = R15 = R16 = R17 = R18 = R19 = R20 = 100 kΩ. R5 = R8 = 68.2 kΩ. R9 = 1 MΩ. R11 = 221 kΩ. R14 = 200 kΩ. C1 = C2 = C3 = C4 = 910 pF. R12 = R_v = variable. Resistor tolerances are 1%, and all capacitors are 5% mica capacitors. All op amps are type 741, and all diodes are type 1n485B.

The chaotic circuit[6] built from these nonlinear elements is shown in Fig. 17. This circuit contains 4 integrators, and may be described by 4 equations:

$$\frac{dx_1}{dt} = -1098[\,1.466A_1x_1 - 2.466x_2 + x_3 - x_4 + g_1(x_4)] \tag{4.2.1}$$

$$\frac{dx_2}{dt} = -10980x_1 \tag{4.2.2}$$

$$\frac{dx_3}{dt} = -4972x_2 \tag{4.2.3}$$

$$\frac{dx_4}{dt} = -10{,}980\left(\frac{-100{,}000}{R_v}\,x_1 + \frac{1}{2}A_4x_4 + g_2(x_4)\right) \tag{4.2.4}$$

$$g_1(x) = 0.5(|x-2.5| - |x+2.5|) \tag{4.2.5}$$

$$g_2((x)) = 0.5x + 0.164\,(|x-1.3| - |x+1.3|) + 0.361(|x-2.6| - |x+2.6|) \tag{4.2.6}$$

The parameter A_4 is set at 1.0, and the control parameter is R_V. Figure 18 shows a typical attractor for this circuit, for $R_V = 47.8$ kΩ. Because of the symmetry of the g_2 function, there is an identical attractor which is a mirror image of this one. The existence of two attractors causes practical problems for synchronization. For smaller values of R_V, these two attractors are merged, as can be seen in Fig. 19, for $R_V = 46.9$ kΩ.

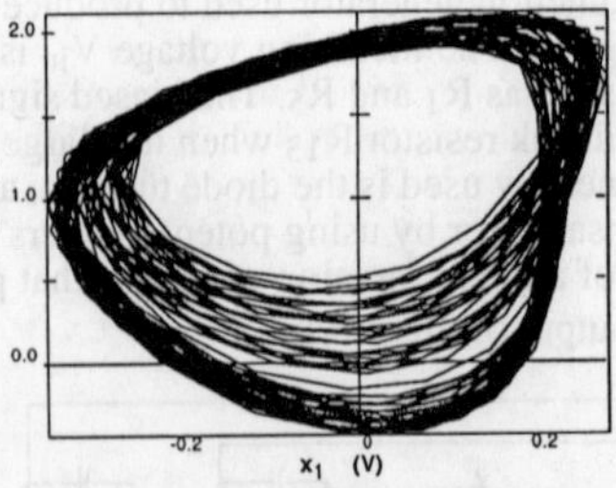

Figure 18. Attractor for cascading circuit when $R_V = 47.8$ kΩ.

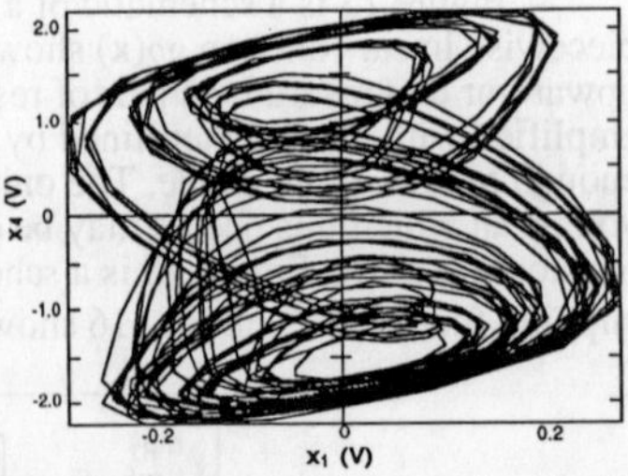

Figure 19. Attractor for the cascading circuit when R_V = 46.9 kΩ.

The x_1, x_2, x_3 subsystem forms one stable subsystem of this circuit, while the x_4 subsystem forms another. We have cascaded these subsystems by driving the x_4 subsystem with the x_1 signal. The synchronized x_4 signal (x_4') then drives the x_1, x_2, x_3 subsystem, producing the synchronized signals x_1'', x_2'', x_3''. When the drive and response circuits are well matched (by using 1% tolerance resistors and 5% or better tolerance capacitors), x_1'' synchronizes to within 2% of x_1, as can be seen in Fig. 20. This synchronization assumes that all initial conditions for the drive and response circuits start in the same basin of attraction; if one circuit is in a different basin of attraction then the other, the circuits will not synchronize, as is seen in Fig. 21.

Although the global conditional Lyapunov exponents for the response circuit are all negative, it is possible for the local Lyapunov exponents to be positive somewhere on the attractor. The effect of these local instabilities depends on how different the two circuits are and how long they spend in the region of instability. Changing the circuit control R_V in both circuits to 46.9 kΩ (as in Fig. 19) puts the circuit into a state where the response circuit spends much more time in a region of local instability. As a result, synchronization is not as good for this state[7]. Figure 22 shows x_1'' vs. x_1 for $R_V = 46.9$ kΩ, while Fig. 23 shows a time series for x_4 and $x_4 - x_4'$. The synchronization appears to degrade when x_4 crosses zero going in the negative direction, because x_4' spends much more time near zero, in a region of local instability. The asymmetry in the $x_4 - x_4'$ signal is caused by asymmetry and mismatch in the drive and response circuits. This synchronization degradation does not show up in numerical simulations of the circuit, where noise and mismatch do not exist.

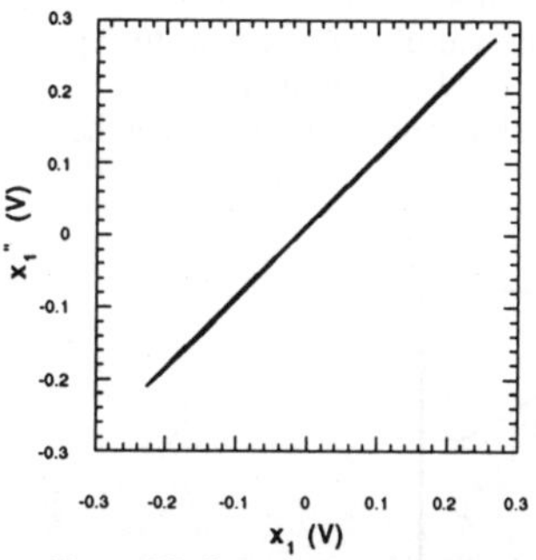

Figure 20. Output of the cascaded response circuit ($x_1{}''$) vs. driving signal (x_1) when the drive and response circuits are synchronized.

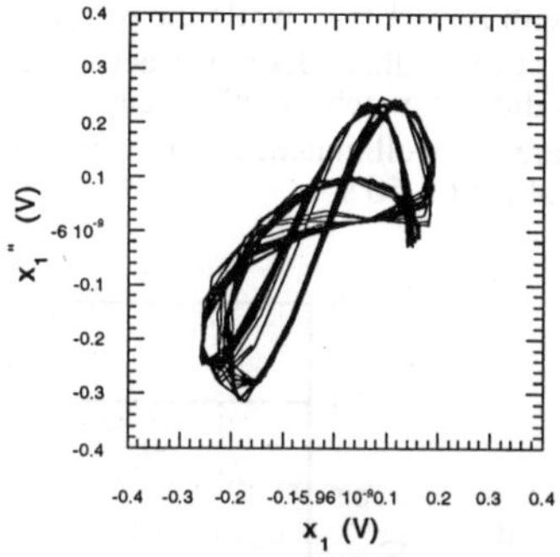

Figure 21. Output of the cascaded response circuit ($x_1{}''$) vs. driving signal (x_1) when the drive and response circuits are synchronized but the response and drive are in different basins of attraction.

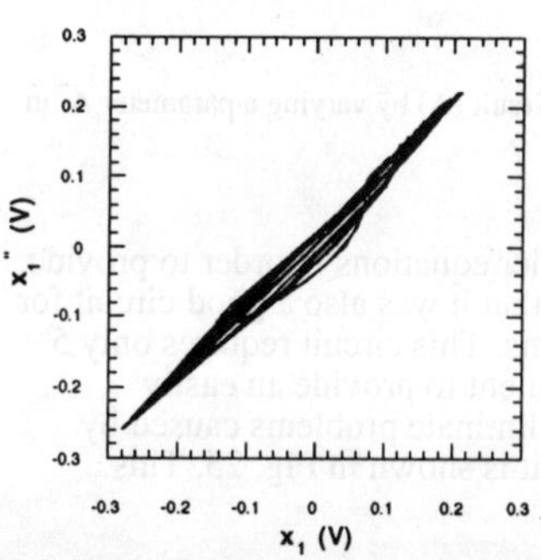

Figure 22. Output of the cascaded response circuit x'' vs. drive signal x when the control parameter R_v is 46.9 kΩ, increasing the effect of local instabilities.

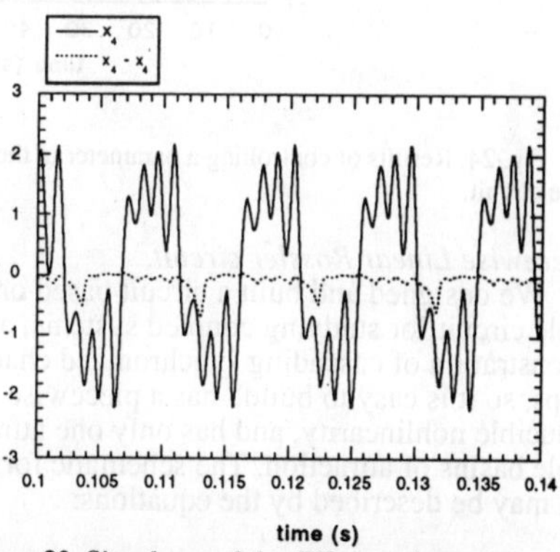

Figure 23. Signal x_4 and the difference between x_4 and $x_4{}''$ showing the effect of local instabilities.

With the cascaded synchronization, it is possible to detect a parameter mismatch between the drive and response circuits and correct this mismatch using only the one drive signal[7]. Comparing the drive signal x_1 to the output of the response system, $x_1{}''$, shows if there is a parameter mismatch between the two circuits. In the actual circuits, the parameter A_1 was changed by using an Analog Devices AD632 multiplier chip. The comparison between x_1 and $x_1{}''$ was made by using the negative going zero crossing of $x_1{}''$ to generate a strobe signal. The value of x_1 was sampled and held at intervals determined by this strobe signal. The result was a step time series consisting of the values of x_1 at the preceding negative going zero crossing of $x_1{}''$. This time series was averaged by an analog integration circuit to produce a correction signal Δ. If the circuits were matched, x_1 and x_1'' crossed zero at the same time, and Δ was zero. If the circuits were not matched, Δ had a finite value roughly proportional to the magnitude and sign of the parameter mismatch. The signal Δ was then used to correct the value of the parameter A_1'' in the response circuit. Figure 24 shows how well this control worked. The drive circuit was started with $A_1 = 1.0$ V, while the response circuit was started with $A_1'' = 0.6$ V. At t = 12 s, the control feedback loop was activated, bringing the value of A_1'' closer to A_1. These values did not match exactly because the multipliers used to modulate the parameter were not well matched. The value of A_1'' did track changes in A_1, as can be seen in Fig. 24 for t > 50 s.

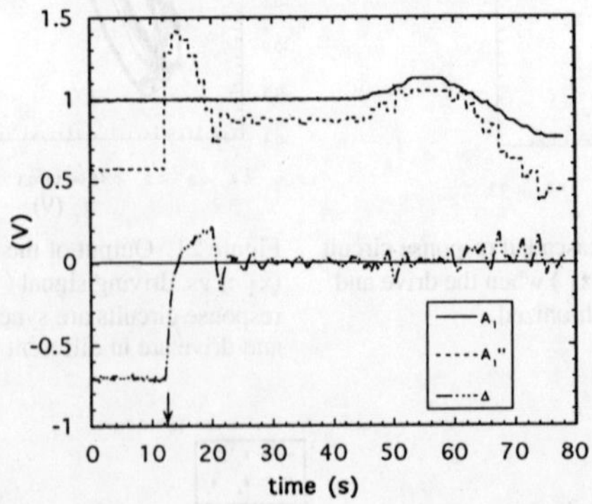

Fig. 24. Results of controlling a parameter in the response circuit (A) by varying a parameter A" in the drive circuit.

4.3 Piecewise Linear Rossler circuit.

We designed and built a circuit based on the Rossler equations in order to provide a simple circuit for studying coupled systems, and found that it was also a good circuit for a demonstration of cascading synchronized chaotic systems. This circuit requires only 5 op amps, so it is easy to build, has a piecewise linear element to provide an easily reproducible nonlinearity, and has only one attractor, to eliminate problems caused by multiple basins of attraction. The schematic for this circuit is shown in Fig. 25. This circuit may be described by the equations:

$$\frac{dx}{dt} = -\alpha\left(\Gamma x + \beta y + \lambda z\right) \tag{4.3.1}$$

$$\frac{dy}{dt} = -\alpha\left(-x - \gamma y + 0.02y\right) \tag{4.3.2}$$

$$\frac{dz}{dt} = -\alpha\left(-g(x) - z\right) \tag{4.3.3}$$

$$g(x) = \begin{cases} 0 & x \le 3 \\ \mu x & x > 3 \end{cases} \tag{4.3.4}$$

where the time factor α is 10^4 s^{-1}, Γ is 0.05, β is 0.5, λ is 1.0, γ is 0.133 and μ is 15. The schematic for the response circuit is also shown in Fig. 25. The y signal drives the x,z subsystem, after which the y subsystem is driven by x' and y to produce y". Only the y term in eq. (4.3.2) with a coefficient of γ is replaced by the incoming y signal; the term with the coefficient of 0..02 is replaced by y" in order to stabilize the integrator circuit. It may be necessary to replace resistors R9 and R10 by trimpots in order to adequately match the nonlinear function. When built, this circuit provides a simple circuit to demonstrate cascaded synchronization of chaotic circuits[60].

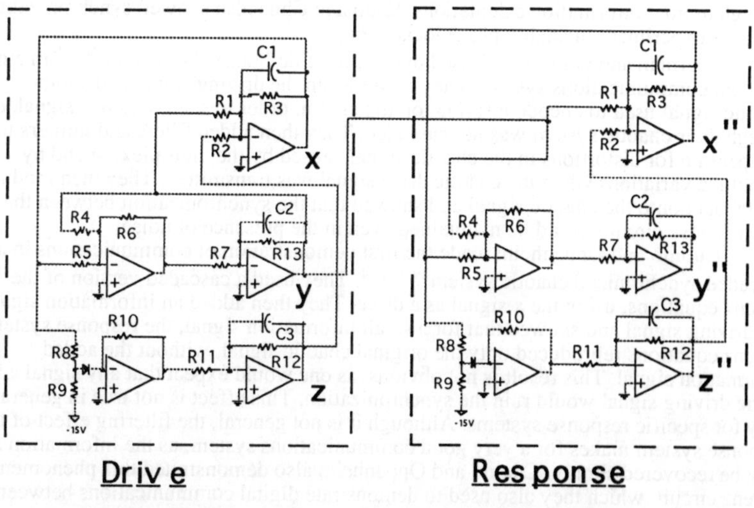

Fig. 25. Piecewise linear Rossler circuits arranged for cascaded synchronization. R1=100 kΩ, R2=200 kΩ, R3=R13 =2MΩ, R4=75 kΩ, R5=10 kΩ, R6=10 kΩ, R7=100 kΩ, R8=10 kΩ, R9=68 kΩ, R10=150 kΩ, R11=100 kΩ, R12=100 kΩ, C1=C2=C3=0.001 μF, and the diode is a type MV2101

5. Other Experiments and Applications

A number of other groups have done actual experiments with synchronized chaotic circuits, and some have demonstrated simple applications of synchronized chaos. De Sousa Vieira et al.[36, 61] demonstrated chaotic synchronization in digital phase locked loop (DPLL) circuits. A digital phase locked loop is used to lock the frequency of a local oscillator to an incoming signal from a remote oscillator. The DPLL samples the signal from the remote oscillator to generate a correction signal which sets the frequency of a voltage controlled oscillator (VCO). The VCO signal in turn determines the interval at which the incoming remote oscillator signal is sampled. If the incoming signal is sinusoidal and the VCO frequency is linearly related to the sampled value of the input signal $v(t_i)$ by the relation $f = f^0 + b(v(t_i) + v^{off})$, where f^0 is the center frequency of the VCO, v^{off} is an offset voltage and b is the loop gain, then the phase difference between the VCO and the incoming signal is described by a circle map. The output of the VCO's used by de Sousa Vieira et al. was a triangle wave given by $v(t) = A\Lambda(\phi(t))$, where $\phi(t) = ft$, t goes from 0 to 1/f, and $\Lambda(t)$ is a triangle wave. Their chaotic circuit consisted of two coupled DPLL's, where the output from one VCO formed the input for the other DPLL, a situation described by:

$$f_1' = f_1^0 + b_1\left[A_2\Lambda(\phi_2) + v_1^{off}\right] \quad (\phi_1 = 0)$$

$$f_2' = f_2^0 + b_2\left[A_1 \Lambda(\phi_1) + v_2^{off}\right] \quad (\phi_2 = 0) \tag{5.1}$$

where the frequency of each loop is changed only when its triangle wave is zero with positive slope. The signal Λ_2 from the output of the second VCO is used as a driving signal for a third DPLL that synchronizes to the first DPLL. De Sousa Vieira et al. [36, 61]demonstrated synchronization of these chaotic circuits and confirmed synchronization through mutual information calculations. Endo and Chua also showed synchronization in a similar experiment in analog phase locked loops.

Sherman and Gullicksen[62]used the circuit of de Sousa Vieira et al.[36, 61] in a noise resistant communications system. They multiplexed the driving signal and another chaotic signal used to encode information together in order to send only one signal, even though their chaotic system was not cascaded. They then added filters and limiters to compensate for distortions in the chaotic signal caused by the multiplexing and by amplitude variations when the chaotic drive signal was transmitted. They then modulated digital data onto the chaotic signal and showed that the synchronization between the drive and response circuits could be maintained even in the presence of noise.

Cuomo and Oppenheim made the first demonstration of communications in a cascaded synchronized chaotic system[29, 63, 64]. They used a cascaded version of the Lorenz equations, using the x signal as a drive. They then added an information signal to the driving signal and showed that for a small information signal, the response system Lorenz equations reproduced only the original chaotic signal, without the added information signal. This result is not obvious, as one would expect that any signal added to the driving signal would ruin the synchronization. This effect is not true in general, but only for specific response systems. Although it is not general, the filtering effect of the response system makes for a very good communications system, as the information signal may be recovered simply. Cuomo and Oppenheim also demonstrated this phenomena in a Lorenz circuit, which they also used to demonstrate digital communications between synchronized chaotic circuits.

Synchronization of chaotic circuits was also studied by Chua et al.[65] in the circuit known as Chua's circuit[66, 67]. This circuit may be described in a dimensionless form by the equations:

$$\begin{aligned} \frac{dx}{dt} &= \alpha(y - x - f(x)) \\ \frac{dy}{dt} &= x - y + z \\ \frac{dz}{dt} &= -\beta y \\ f(x) &= bx + \frac{1}{2}(a - b)\left[\,|x - 1| - |x + 1|\,\right] \end{aligned} \tag{5.2}$$

where the function f(x) is a negative impedance function formed by an op amp using both positive and negative feedback, an arrangement known as a negative impedance converter[58, 68]. After first examining mutual diffusive coupling between two circuits, Chua et al. considered one way driving using either the x, y, or z variables as driving signals. For the x and y driving signals, they found that the response systems had all negative Lyapunov exponents, and they found that synchronization took place. For the z driving, they found a positive Lyapunov exponent for the response system, and no synchronization was seen.

The same group also studied cascaded synchronization of chaotic circuits. They first reproduced the scheme of Cuomo and Oppenheim[29, 63, 64] in cascaded Chua's circuits[69], transmitting a sine wave as the information signal. Later, they modulated the chaotic drive signal with a digital signal by switching a parameter in the drive circuit[70].

6. Additions to Synchronization

6.1 Synchronizing Non-Autonomous Circuits

One drawback of the synchronizing circuits described so far is that the synchronization degrades quickly when noise is added to the chaotic driving signal. It is still possible to get information from the noisy chaotic signal if the signal comes from a nonautonomous system. The same conditions for synchronization that were found for autonomous circuits also hold for nonautonomous circuits, with the additional condition that the phases of the periodic forcing signals in the nonautonomous circuits must also match. The phase of the periodic driving source represents a zero Lyapunov exponent which must be made less than zero by some additional control device before synchronization is possible. We have found that there is enough phase information in the chaotic driving signal to synchronize the phase of the periodic forcing source in the response circuit, so that synchronization between nonautonomous chaotic systems is possible[8].

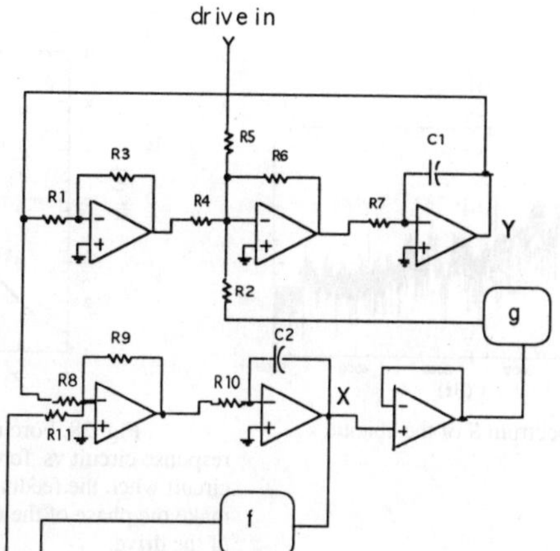

Fig. 26 Circuit described by eqs. (6.1.1-6.1.5). R1= R3=R4= R5= R6=10kΩ, R2= 39.2 kΩ, R7=R10=R12=R13=R14=R16= R17= R18 =100 kΩ, R8=R9=R18 =1 MΩ, R15=5.2 kΩ , C1=C2= C3= 0.001 μF. The box labeled f corresponds to the nonlinear function f(x), while the box labeled g corresponds to the nonlinear function g(x). All op amps are type 741 or equivalents.

The circuit of Fig. 27 is a nonautonomous chaotic circuit that may be synchronized[8]. This circuit is based on the well known Duffing equations[56] with one extra dimension added. The boxes labeled f and g are piecewise linear diode function generators, similar to those described above. This circuit may be described by the equations

$$\frac{dx}{dt} = \beta(y - z) \tag{6.1.1}$$

$$\frac{dy}{dt} = \beta\left(-\Gamma_y y - g(x) + \alpha cos(\omega t)\right) \qquad (6.1.2)$$

$$\frac{dz}{dt} = \beta\left(f(x) - \Gamma_z z\right) \qquad (6.1.3)$$

$$g(x) = -3.8 + \left(\frac{1}{2}\right)\left(|x + 2.6| + |x - 2.6| + |x + 1.2| + |x - 1.2|\right) \qquad (6.1.4)$$

$$f(x) = \frac{1}{2}x + |x - 1| - |x + 1| \qquad (6.1.5)$$

where $\alpha = 2.0$, $\Gamma_y = 0.2$, $\Gamma_z = 0.1$; the time factor β is 10^4 s^{-1}; and the frequency ω is $2\pi f_d$, where the forcing frequency f_d -s 769 Hz. The power spectrum of the chaotic x signal is shown in Fig. 27.

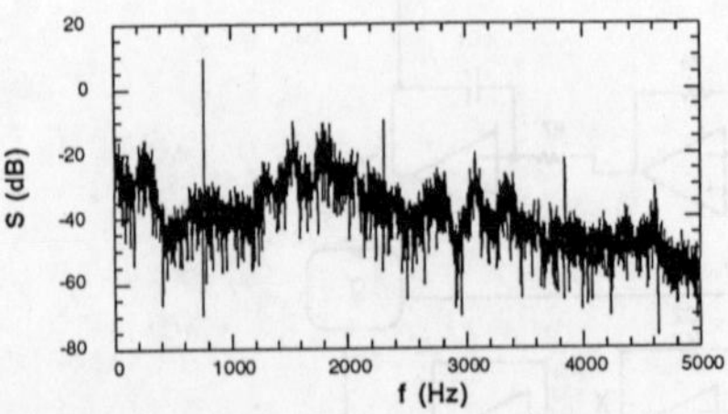

Fig 27. Power spectrum S of the chaotic x driving signal

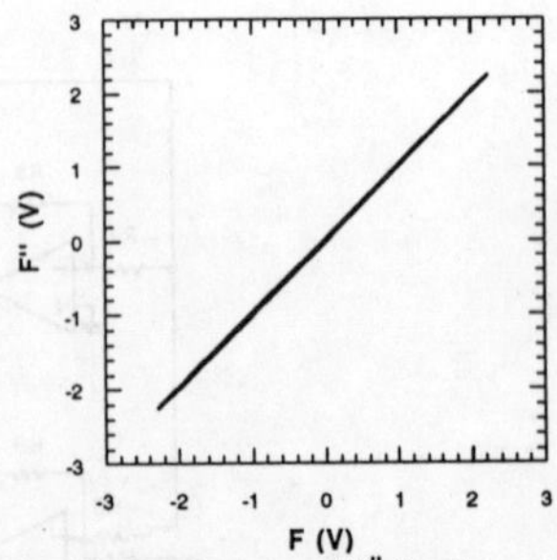

Fig. 28. Forcing signal F" for the response circuit vs. forcing signal F for the drive circuit when the feedback loop has been used to make the phase of the response match the phase of the drive.

This circuit has two stable subsystems, the x,z subsystem and the y subsystem. We used the x signal as a driving signal and cascaded the x,z subsystem, which then provided an x driving signal for the y subsystem. In order to synchronize the response to the drive, the phase of the response periodic forcing source F" had to be matched to the phase of the drive periodic forcing F. To extract the phase information, we treated the phase as we would any other parameter and used the method described above[7, 8] to extract the value of the phase from the chaotic driving signal. We used the output x" of the response circuit to generate a strobe pulse when it crossed zero in the positive direction. This strobe pulse was used to sample the drive signal x. When the circuits were synchronized, the drive signal x was zero at the time of the strobe pulse; if the circuits were not synchronized, the drive signal had some other value. The sequence of values of x during the strobe pulse was averaged with an analog integrator to produce a correction signal Δ that was proportional to the parameter difference between drive and response circuits. In this case, the parameter was the phase of the periodic forcing, so the error signal Δ was applied to the FM input of the function generator providing the periodic forcing signal for the response circuit in order to correct the phase of the forcing F". Fig. 28. is a plot of the forcing F" in the response vs. the forcing F in the drive, showing that the two phases are matched.

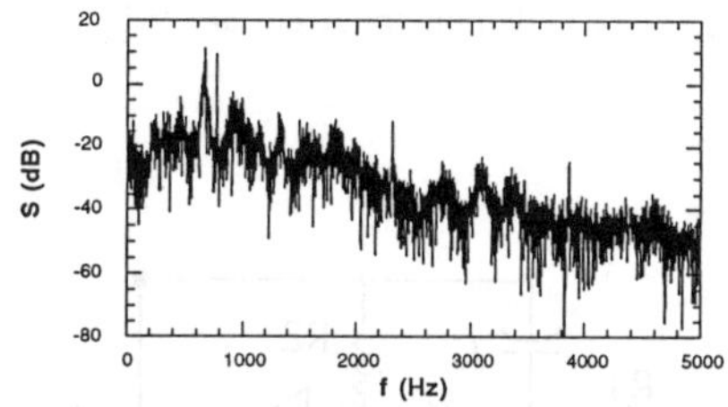

Fig. 29. Power spectrum of the driving signal x when a chaotic signal has been added.

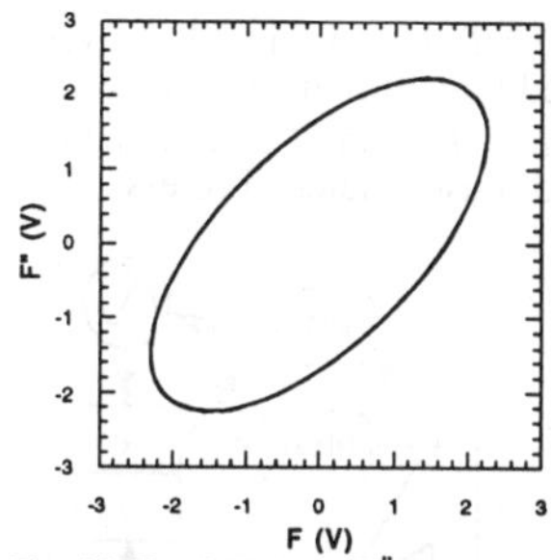

Fig. 30. Forcing signal F'' for the response circuit vs. forcing signal F for the drive circuit when chaos has been added to the driving signal.

This phase matching was stable even when a contaminating signal was added to the drive signal x. The most difficult contaminating signal to separate from a chaotic signal is another chaotic signal with components in the same frequency range. It is difficult to separate signals occupying the same frequency band with a filter, while the deterministic nature of the contaminating chaotic signal makes it impossible to use many recent signal separation algorithms that depend on the contaminating signal being random. The cascaded synchronized chaotic circuits are less sensitive to noise because the incoming signal is run through the cascaded response system before being compared with itself. If the incoming signal is not the correct chaotic signal, then the response system output will not be simply related to the input, and an average of the input strobed with the output will not respond to changes in the input signal. The correct chaotic signal will yield an average that responds to changes in the drive parameters, Adding a contaminating signal that is not too large and not too similar will at most add a constant to this average, so that the error signal Δ will only be responsive to parameter differences when the correct chaotic signal is present.

Using an analog adder circuit, we used chaos from the circuit of Fig. 17 as a contaminating signal which was added to the x driving signal with about twice the amplitude. Figure 29 is a power spectrum of this sum. The response forcing source was still able to lock to the phase of the drive forcing signal, as is seen in Fig. 30, but a constant phase offset was present. This phase offset is not a problem, as a constant phase relationship is good enough for applications such as communications. The response chaotic circuit is not actually synchronized to the drive, but the added noise makes it impossible to detect synchronization anyway. The solution to the noise sensitivity problem is to use a chaotic signal to synchronize periodic signals, rather than looking for the synchronization of the chaotic signals.

6.2 Filtering the Driving Signal

The power spectrum of Fig. 27 contains large peaks at the forcing frequency and odd harmonics. These peaks make it easy to recognize this as the spectrum of a deterministic system, and transmitting a signal with these large peaks requires a large amount of power. These peaks could be removed by filtering, but this would change the driving signal, so that synchronization no longer occurred. There may be other reasons for filtering the chaotic drive signal, such as limiting the bandwidth before transmission. In general, filtering alters the chaotic drive signal, making synchronization impossible. There are some types of transformation, however, that may be undone by using the synchronized response circuit. The undoing of these transformations (or reconstruction of the driving signal) does not involve inverting a function, so there are no problems with ill-conditioned transformations. There are probably many types of transformation that can be undone without being ill-conditioned; only one has been demonstrated so far[9, 10] by one of us, and it uses matched filters.

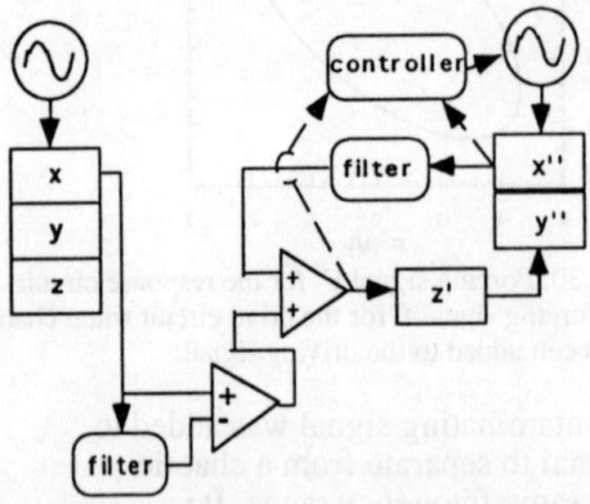

Figure 31. Block diagram of filtering and reconstruction of chaotic driving signal for a nonautonomous chaotic circuit.

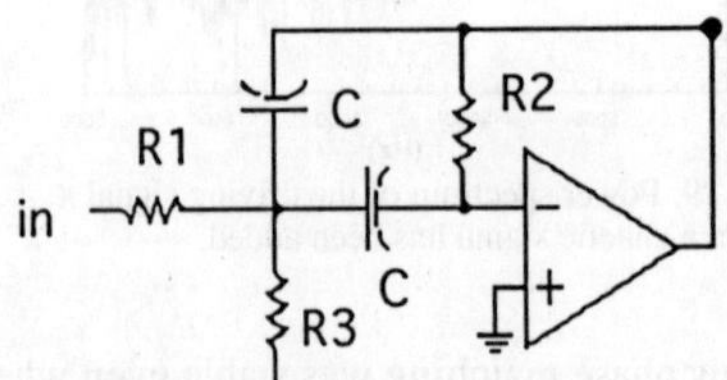

Figure 32. Schematic of one element of the band pass filter used to filter the chaotic driving signal. The resistor values are given in Table I; the capacitor value is C = 0.001 μF.

The general method (shown in block diagram form in Fig. 31) is to use a filtering operator F and operate on the driving signal x with (1-F) to produce a transmitted signal x_t. To remove the peaks from the power spectrum of Fig. 27, for example, a set of bandpass filters was used that each passed one of the harmonics of the forcing frequency. The sum of the signals from the filters was then subtracted from the driving signal x to produce x_t. At the receiver, a reconstructed driving signal x_d is produced by filtering the output x" of the response circuit and adding the filter output to x_t. The filtering and reconstruction scheme is described by:

$$w = \frac{d(x - \alpha\cos(\omega t))}{dt} \tag{6.2.1}$$

$$\frac{du_i}{dt} = \frac{-2.0}{R_{i2}C}u_i - \frac{1}{R_{i2}C}\left(\frac{1}{R_{i3}C} + \frac{1}{R_{i1}C}\right)v_i - \frac{1}{R_{i1}C}w \tag{6.2.2}$$

$$\frac{dv_i}{dt} = u_i \tag{6.2.3}$$

$$x_t = x + \sum_{i=1}^{5} v_i \tag{6.2.4}$$

$$x_d = x_t - \sum_{i=1}^{5} r_i \tag{6.2.5}$$

$$\frac{dq_i}{dt} = \frac{-2.0}{R_{i2}C}q_i - \frac{1}{R_{i2}C}\left(\frac{1}{R_{i3}C} + \frac{1}{R_{i1}C}\right)r_i - \frac{1}{R_{i1}C}\frac{dx''}{dt} \tag{6.2.6}$$

$$\frac{dr_i}{dt} = q_i \tag{6.2.7}$$

where x is the x variable in eq. (6.1.1) and x_d replaces x as the chaotic driving signal for the response system. The individual band pass filters are second order band pass filters, with the resistance values defined in Table I. Figure 32 shows a single filter.

Table I. Resistor values for band pass filters

center frequency (Hz)	R_1 (ohms)	R_2 (ohms)	R_3 (ohms)
780	204,000	408,000	1026
1560	102,000	204,000	513
2340	68,000	136,000	342
3120	51,000	102,000	256
3900	40,800	82,000	205

If the drive and response circuits are synchronized, the filter in the response circuit will just put back what the filter in the drive circuit takes out. Synchronization can occur if this synchronized state is stable and the response system does not fall into some other stable state. For the circuit described here, the synchronized state is stable; numerical simulations (using only a filter at the fundamental frequency to speed up the computation) show that the response circuit has a largest Lyapunov exponent of $-10.0\ s^{-1}$, which may be compared with the largest Lyapunov exponent for the driven response circuit without filtering of $-780\ s^{-1}$.

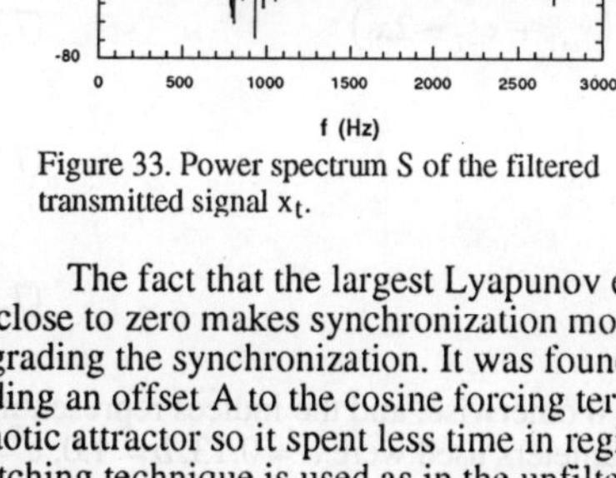

Figure 33. Power spectrum S of the filtered transmitted signal x_t.

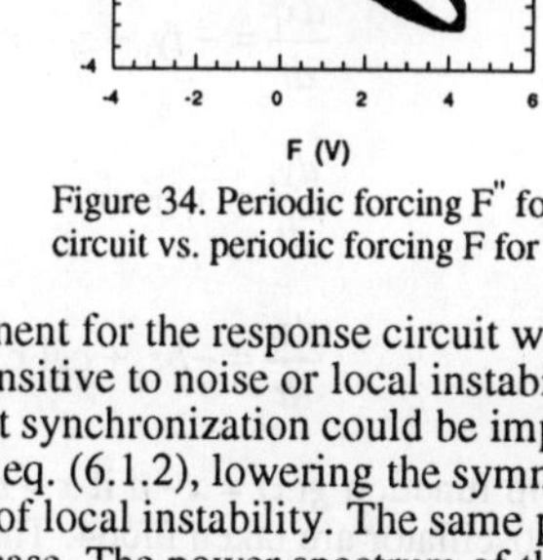

Figure 34. Periodic forcing F" for the response circuit vs. periodic forcing F for the drive circuit.

The fact that the largest Lyapunov exponent for the response circuit with a filter is so close to zero makes synchronization more sensitive to noise or local instabilities, degrading the synchronization. It was found that synchronization could be improved by adding an offset A to the cosine forcing term in eq. (6.1.2), lowering the symmetry of the chaotic attractor so it spent less time in regions of local instability. The same phase matching technique is used as in the unfiltered case. The power spectrum of the filtered signal x_t is shown in Fig. 33, while the forcing F" for the response system with filter vs. the forcing F for the drive system is shown in Fig. 34. The synchronization is not as good as in the unfiltered case, probably due to filter mismatch.

This filtered synchronization still maintains some robustness in the presence of noise. As a contaminating signal, a chaotic signal from another Duffing circuit with different parameters was used. The second Duffing circuit was driven by an independent forcing source at the same frequency as the original Duffing circuit to produce a contaminating signal which was filtered in the same way as the signal x_t. Figure 35 is the power spectrum of the contaminating signal, which was added to the x_t signal. Synchronization was maintained with the contaminating signal added.

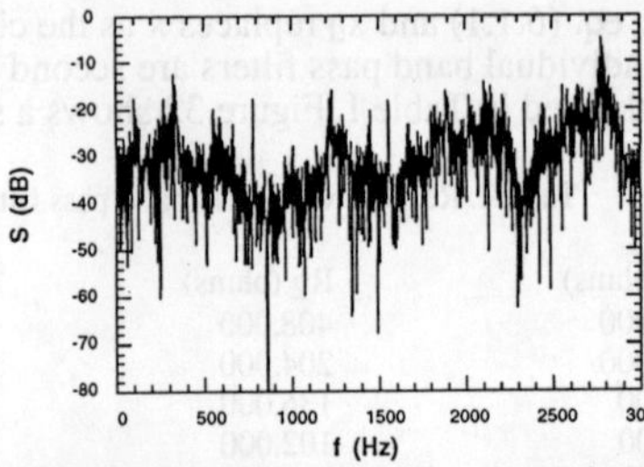

Figure 35. Power spectrum of the filtered contaminating signal from a different Duffing circuit that was added to the transmitted signal x_t

These circuits were also used to demonstrate communications and signal separation[9]. The offset A that was added to eq. (6.1.2) could be switched between two values, and this switching could be detected in the controller error signal Δ, even when the contaminating Duffing signal was present. As a further test, an offset was added to the periodic forcing in the contaminating Duffing circuit. The switching of this offset in the contaminating circuit could not be detected in the error signal Δ.

7. Riddled Basins

As mentioned above in section 3.1, it is possible to meet all the conditions for synchronization and still not see synchronization in two circuits. The initial conditions for the response circuit must be in the proper basin of attraction for synchronization to occur. It is possible in some cases for two basins of attraction to be so finely intermingled with each other that there is always a possibility of going to the asynchronous attractor no matter how close it is to the synchronous attractor. Such basins of attraction are known as riddled basins[71].

We have observed riddled basins in a set of four bi-directionally coupled circuits[53] similar to the Rossler system[72]. Each circuit is described by[60]:

$$\frac{dx_i}{dt} = -fy_i - z_i - rx_i + k(x_{i+1} + x_{i-1} - 2x_i) \tag{7.1}$$

$$\frac{dy_i}{dt} = x_i + ay_i \tag{7.2}$$

$$\frac{dz_i}{dt} = -bz_i + cg(x_i) \tag{7.3}$$

where the ramp function $g(x) = x - d$ if $x > d$ and 0 otherwise, and the indices representing the particular oscillator are taken mod4. The parameters used were $a = 0.13$, $b = 1.0$, $c =$ 15.0, $d = 3.0$, $f = 0.5$, and $r = 0.05$. The coupling k was set to 0.935 in the equations or 0.805 in the circuit.

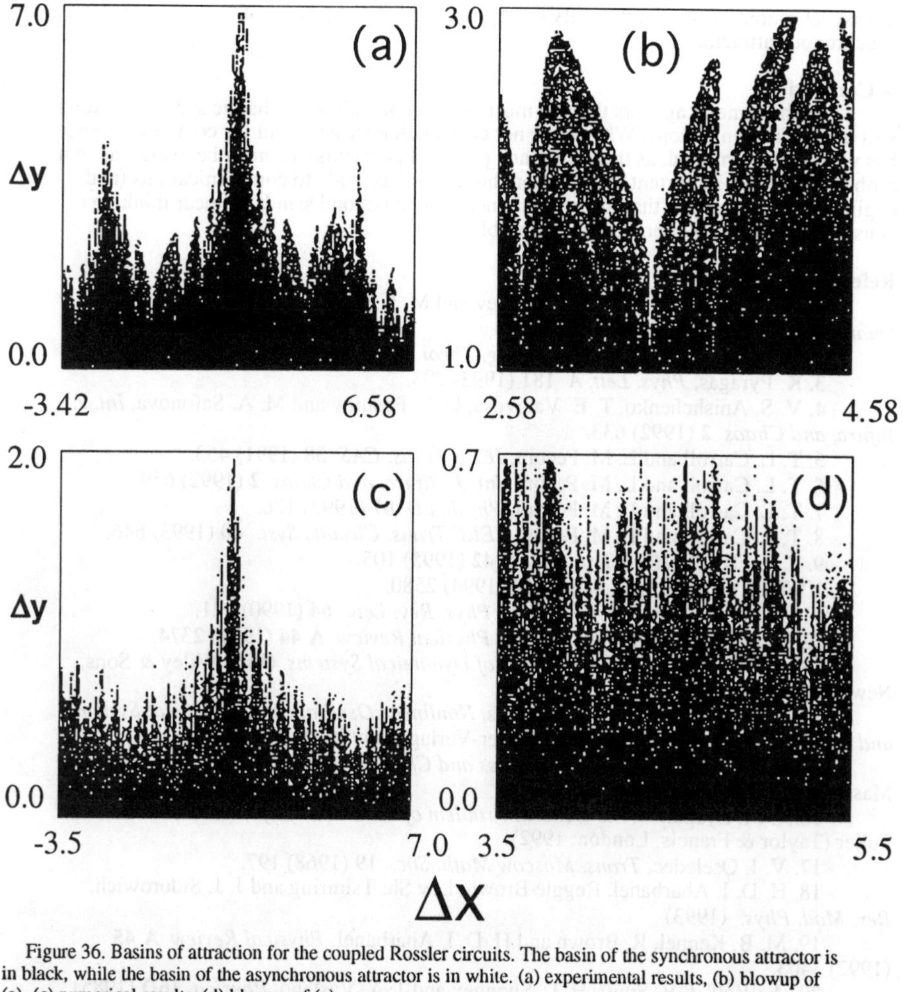

Figure 36. Basins of attraction for the coupled Rossler circuits. The basin of the synchronous attractor is in black, while the basin of the asynchronous attractor is in white. (a) experimental results, (b) blowup of (a), (c) numerical results, (d) blowup of (c).

The riddled basins could be seen by using a computer to set a grid of 40,000 initial conditions for the circuit and recording which attractor each initial condition led to. A point on the synchronous attractor (x_S, y_S, z_S) was selected for a reference, and a variation Δx was the variation of the x coordinate from this point. The y component of the variation was chosen to have the symmetry of the asynchronous attractor, so that for the i'th circuit, the initial condition was $y_i = y_S + (-1)^i \Delta y$. The resulting basin of attraction plots are shown in Fig. 36, where the basin of the synchronous attractor is in black and the basin of the asynchronous attractor is white. It can be seen for both the numerical and experimental results that the basin of attraction is riddled with points from the basin of the asynchronous attractor. In a system with riddled basins, it is not possible to predict if the system will be synchronized because there are points from the basin of attraction of the asynchronous attractor arbitrarily close to all points in the basin of attraction of the synchronous attractor.

8. Conclusions

At this time, it appears that the most likely application for chaotic synchronization will be in communications. Whether or not chaos will ever be useful for communications has yet to be determined, as their are many practical problems that must be overcome, but probably the most important contribution that chaos has made to communications (and engineering in general is that it forces engineers to go beyond standard linear thinking to consider new ways of approaching old problems.

References

1. V. S. Afraimovich, N. N. Verichev and M. I. Rabinovich, *Radiophys. and Quant. Elect.* **29** (1986) 795.
2. H. Fujisaka and T. Yamada, *Prog. Theor. Phys.* **69** (1983) 32.
3. K. Pyragas, *Phys. Lett. A* **181** (1993) 203.
4. V. S. Anishchenko, T. E. Vadivosa, D. E. Postnov and M. A. Safonova, *Int. J. Bifurc. and Chaos* **2** (1992) 633.
5. T. L. Carroll and L. M. Pecora, *IEEE Trans. CAS* **38** (1991) 453.
6. T. L. Carroll and L. M. Pecora, *Int. J. Bifurc. and Chaos* **2** (1992) 659.
7. T. L. Carroll and L. M. Pecora, *Physica D* **67** (1993) 126.
8. T. L. Carroll and L. M. Pecora, *IEEE Trans. Circuits Syst.* **40** (1993) 646.
9. T. L. Carroll, *IEEE Trans. CAS* **42** (1995) 105.
10. T. L. Carroll, *Phys Rev. E* **50** (1994) 2580.
11. L. M. Pecora and T. L. Carroll, *Phys. Rev. Lett.* **64** (1990) 821.
12. L. M. Pecora and T. L. Carroll, *Physical Review* **A 44** (1991) 2374.
13. J. L. Willems, *Stability Theory of Dynamical Systems* (John Wiley & Sons, New York, 1970).
14. J. Guckenheimer and P. Holmes, *Nonlinear Oscillations, Dynamical Systems, and Bifurcations of Vector Fields* (Springer-Verlag, New York, 1983).
15. S. Strogatz, *Nonlinear Dynamics and Chaos* (Addison-Wesley, Reading Massachusettes, 1994).
16.A. M. Lyapunov,*The General Problem of the Stability of Motion* , ed. A. T. Fuller (Taylor & Francis, London, 1992)
17. V. I. Oseledec, *Trans. Moscow Math. Soc.* **19** (1968) 197.
18. H. D. I. Abarbanel, Reggie Brown, Lev Sh. Tsimring and J. J. Sidorowich, *Rev. Mod. Phys.* (1993)
19. M. B. Kennel, R. Brown and H. D. I. Abarbanel, *Physical Review* **A 45** (1992) 3403.
20. A. Wolf, J. B. Swift, H. L. Swinney and J. A. Vastano, *Physica* **16D** (1985) 285.
21. R. Brown and P. Bryant, *Physical Review* **A 43** (1991) 2787.
22. H. D. I. Abarbanel, R. Brown and M. B. Kennel, *International Journal of Modern Physics B* **5** (1991) 1347.
23. R. Brown, *Physical Review* **E 47** (1993) 3962.
24. M. Sano and Y. Sawada, *Phys. Rev. Lett.* **55** (1985) 1082.
25. M. T. Rosenstein, J. J. Collins and C. J. D. Luca, *Physica* **D 65** (1993) 117.
26. J.-P. Eckmann, S. O. Kamphorst, D. Ruelle and S. Ciliberto, *Phys. Rev. A* **34** (1986) 4971.
27. J.-P. Eckmann and D. Ruelle, *Rev. Mod. Phys.* **57** (1985) 617.
28.E. Cawley, personal communication (1991)
29. K. M. Cuomo and A. V. Oppenheim, *Phys. Rev. Lett.* **71** (1993) 65.
30.Rabinovich, personal communication (1993)

31. K. Pyragas, *Phys. Lett. A* **170** (1992) 421.

32. K. M. Cuomo, *Int. J. Bifurc. and Chaos* **3** (1993) 1327.

33. N. F. Rul'kov, A. R. Volkovskii, A. Rodriguez-Lozano, E. d. Rio and M. G. Velarde, *Chaos, Solitons & Fractals* **4** (1994) 201.

34.N. F. Rulkov and A. R. Volkovskii, in *proceedings of Chaos in Communications (SPIE)*, ed. L. M. Pecora (The International Society fo Optical Engineering, San Diego, California, 1993).

35. A. R. Volkovskii and N. F. Rul'kov, *Sov. Phys. Tech. Phys. Lett.* **19** (1993) 97.

36. M. d. S. Viera, P. Khoury, A. J. Lichtenberg, M. A. Lieberman, W. Wonchoba, J. Gullicksen, J. Y. Huang, R. Sherman and M. Steinberg, *Int. J. Bifurc. and Chaos* **2** (1992) 645.

37. C. W. Wu and L. O. Chua, *Int. J. Bifurc. and Chaos* **4** (1994) 979.

38. R. Brown, N. F. Rul'kov and N. B. Tufillaro, *Phys. Rev. E* **50** (1994) 4488.

39. V. S. Afraimovich, N. N. Verichev and M. I. Rabinovich, *Inv. VUZ. Radiofiz. RPQAEC* **29** (1986) 795.

40. V. S. Anishchenko, T. E. Vadivasova, D. E. Postnov and M. A. Safonova, *Int. J. Bifurc. and Chaos* **2** (1992) 633.

41. P. Ashwin, J. Buescu and I. Stewart, *Physics Letters* **A 193** (1994) 126.

42. N. F. Rul'kov, A. R. Volkovskii, A. Rodriguez-Lozano, E. d. Rio and M. G. Velarde, *International Journal of Bifurcation and Chaos* **2** (1992) 669.

43. R. Roy and K. S. T. Jr., *Phys. Rev. Lett.* **72** (1994) 2009.

44. A. R. Volkovskii and N. F. Rul'kov, *Sov. Tech. Phys. Lett.* **15** (1989) 249.

45. L. O. Chua, L. Kocarev, K. Eckert and M. Itoh, *International Journal of Bifurcation and Chaos* **2** (1992) 705.

46. T. Kapitaniak, *Physical Review* **E 47** (1993) R2975.

47. T. Yamada and H. Fujisaka, *Progress in Theoretical Physics* **70** (1983) 1240.

48. T. Yamada and H. Fujisaka, *Progress in Theoretical Physics* **72** (1984) 885.

49. H. Fujisaka and T. Yamada, *Progress in Theoretical Physics* **74** (1985) 918.

50. J. M. Kowalski, G. L. Albert and G. W. Gross, *Physical Review* **A 42** (1990) 6260.

51. H. G. Winful and L. Rahman, *Phys. Rev. Lett.* **65** (1990) 1575.

52. J. F. Heagy, T. L. Carroll and L. M. Pecora, *Physical Review* **E 50** (1994) 1874.

53. J. F. Heagy, T. L. Carroll and L. M. Pecora, *Phys. Rev. Lett.* **73** (1994) 3528.

54. R. W. Newcomb and N. El-Leithy, *Circ. Syst. Sig. Proc.* **5** (1986) 321.

55. R. W. Newcomb and S. Sathyan, *IEEE Trans. CAS* **30** (1983) 54.

56. P. Berge, Y. Pomeau and C. Vidal, *Order Within Chaos* (Wiley, New York, 1984).

57. C. L. Johnson, *Analog Computer Techniques* (McGraw-Hill, New York, 1963).

58. E. S. o. P. N. Research, *Applications Manual for Operational Amplifiers* (Philbrick/Nexus Research, Dedham, MA, 1968).

59. D. H. Sheingold, *Nonlinear Circuits Handbook* (Analog Devices, Norwood, MA, 1976).

60. T. L. Carroll, *Am. J. Phys.* **63** (1995) 377.

61. M. d. S. Viera, A. J. Lichtenberg and M. A. Lieberman, *Phys. Rev. A* **46** (1992) R7359.

62.R. H. Sherman and J. Gullicksen, in *proceedings of SPIE Conference on Chaos in Communications*, ed. L. M. Pecora (SPIE, San Diego, CA, 1993).

63.A. V. Oppenheim, G. W. Wornell, S. H. Isabelle and K. M. Cuomo, in *proceedings of International Conference on Acoustics, Speech and Signal Processing*, ed. (IEEE, San Francisco, CA, 1992).

64. K. M. Cuomo, A. V. Oppenheim and S. H. Strogatz, *IEEE Trans. CAS* **40** (1993) 626.

65. L. O. Chua, L. Kocarev, K. Eckart and M. Itoh, *Int. J. Bifurc. and Chaos* **2** (1992) 705.

66. M. Shinriki, M. Yamaoto and S. Mori, *Proc. IEEE* **69** (1981) 394.

67. T. Matsumoto, L. O. Chua and M. Komuro, *IEEE Trans. CAS* **32** (1985) 798.

68. P. Horowitz and W. Hill, *The Art of Electronics* (Cambridge Press, New York, 1980).

69. L. Kocarev, K. S. Halle, K. Eckert, L. O. Chua and U. Parlitz, *Int. J. Bifurc. and Chaos* **2** (1992) 709.

70. U. Parlitz, L. O. Chua, L. Kocarev, K. S. Halle and A. Shang, *Int. J. Bifurc. and Chaos* **2** (1992) 973.

71. J. C. Alexander, J. A. Yorke, Z. You and I. Kan, *Int. J. Bifurc. and Chaos* **2** (1992) 795.

72. O. E. Rossler, *Z. Naturforsch.* **38a** (1983) 788.

Analysis, Synthesis, and Applications of Self-Synchronizing Chaotic Systems

KEVIN M. CUOMO

Research Laboratory of Electronics and the Department of Electrical Engineering and Computer Science, MIT, Cambridge, MA 02139, USA

and

ALAN V. OPPENHEIM

Research Laboratory of Electronics and the Department of Electrical Engineering and Computer Science, MIT, Cambridge, MA 02139, USA

Abstract

Chaotic systems provide a rich mechanism for signal design and generation, with potential applications to communications and signal processing. Of practical significance are chaotic systems that possess the self-synchronization property. This property allows two identical chaotic systems to synchronize when the second system is driven by the first. In certain communication contexts, the first system can be viewed as the transmitter and the second system as the receiver. A potential drawback to utilizing self-synchronizing chaotic systems in applications is that the analysis and synthesis of these systems is not well-understood due to their highly nonlinear nature. This chapter summarizes work that we have recently done in both of these areas.

1. Introduction

In 1990, it was discovered that a certain class of dissipative chaotic systems possess a self-synchronization property[1,2,3]. This property allows two identical chaotic systems to synchronize when the second system (receiver) is driven by the first (transmitter). Because chaotic signals are typically broadband, noise-like, and difficult to predict, their use in various communications contexts, *e.g.*, as masks for information-bearing waveforms and as modulating waveforms in spread spectrum systems has been proposed[4]. These proposed applications exploit the self-synchronization property to faithfully recover the information at the receiver. A major drawback to utilizing self-synchronizing chaotic systems in communication applications is that the analysis and synthesis of these systems is not well-understood due to their highly nonlinear nature. This chapter summarizes some of our recent work directed at analysis, synthesis, and applications of self-synchronizing chaotic systems.

In Sec.2, we summarize our approach to analysis of the self-synchronization property of the Lorenz system from the viewpoint of nonlinear stability theory. In this section we also describe our results on the sensitivity of synchronization to perturbation of the drive signal.

In Sec. 3, we describe a systematic procedure for synthesizing a class of high-dimensional dissipative chaotic systems that possess the self-synchronization property. We refer to this class of systems as "Linear Feedback Chaotic System" (LFBCSs)[5]. LF-BCSs are composed of a low-dimensional chaotic system and a linear feedback system. We focus on LFBCSs that utilize the Lorenz system as the chaotic system component. A generalization of this concept allowed for multiple Lorenz systems and a linear system to be combined into a chaotic array; a systematic procedure for synthesizing this class of systems has been recently reported[6]. A further generalization of these concepts eliminates the necessity of the linear system and allows the entire system to be nonlinear. A systematic synthesis capability allows high-dimensional non-Lorenz self-synchronizing chaotic systems to be designed[7]. The synthesis techniques vary in the number of drive signals required for synchronization and the resulting complexity of the system dynamics. The flexibility afforded by the various synthesis techniques enhances the potential usefulness of synchronized chaotic systems for communications and signal processing.

In Sec. 4, we summarize our work on some applied aspects of self-synchronizing systems. Specifically, we describe the implementation of the Lorenz transmitter and receiver systems as simple analog circuits using commercially available hardware[8,9]. We also describe two techniques for embedding an information-bearing waveform in the chaotic drive signal, and for recovering the information with the synchronizing receiver. With the first approach, low-level speech signals can be privately transmitted and recovered with the receiver circuit. The second approach allows binary-valued bit streams to privately transmitted and recovered. While these approaches do not represent the ultimate in privacy or practicality, they perhaps suggest important directions for applications. In Sec. 5, we summarize our analysis, synthesis, and applications efforts.

2. Analysis of Self-Synchronization in the Lorenz System

The Lorenz system is a well-known chaotic system that possesses the self-synchronization property. The Lorenz equations, first introduced by E. N. Lorenz as a simplified model of fluid convection[10], are given by

$$\begin{aligned} \dot{x} &= \sigma(y - x) \\ \dot{y} &= rx - y - xz \\ \dot{z} &= xy - bz \ , \end{aligned} \tag{1}$$

where σ, r, and b are positive parameters. In signal processing applications, it is typically of interest to adjust the time scale of the chaotic signals. This is accomplished in a straightforward way by establishing the convention that $\dot{x}, \dot{y}$, and $\dot{z}$ denote $dx/d\tau, dy/d\tau$, and $dz/d\tau$ respectively where $\tau = t/T$ is normalized time and T is a time scale factor. It is also convenient to define the normalized frequency $\omega = \Omega T$, where Ω denotes the angular frequency in units of rad/s. In our circuit implementation of Eq. 1, the parameter values used were $T = 400\ \mu sec, \sigma = 16, r = 45.6$, and $b = 4$.

Viewing the Lorenz system (1) as a set of transmitter equations, a dynamical receiver system that will synchronize to the transmitter is given by

$$\begin{aligned}\dot{x}_r &= \sigma(y_r - x_r) \\ \dot{y}_r &= rx(t) - y_r - x(t)z_r \\ \dot{z}_r &= x(t)y_r - bz_r \ .\end{aligned} \qquad (2)$$

In this case, the chaotic signal $x(t)$ from the transmitter is used as the driving input to the receiver system. In Sec. 2.1, we show that the synchronization of the transmitter and receiver is global, *i.e.*, the receiver can be initialized in any state and the synchronization still occurs. In Sec. 2.2, we briefly examine the sensitivity of synchronization to perturbation of the drive signal.

2.1 Self-Synchronization and Asymptotic Stability

A close relationship exists between the concepts of self-synchronization and asymptotic stability. Specifically, self-synchronization in the Lorenz system is a consequence of globally stable error dynamics. To show this, we define the error signals as

$$\begin{aligned}e_x(t) &= x(t) - x_r(t) \\ e_y(t) &= y(t) - y_r(t) \\ e_z(t) &= z(t) - z_r(t) \ .\end{aligned}$$

Assuming that the Lorenz transmitter and receiver parameters are identical, a set of equations which govern their error dynamics is given by

$$\begin{aligned}\dot{e}_x &= \sigma(e_y - e_x) \\ \dot{e}_y &= -e_y - x(t)e_z \\ \dot{e}_z &= x(t)e_y - be_z \ .\end{aligned} \qquad (3)$$

A sufficient condition for the error equations to be globally asymptotically stable at the origin can be determined by considering a Lyapunov function of the form

$$E(\mathbf{e}) = \frac{1}{2}(\frac{1}{\sigma}e_x^2 + e_y^2 + e_z^2) \ .$$

Note that $E(\mathbf{e})$ is positive definite provided that $\sigma > 0$. The time rate of change of $E(\mathbf{e})$ along trajectories is given by

$$\begin{aligned}\dot{E}(\mathbf{e}) &= \frac{1}{\sigma}e_x\dot{e}_x + e_y\dot{e}_y + e_z\dot{e}_z \\ &= -(e_x - \frac{1}{2}e_y)^2 - \frac{3}{4}e_y^2 - be_z^2 \ .\end{aligned}$$

Provided that $b > 0$, $\dot{E}(\mathbf{e})$ is negative definite. Since σ and b in the Lorenz equations are both assumed to be positive, E is positive definite and $\dot{E}$ is negative definite. It then follows from Lyapunov's theorem that $\mathbf{e}(t) \to 0$ as $t \to \infty$. Therefore, synchronization occurs as $t \to \infty$ regardless of the initial conditions imposed on the transmitter and receiver systems. A similar Lyapunov argument has also been given for the synchronization of the (y, z) subsystem of the Lorenz equations[11].

2.2. Robustness and Signal Recovery in the Lorenz System

When a message or other perturbation is added to the chaotic drive signal, the receiver does not regenerate a perfect replica of the drive; there is always some synchronization error. By subtracting the regenerated drive signal from the received signal, successful message recovery would result if the synchronization error was small relative to the perturbation itself. An interesting property of the Lorenz system is that the synchronization error is *not* small compared to a narrowband perturbation; nevertheless, the message can be recovered because the synchronization error is nearly *coherent* with the message. This section summarizes experimental evidence for this effect; a more detailed explanation has been given in terms of an approximate analytical model[12].

The series of experiments to demonstrate the robustness of synchronization to white noise perturbations and the ability to recover speech perturbations focus on the synchronizing properties of the transmitter equations (1) and the corresponding receiver equations,

$$\begin{aligned} \dot{x}_r &= \sigma(y_r - x_r) \\ \dot{y}_r &= rs(t) - y_r - s(t)z_r \\ \dot{z}_r &= s(t)y_r - bz_r \ . \end{aligned} \tag{4}$$

In Sec. 2.1, we showed that with $s(t)$ equal to the transmitter signal $x(t)$, the signals x_r, y_r, and z_r will asymptotically synchronize to x, y, and z, respectively. In this section, we experimentally examine the synchronization error when a perturbation $p(t)$ is added to $x(t)$, *i.e.*, when $s(t) = x(t) + p(t)$.

2.2.1 Sensitivity of Synchronization to Additive White Noise

In this experiment, the perturbation $p(t)$ is Gaussian white noise. In Fig. 1, we show the perturbation and error spectra for each of the three state variables vs. normalized frequency ω. Note that at relatively low frequencies, the error in reconstructing $x(t)$ slightly exceeds the perturbation of the drive but that for normalized frequencies above 20 the situation quickly reverses. An analytical model closely predicts and explains this behavior[12]. These figures suggest that the sensitivity of synchronization depends on the spectral characteristics of the perturbation signal. For signals which are bandlimited to the frequency range $0 < \omega < 10$, we would expect that the synchronization errors will be larger than the perturbation itself. This turns out to be the case, although the next experiment suggests there are additional interesting characteristics as well.

2.2.2 Sensitivity of Synchronization to Additive Speech

In this experiment, $p(t)$ is a low-level speech signal (the message to be transmitted

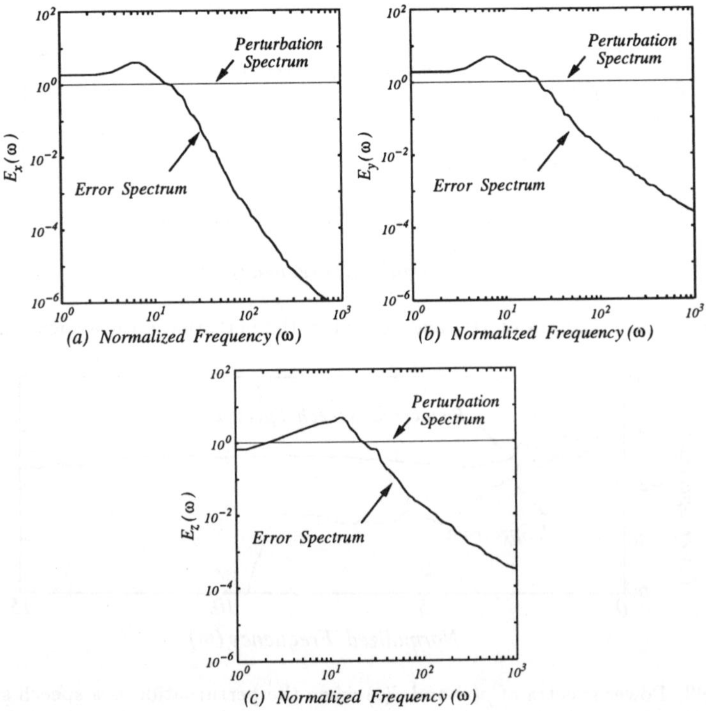

Figure 1: Power spectra of the error signals: (a) $E_x(\omega)$. (b) $E_y(\omega)$. (c) $E_z(\omega)$.

and recovered). The normalizing time parameter is 400 μsec and the speech signal is bandlimited to 4 kHz or equivalently to a normalized frequency ω of 10. Figure 2 shows the power spectrum of a representative speech signal and the chaotic signal $x(t)$. The overall chaos-to-perturbation ratio in this experiment is approximately 20 dB.

To recover the message, we subtract the regenerated drive signal at the receiver from the received signal. In this case, the recovered message is $\hat{p}(t) = p(t) + e_x(t)$. It would be expected that successful message recovery would result if $e_x(t)$ was small relative to the perturbation signal. For the Lorenz system, however, we have shown that although the synchronization error is not small compared to the perturbation the message can be recovered because $e_x(t)$ is nearly coherent with the message. Experimental evidence for this effect is presented below; an explanation has also been given in terms of an approximate analytical model[12].

In Fig. 3, we show the spectrum of $\hat{p}(t)$ for this same example. Notice that $\hat{p}(t)$ includes considerable energy beyond the bandwidth of the speech. Furthermore, $\hat{p}(t)$ resembles a scaled version of the message at low frequencies. These observations turn

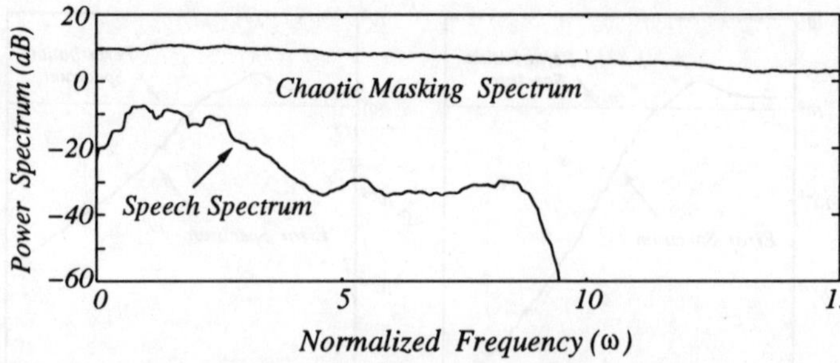

Figure 2: Power spectra of $x(t)$ and $p(t)$ when the perturbation is a speech signal.

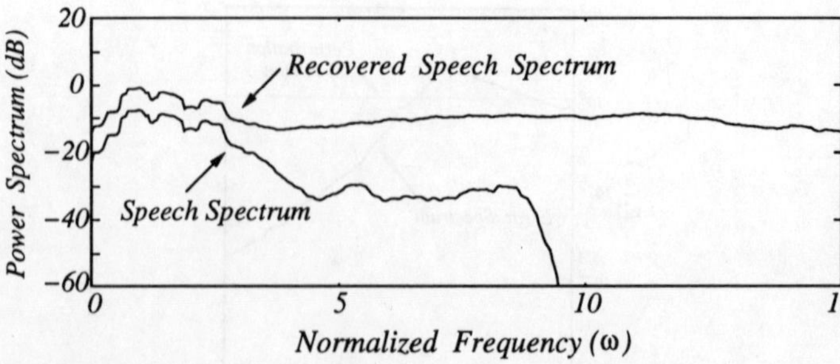

Figure 3: Power spectra of $p(t)$ and $\hat{p}(t)$ when the perturbation is a speech signal.

out to be consistent with the fact that the synchronization error $e_x(t)$ is nearly coherent with the message at low frequencies and noise-like at high frequencies. Consequently, the speech recovery can be improved by lowpass filtering $\hat{p}(t)$. With this lowpass filtering, the message-to-error ratio is approximately 10 dB. In Sec. 4, we utilize an analog circuit implementation of the Lorenz equations to verify that the speech waveform can be recovered.

In summary, we showed that: *(i)* the synchronization is robust to white noise perturbations; and *(ii)* the synchronizing properties of the Lorenz system allow speech perturbations to be faithfully recovered at the receiver. In Sec. 3, we summarize an approach for synthesizing new chaotic systems with the same robust synchronizing properties as the Lorenz system.

3. Synthesizing Self-Synchronizing Chaotic Systems

Our approach to synthesis is based on considering a class of chaotic systems which we refer to as "Linear Feedback Chaotic Systems" (LFBCSs)[5]. LFBCSs are composed of a low-dimensional chaotic system and a linear feedback system as illustrated in Fig.

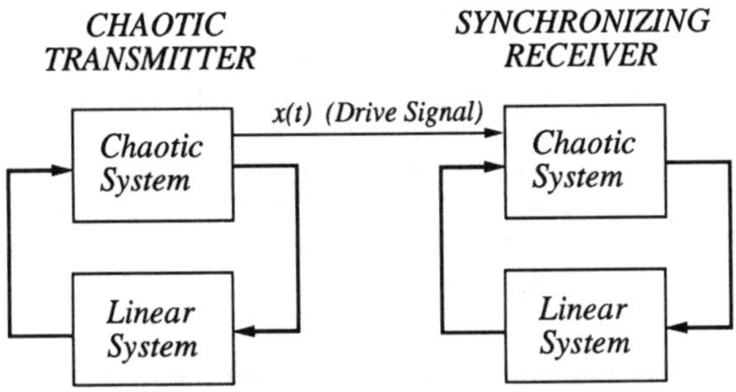

Figure 4: Communicating with linear feedback chaotic systems.

4. Although this approach is applicable to any chaotic system, we will focus our attention on LFBCSs which utilize the Lorenz system. The advantages of applying linear feedback to the Lorenz system are that the resulting high-dimensional chaotic systems are analytically tractable and relatively easy to implement. Furthermore, this approach allows an unlimited number of self-synchronizing LFBCSs to be designed.

While many types of LFBCSs are possible, we have considered two specific cases: *(i)* the chaotic Lorenz signal $x(t)$ drives an N-dimensional linear system and the output of the linear system is added to the equation for $\dot{x}$ in the Lorenz system; and *(ii)* the Lorenz signal $z(t)$ drives an N-dimensional linear system and the output of the linear system is added to the equation for $\dot{z}$ in the Lorenz system. In both cases, a complete set of sufficient conditions were developed such that:

— there exists algebraically similar transmitter and receiver systems which possess the global self-synchronization property; and

— the transmitter system is globally stable.

A detailed linear stability analysis was then performed to determine conditions which ensure that all of the transmitter's fixed points are unstable so that non-trivial motion will occur.

To illustrate the approach, consider the x-input/x-output case, *i.e.*, where the transmitter equations are given by

$$\begin{aligned} \dot{x} &= \sigma(y-x)+\nu \\ \dot{y} &= rx-y-xz \\ \dot{z} &= xy-bz \\ \dot{\mathrm{l}} &= A\mathrm{l}+Bx \\ \nu &= C\mathrm{l}+Dx \ . \end{aligned} \tag{5}$$

The vector l and scalar ν denote the state variables and output of the linear system, respectively. The linear system is N-dimensional, *i.e.*, A is $N \times N$, B is $N \times 1$, C

is $1 \times N$, and D is 1×1. For notational simplicity, we refer to the transmitter state variables collectively by the vector $\mathbf{x} = (x, y, z, \mathbf{l})$, when convenient.

The self-synchronization properties of the Lorenz system suggested a receiver system of the form

$$\begin{aligned} \dot{x}_r &= \sigma(y_r - x_r) + \nu_r \\ \dot{y}_r &= rx(t) - y_r - x(t)z_r \\ \dot{z}_r &= x(t)y_r - bz_r \\ \dot{\mathbf{l}}_r &= A\mathbf{l}_r + Bx(t) \\ \nu_r &= C\mathbf{l}_r + Dx(t) \ . \end{aligned} \tag{6}$$

Algebraically, the receiver system (6) is obtained from the transmitter (5) by renaming variables $\mathbf{x} \rightarrow \mathbf{x}_r$ and substituting the drive signal $x(t)$ for $x_r(t)$ in all state equations except the first.

We can study the self-synchronization properties of the transmitter and receiver equations by forming the error system. The error system is derived by subtracting (6) from (5) to obtain

$$\begin{aligned} \dot{e}_x &= \sigma(e_y - e_x) + C\mathbf{e}_l \\ \dot{e}_y &= -e_y - x(t)e_z \\ \dot{e}_z &= x(t)e_y - be_z \\ \dot{\mathbf{e}}_l &= A\mathbf{e}_l \ . \end{aligned} \tag{7}$$

Since the dynamics of $\mathbf{e}_l$ are independent of e_x, e_y, and e_z, we can see that if A is a stable matrix, then the $\mathbf{e}_l$ subsystem is globally asymptotically stable at the origin. The (e_y, e_z) subsystem is also decoupled from the rest of the system, and was shown previously to be globally asymptotically stable at the origin. The error signal $e_x(t)$ must also go to zero as $t \rightarrow \infty$ because $e_x(t)$ is the output of a stable linear time-invariant system that is driven by $e_y(t)$ and $\mathbf{e}_l(t)$. From this analysis, we conclude that the error system is globally asymptotically stable at the origin if A is a stable matrix. Equivalently, with A as a stable matrix, the transmitter and receiver are guaranteed to synchronize regardless of the initial conditions imposed on these systems.

The next step is to determine an appropriate set of conditions which gaurantee that the transmitter is globally stable. Below, we summarize the various self-synchronization and global stability conditions; a complete development is given elsewhere[5].

Self-Synchronization { 1. A is stable.

Global Stability {
2. $PB + rC^T = \mathbf{0}$, for some $N \times N$ positive definite matrix P.
3. $PA + A^TP$ is negative definite.
4. $\sigma - D > 0$.

Condition 1 implies that there exists a positive definite solution P to the matrix Lyapunov equation

$$PA + A^TP + Q \ = \ 0 \ , \ Q \ > \ 0 \ .$$

By choosing Q to be any symmetric positive definite matrix, condition 3 can always be satisfied. Exploiting this relationship, the following synthesis procedure is suggested.

Synthesis Procedure

1. Choose any stable A matrix and any $N \times N$ symmetric positive definite matrix Q.
2. Solve $PA + A^T P + Q = 0$ for the positive definite solution P.
3. Choose any vector B and set $C = -B^T P/r$.
4. Choose any D such that $\sigma - D > 0$.

For the purpose of demonstration, consider the following five-dimensional x-input/x-output LFBCS.

$$\begin{aligned} \dot{x} &= \sigma(y - x) + \nu \\ \dot{y} &= rx - y - xz \\ \dot{z} &= xy - bz \\ \begin{bmatrix} \dot{l}_1 \\ \dot{l}_2 \end{bmatrix} &= \begin{bmatrix} -\frac{1}{2} & 10 \\ -10 & -\frac{1}{2} \end{bmatrix} \begin{bmatrix} l_1 \\ l_2 \end{bmatrix} + \begin{bmatrix} 1 \\ 1 \end{bmatrix} x \\ \nu &= -\begin{bmatrix} 1 & 1 \end{bmatrix} \begin{bmatrix} l_1 \\ l_2 \end{bmatrix} \end{aligned} \tag{8}$$

It can be shown in a straightforward way that the linear system satisfies the self-synchronization and global stability conditions for suitable choices of P, Q, and R. For the numerical demonstrations presented below, the Lorenz parameters chosen are $\sigma = 16$ and $b = 4$; the bifurcation parameter r will be varied.

In Fig. 5, we show the computed Lyapunov dimension[13] as r is varied over the range, $20 < r < 100$. This figure demonstrates that the LFBCS achieves a greater Lyapunov dimension than the Lorenz system without feedback. The Lyapunov dimension could be increased by using more states in the linear system. However, numerical experiments suggest that a limit will be reached since the Lyapunov dimension depends more heavily on the most positive exponents and stable linear feedback creates only negative exponents.

In Fig. 6, we demonstrate the rapid synchronization between the transmitter and receiver systems. The curve measures the distance in state space between the transmitter and receiver trajectories when the receiver is initialized from the zero state. Synchronization is maintained indefinitely.

The synthesis results presented in this section have further enhanced our understanding of self-synchronization in a class of chaotic systems. The development of a systematic procedure for synthesizing high-dimensional chaotic systems which possess the self-synchronization property may serve a useful purpose for future communications applications. In the next section, we consider some applied aspects of synchronized chaotic systems.

4. Applications of Self-Synchronizing Chaotic Systems

In Sec. 2, we showed that, theoretically, a low-level speech signal could be added to the synchronizing drive signal and approximately recovered at the receiver. These results

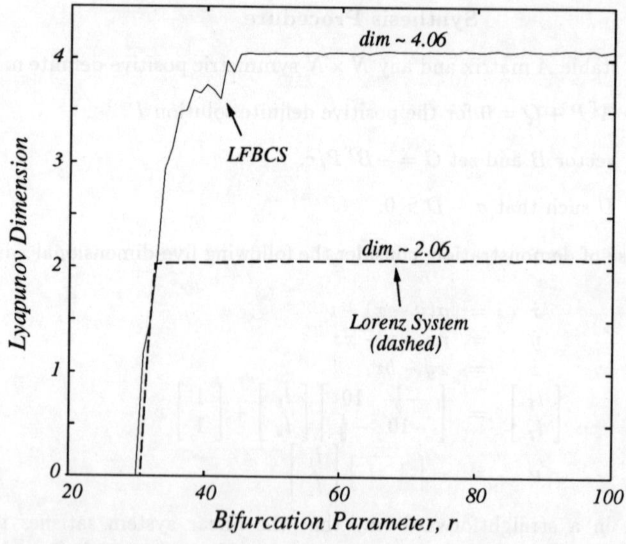

Figure 5: Lyapunov dimension for a 5-dimensional x-input/x-output LFBCS.

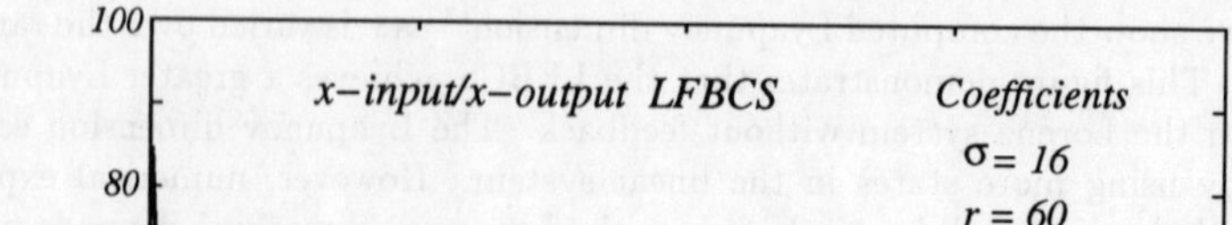

Figure 6: Self-synchronization in a 5-dimensional x-input/x-output LFBCS.

were based on an analysis of the *exact* Lorenz transmitter and receiver equations. When implementing synchronized chaotic systems in hardware, the limitations of available circuit components result in approximations of the defining equations. The resulting system performance needs to be re-evaluated to assess any hardware-induced effects or limitations and to ensure that the system is performing within the desired specifications. To address this issue, we implemented the Lorenz transmitter and receiver equations with simple analog circuits[8,9]. The resulting system performance was assessed and shown to be in excellent agreement with numerical and theoretical predictions. Some potential implementation difficulties were avoided by scaling the Lorenz state variables according to $u = x/10, v = y/10$, and $w = z/20$. With this scaling, the Lorenz equations are transformed to

$$\begin{aligned} \dot{u} &= \sigma(v-u) \\ \dot{v} &= ru - v - 20uw \\ \dot{w} &= 5uv - bw \ . \end{aligned} \tag{9}$$

This system, which we refer to as the circuit equations, can be readily implemented with an analog circuit; the state variables all have similar dynamic range and circuit voltages remain well within the range of typical power supply limits. Below, we discuss and demonstrate some applied aspects of the Lorenz circuits.

4.1 Chaotic Signal Masking and Recovery

In Fig. 7, we illustrate a communication scenario that is based on chaotic signal masking and recovery[4,8,9]. In this figure, a chaotic masking signal $u(t)$ is added to the information-bearing signal $p(t)$ at the transmitter, and at the receiver the masking is removed. By subtracting the regenerated drive signal $u_r(t)$ from the received signal $s(t)$ at the receiver, the recovered message is

$$\hat{p}(t) = s(t) - u_r(t) = p(t) + [u(t) - u_r(t)] \ .$$

In this context, $e_u(t)$, the error between $u(t)$ and $u_r(t)$, corresponds directly to the error in the recovered message. As pointed out in Sec. 2.2, although $e_u(t)$ is not small compared to the message, the message can be recovered because $e_u(t)$ is nearly coherent with the message at low frequencies. Below, we demonstrate this effect and the ability to recover the message with the Lorenz receiver circuit.

For this experiment, $p(t)$ is a low-level speech signal (the message to be transmitted and recovered). The normalizing time parameter is 400 μsec and the speech signal is bandlimited to 4 kHz or, equivalently, to a normalized frequency ω of 10. In Fig. 8, we show the power spectrum of $p(t)$ and $\hat{p}(t)$, where $\hat{p}(t)$ is obtained from both a simulation and from the circuit. The two spectra for $\hat{p}(t)$ are in excellent agreement, indicating that the circuit performs very well. Because $\hat{p}(t)$ includes considerable energy beyond the bandwidth of the speech, the speech recovery can be improved by lowpass filtering $\hat{p}(t)$. We denote the lowpass filtered version of $\hat{p}(t)$ by $\hat{p}_f(t)$. In Fig. 9(a) and (b), we show a comparison of $\hat{p}_f(t)$ from both a simulation and from the circuit, respectively. Clearly, the circuit performs well and, in informal listening tests, the recovered message

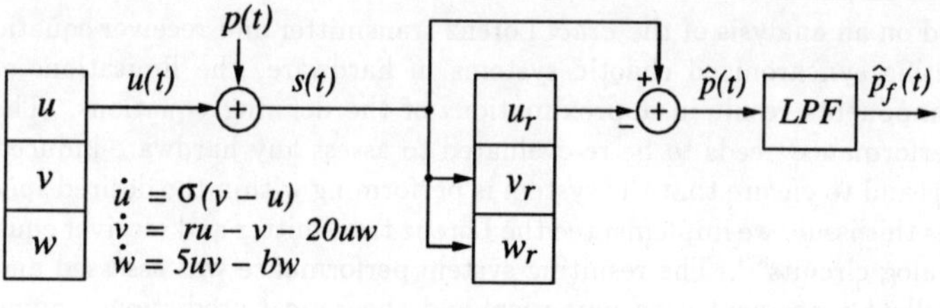

Figure 7: Chaotic signal masking and recovery system.

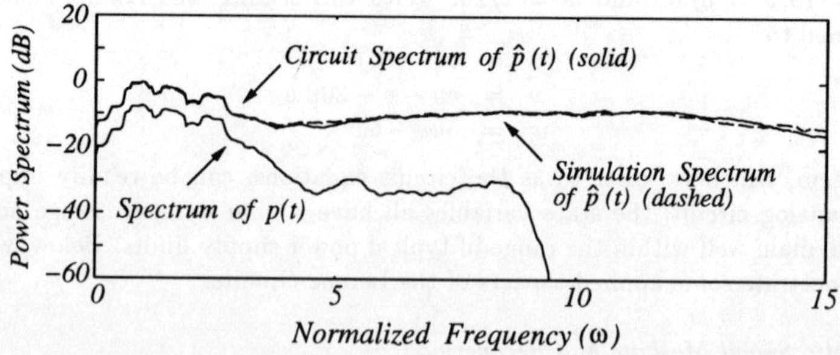

Figure 8: Power spectra of $p(t)$ and $\hat{p}(t)$ when the perturbation is a speech signal.

is of reasonable quality. This signal masking concept was also shown to work using Chua's circuit[14].

Although $\hat{p}_f(t)$ is of reasonable quality in this experiment, the presence of additive channel noise will produce message recovery errors that cannot be completely removed by lowpass filtering; there will always be some error in the recovered message. Because the message and noise are directly added to the synchronizing drive signal, the message-to-noise ratio should be large enough to allow a faithful recovery of the original message. This requires a communication channel which is nearly noise free.

4.2 Chaotic Binary Communications

In this section, we discuss a private communication technique called chaotic binary communications[8,9]. The basic idea behind this technique is to modulate a transmitter parameter with the information-bearing waveform and to transmit the chaotic drive signal. At the receiver, the parameter modulation will produce a synchronization error between the received drive signal and the receiver's regenerated drive signal with an error signal amplitude that depends on the modulation. Using the synchronization error, the modulation can be detected.

This modulation/detection process is illustrated in Fig. 10. To test the approach, we use a periodic square-wave for $p(t)$ as shown in Fig. 11(a). The square-wave has a

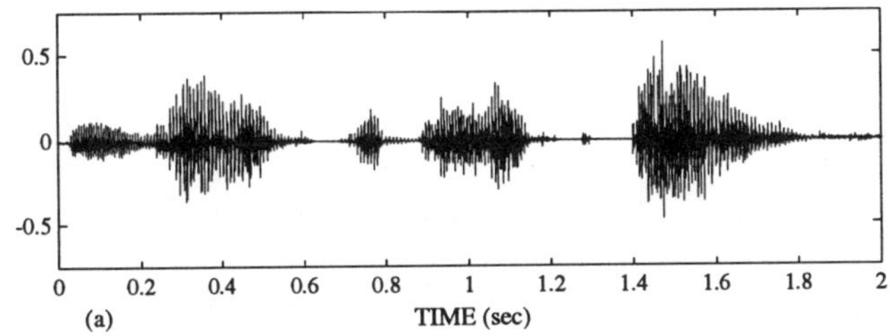

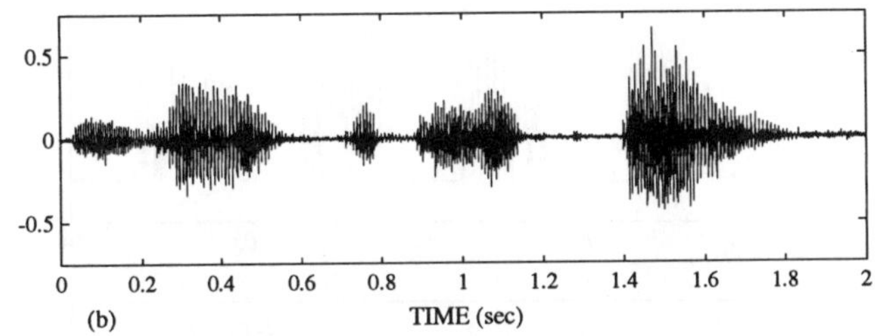

Figure 9: (a) Recovered speech (simulation) (b) Recovered speech (circuit).

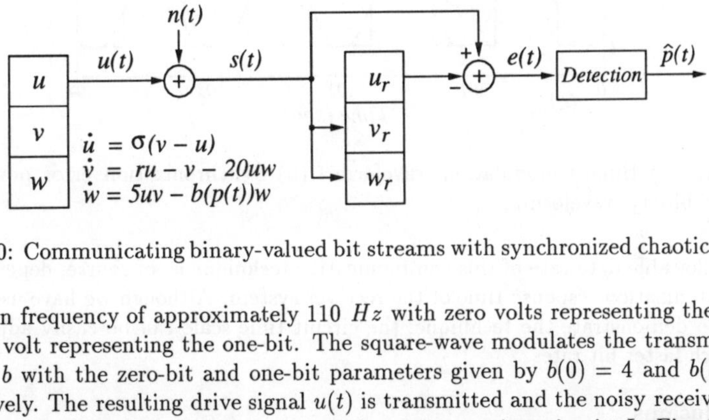

Figure 10: Communicating binary-valued bit streams with synchronized chaotic systems.

repetition frequency of approximately 110 Hz with zero volts representing the zero-bit and one volt representing the one-bit. The square-wave modulates the transmitter parameter b with the zero-bit and one-bit parameters given by $b(0) = 4$ and $b(1) = 4.4$, respectively. The resulting drive signal $u(t)$ is transmitted and the noisy received signal $s(t)$ is used as the driving input to the synchronizing receiver circuit. In Fig. 11(b), we show the synchronization error power $e^2(t)$. The parameter modulation produces significant synchronization error during a "1" transmission and very little error during a "0" transmission. It is plausible that a detector based on the average synchronization error power, followed by a threshold device, could yield reliable performance. We illustrate in Fig. 11(c) that the square-wave modulation can be reliably recovered by lowpass filtering the synchronization error power waveform and applying a threshold test. The threshold device used in this experiment consisted of a simple analog comparator circuit.

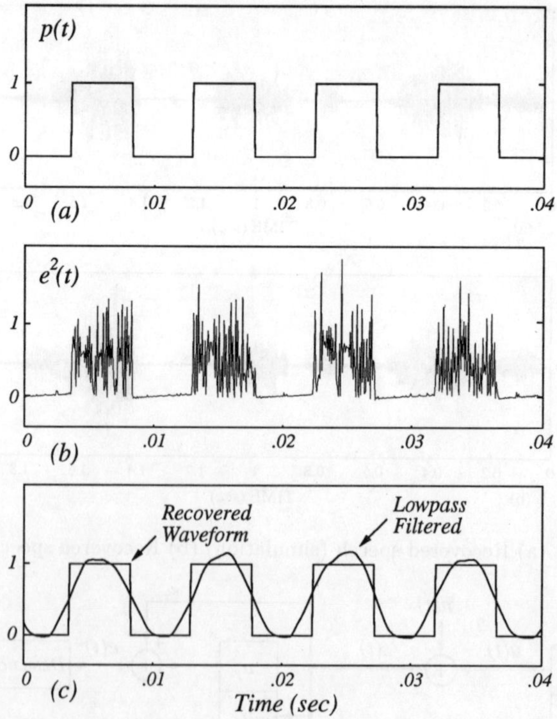

Figure 11: (a) Binary modulation waveform. (b) Synchronization error power. (c) Recovered binary waveform.

The allowable data rate of this communication technique is, of course, dependent on the synchronization response time of the receiver system. Although we have used a low bit rate to demonstrate the technique, the circuit time scale can be easily adjusted to allow much faster bit rates.

5. Conclusions

Since its discovery in 1990, self-synchronization of chaotic systems has become an extremely active research area. The work presented in this chapter has focused on the development of general procedures for analyzing and synthesizing chaotic systems which possess the self-synchronization property. We also investigated and demonstrated some applied aspects of these systems. While these results appear promising and represent a starting point for utilizing self-synchronizing chaotic systems in communications, much work remains before these approaches can be considered truly practical.

6. Acknowledgements

The authors thank S. H. Strogatz and A. S. Willsky for many helpful discussions during various stages of this work. This work was sponsored in part by the Air Force Office of Scientific Research under Grant Number AFOSR-91-0034-A, and in part by the Department of the Navy, Office of the Chief of Naval Research, contract number N00014-93-1-0686 as part of the Advanced Research Projects Agency's RASSP program. K.M.C. gratefully acknowledges support from the M.I.T. Lincoln Laboratory Staff Associate Program.

7. References

1. L. M. Pecora and T. L. Carroll, *Phys. Rev. Lett.* **64** (1990) 821.
2. L. M. Pecora and T. L. Carroll, *Phys. Rev.* **A44** (1991) 2374.
3. T. L. Carroll and L. M. Pecora, *IEEE Trans. Circuits Systems* **38** (1991) 453.
4. A. V. Oppenheim, G. W. Wornell, S. H. Isabelle, and K. M. Cuomo, in *Proc. 1992 IEEE ICASSP* **IV** (1992) 117.
5. K. M. Cuomo, Ph.D. Thesis, Massachusetts Institute of Technology, *Res. Lab. of Elect. TR 582* (1993).
6. K. M. Cuomo, *Int. J. Bifurcation and Chaos* **4**(3) (1994) 727.
7. K. M. Cuomo, *Int. J. Bifurcation and Chaos* **3**(5) (1993) 1327.
8. K. M. Cuomo and A. V. Oppenheim, *Phys. Rev. Lett.* **71** (1993) 65.
9. K. M. Cuomo, A. V. Oppenheim, and S. H. Strogatz, *IEEE Trans. Circuits and Syst.* **40** (1993) 626.
10. E. N. Lorenz, *J. Atmospheric Sciences* **20** (1963) 130.
11. R. He and P. G. Vaidya, *Phys. Rev.* **A46** (1992) 7387.
12. K. M. Cuomo, A. V. Oppenheim, and S. H. Strogatz, *Int. J. Bifurcation and Chaos* **3**(6) (1993) 1629.
13. P. Frederickson, J. L. Kaplan, E. D. Yorke, and J. A. Yorke, *J. Diff. Eqns.* **49** (1983) 185.
14. Lj. Kocarev, K. Halle, K. Eckert, and L. Chua, *Int. J. Bifurcation and Chaos* **2**(3) (1992) 709.

CHUA'S CIRCUIT: CHAOTIC PHENOMENA AND APPLICATIONS

LADISLAV PIVKA, CHAI WAH WU, and LEON O. CHUA

Electronics Research Laboratory and
Department of Electrical Engineering and Computer Sciences
University of California, Berkeley
CA 94720, U. S. A.

Abstract

This paper reports on some recent results on Chua's circuit, including the fabrication of a 2.5mm $\times$ 2.8mm monolithic CMOS chip of Chua's circuit, as well as some recent applications on stochastic resonance, secure communication, image processing, and music. It will also summarize several recent generalizations of Chua's circuit, including a globally unfolded version with some general results on synchronization and control, and an extension to 1- and 2-dimensional arrays which exhibit spatio-temporal chaos, autowaves, spiral waves, concentric waves, and Turing patterns.

1. Introduction

In 1963, the meteorologist E.N. Lorenz proposed the first three-dimensional autonomous system of equations[1] exhibiting chaotic behavior, which has since become a subject of intense research. It was not until 20 years later that a real physical object was discovered[2] and built[3], which is capable of reproducing chaotic phenomena known from the theory. Chua's circuit has since become the most widely studied paradigm for different types of dynamical, especially chaotic, behaviors. Fig. 1 shows the diagram of the circuit whose state equations, in a dimensionless form, are as follows:

$$\begin{aligned} \dot{x} &= \alpha\,(y - x - f(x)) \\ \dot{y} &= x - y + z \\ \dot{z} &= -\beta y \end{aligned} \tag{1}$$

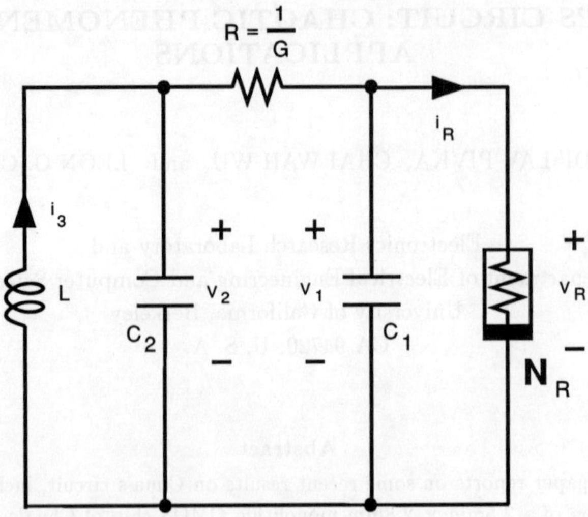

Fig. 1. Circuit diagram of Chua's circuit.

where f is a function obtained from the $v - i$ characteristic of the nonlinear resistor through scaling.

The history of the inception and evolution of Chua's circuit has been covered extensively in review papers *The genesis of Chua's circuit*[4] and *Chua's circuit: Ten years later*[5] where an overview of the most important recent results is given along with extensive bibliographies.

Given the large volume of publications (more than 200 papers, including two special issues in the Journal of Circuits, Systems, and Computers, several papers in the Chaos in Nonlinear Electronic Circuits special issues of the IEEE Transactions on Circuits and Systems, and a book containing papers on Chua's circuit have been written during the past decade) it is rather difficult to present, in a limited space, an overview of all major aspects of current research.

More technical details will be given in this paper than in the above reviews, although we will avoid using involved mathematical concepts and details. In addition, we will report on some recent advances not yet published at the time of this writing, namely, an implementation of Chua's circuit with a cubic nonlinearity (Sec. 2.), resonance phenomena in the driven Chua's circuit (Sec. 4.6.2.), the generation of Turing patterns in two-dimensional arrays of Chua's circuits (Sec. 5.3.), spatio-temporal chaos in chains of Chua's circuits (Sec. 5.4.), and a unified framework for synchronization and control (Sec. 7.).

2. Robust Implementations of Chua's Circuit

One of the distinguishing features of Chua's circuit in contrast to other chaotic systems is that it can be easily constructed as a physical electronic circuit. More precisely, simple electronic circuit implementations of Chua's circuit exist whose behaviors can be accurately modeled by Eq. 1. Furthermore, electrical quantities in electronic circuits are easy to measure accurately, in contrast to quantities in chemical or mechanical systems.

The linear components of Chua's circuit are readily available as off-the-shelf components. The nonlinear resistor (Chua's diode) has been implemented in the literature using diodes, operational amplifiers, transistors, and operational transconductance amplifiers. We reproduce here the inexpensive and robust implementation using operational amplifiers as given in Ref. 6 which is easy to construct and implements Chua's diode with a piecewise-linear $v-i$ characteristic.

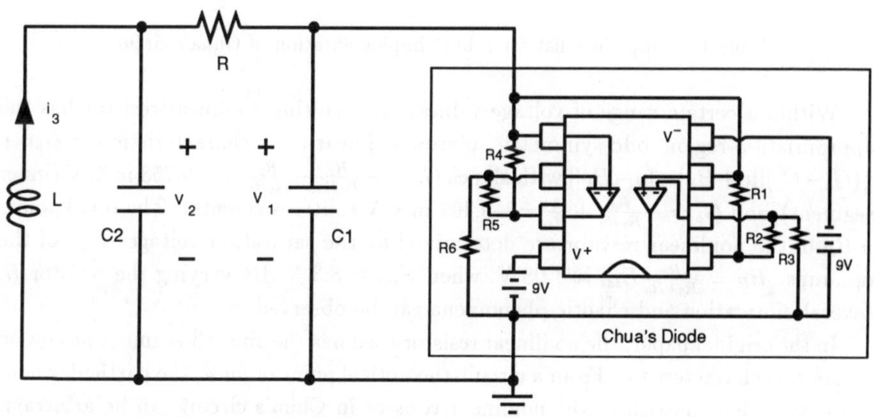

Fig. 2. Diagram of Chua's circuit and Chua's diode.

The circuit diagram is shown in Fig. 2, where an 8-pin dual op-amp package is used to implement the negative resistor. The component list is given in Table 1.

Element	Description	Value	Tolerance
A_1	Op amp ($\frac{1}{2}$ AD712, TL082, or equivalent)		
R_1	$\frac{1}{4}$W Resistor	220 Ω	±5%
R_2	$\frac{1}{4}$W Resistor	220 Ω	±5%
R_3	$\frac{1}{4}$W Resistor	2.2 kΩ	±5%
A_2	Op amp ($\frac{1}{2}$ AD712, TL082, or equivalent)		
R_4	$\frac{1}{4}$W Resistor	22 kΩ	±5%
R_5	$\frac{1}{4}$W Resistor	22 kΩ	±5%
R_6	$\frac{1}{4}$W Resistor	3.3 kΩ	±5%
C_1	Capacitor	10 nF	±5%
R	Potentiometer	2 kΩ	
C_2	Capacitor	100 nF	±5%
L	Inductor (TOKO type 10RB or equivalent)	18 mH	± 10%

Table 1. Component list for robust implementation of Chua's circuit.

Within a certain range of voltage values, the resulting nonlinear resistor has the appropriate 3-region, odd-symmetric, piecewise-linear $v-i$ characteristic $i = G_b v + \frac{1}{2}(G_a - G_b)[|v+B_P| - |v-B_P|]$ with slopes $G_a = -\frac{R_2}{R_1 R_3} - \frac{R_5}{R_4 R_6} = -0.758$ mA/V (inner segment), and $G_b = -\frac{R_2}{R_1 R_3} + \frac{1}{R_4} = -0.409$ mA/V (outer segments). The breakpoints $\pm B_P$ of the nonlinear resistor are determined by the saturation voltages E_{sat} of the op amps. $B_P = \frac{R_6}{R_5+R_6} E_{sat} \approx 1.08$ V, when $E_{sat} \approx 8.3$ V. By varying the resistor R, several bifurcation and chaotic phenomena can be observed.

In the original paper, the nonlinear resistor used has the above 3-segment piecewise-linear $v-i$ characteristic. From a circuit-theoretical point of view, the particular nonlinearity which constitutes the nonlinear resistor in Chua's circuit can be arbitrary, yet analysis of the circuit is essentially the same as it has the same circuit topology. Using a piecewise-linear nonlinearity with many segments, Suykens *et al.*[7] were able to synthesize Chua's circuit which generates n-double scrolls. Altman[8] and Khibnik *et al.*[9] consider Chua's circuit with a smooth polynomial nonlinearity. We give here an implementation of a cubic nonlinearity due to G.-Q. Zhong[10]. It consists of two four-quadrant analog multipliers in conjunction with a negative linear resistor, which is implemented using an op-amp. The circuit diagram is shown in Fig. 3, while the $v-i$ characteristic is shown in Fig. 4. Several periodic orbits and strange attractors that were obtained using this Chua's diode are shown in Fig. 5.

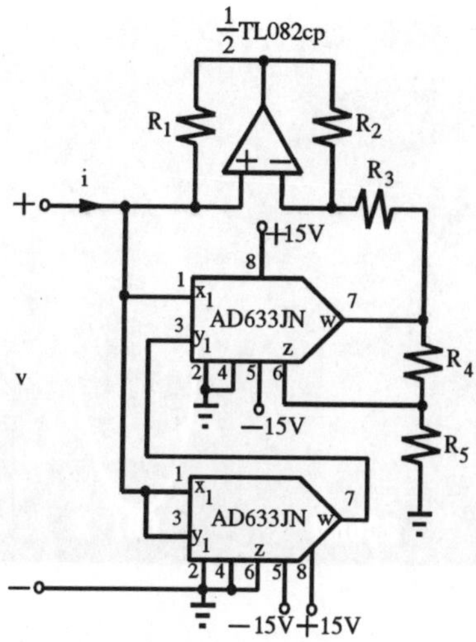

Fig. 3. Circuit diagram of cubic Chua's diode. Two analog multipliers (AD 633) are used along with a negative linear resistor which is implemented here using an op-amp. $R_1 = R_2 = 2\text{k}\Omega$, $R_3 = 1.668\text{k}\Omega$, $R_4 = 3.01\text{k}\Omega$, $R_5 = 7.91\text{k}\Omega$.

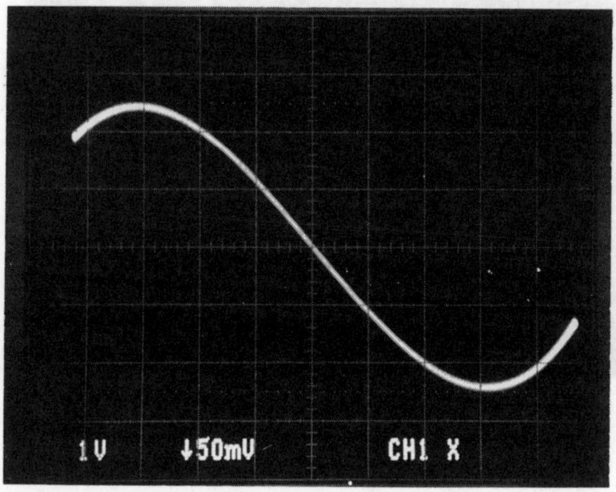

Fig. 4. Measured $v - i$ characteristic of the cubic polynomial Chua's diode. The vertical scale is 0.5mA/div and the horizontal scale is 1V/div. The $v - i$ characteristic has the form $i = av + cv^3$, where $a = -0.59\text{mS}$, $c = 0.02\text{mS/V}^2$.

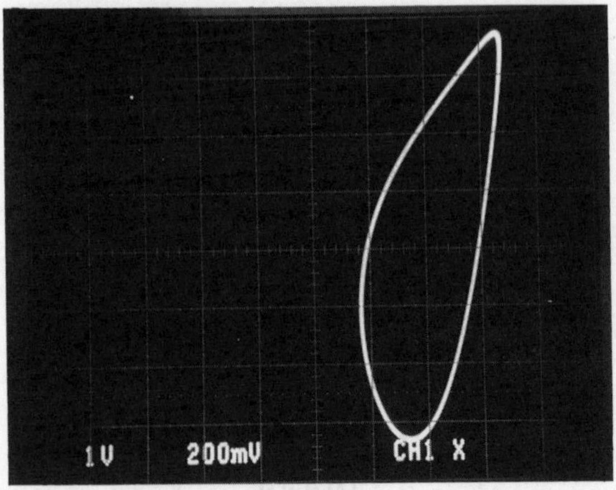

(a)

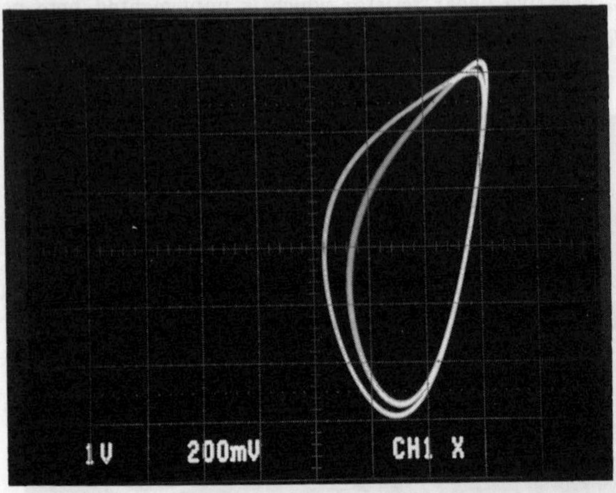

(b)

Fig. 5. Attractors obtained from Chua's circuit with cubic polynomial Chua's diode. The component values of the linear elements are C_1 = 7nF, C_2 = 78nF, L = 18.92mH. The inductor has a series resistance of 14.99Ω. (a) $R = 2200\Omega$, (b) $R = 2103\Omega$, (c) $R = 2033\Omega$, (d) $R = 1964\Omega$.

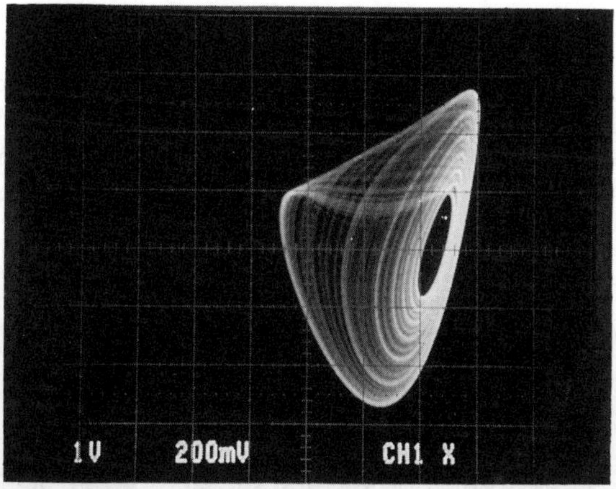

(c)

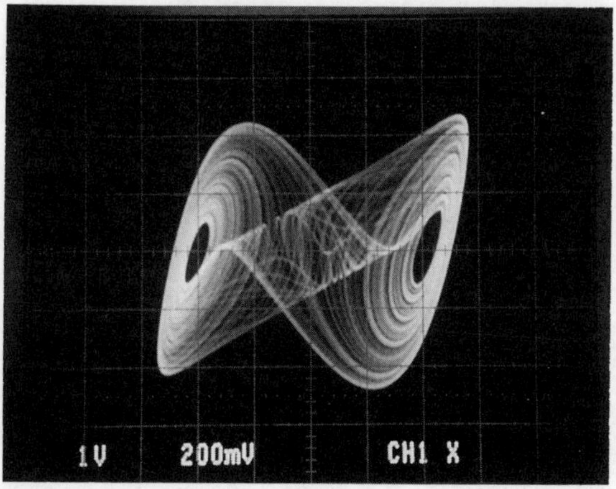

(d)

Fig. 5. (continued).

3. Monolithic IC Chips of Chua's Circuit

We have seen in the last section how Chua's circuit can be built using standard off-the-shelf components[6]. The nonlinear resistor, also called Chua's diode, has been integrated into a single chip[11] as a piecewise-linear function. The nonlinear resistor is implemented in CMOS technology and is based on operational transconductance amplifiers.

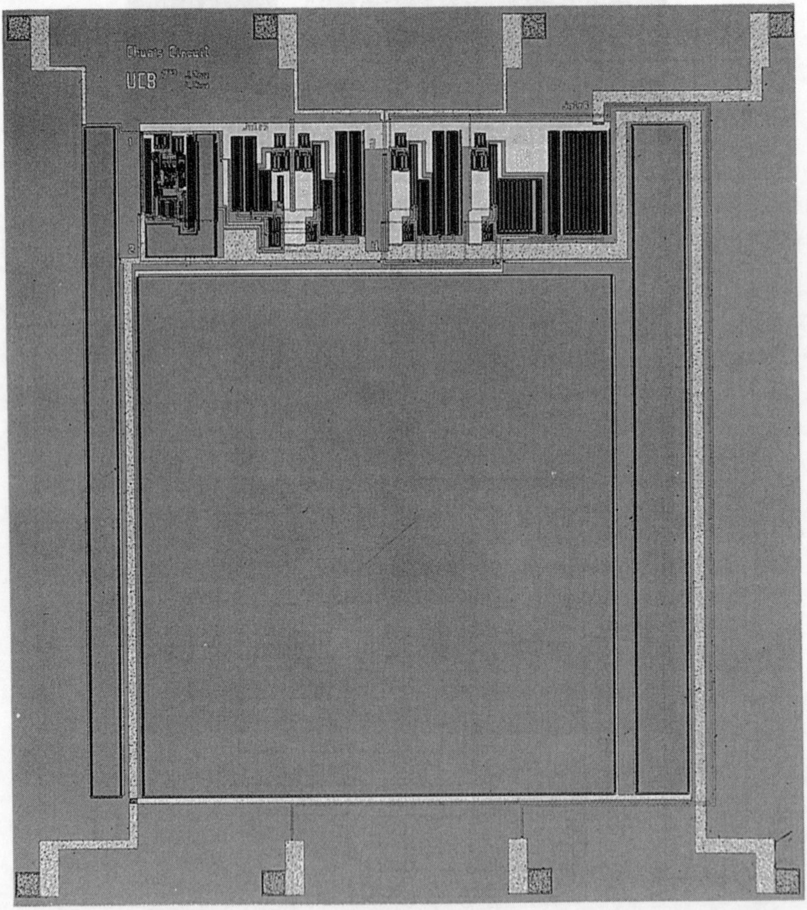

Fig. 6. Micrograph of the fabricated circuit.

For applications which require a large number of Chua's circuits such as arrays of Chua's circuits (Sec. 5.), it is necessary to implement one or several Chua's circuits in integrated form. In Ref. 12, the inductor, implemented as a gyrator-capacitor pair, is integrated on a single chip along with a piecewise-linear Chua's diode and two linear capacitors. Two external linear resistors complete the circuit. One resistor is the linear resistor R accross the two capacitors and the other resistor controls the inductance value L that is being realized. These two external resistors can serve as independent tuning or bifurcation parameters. The chip has been fabricated in 2μm CMOS technology and in an experimental setup generates chaotic signals from all three state variables. The circuit occupies a silicon area of $2.5mm \times 2.8mm$, with most of the area devoted to implementing the capacitors. A photomicrograph of this chip is shown in Fig. 6.

Using a systematic state variables approach[13], the *entire* Chua's circuit has been implemented in 2.4μm CMOS technology.

4. Dynamical Phenomena From Chua's Circuit

Since its inception in 1983, Chua's circuit has become a vehicle for modeling a large number of dynamical phenomena. In this section, we summarize those phenomena, both classical and recently observed, from the original Chua's circuit[2].

4.1. Routes to Chaos and Coexistence of Attractors

Period-doubling route to chaos is a classical example of transition to chaos and was also observed when the autonomous Chua's circuit was first simulated[2] on a digital computer. Among the other modes of transition to chaos, torus breakdown route to chaos has also been observed and both routes can be conveniently interpreted and explained in terms of characteristic multipliers of the corresponding Poincaré map[14]. Extensive computer simulations and physical experiments were performed in a two-parameter study[15] to describe several types of transition to chaos in the nonautonomous Chua's circuit driven by an external sinusoidal signal.

Coexistence of attractors is an interesting phenomenon in which the interaction of attractors can give rise to different dynamical phenomena described in the following two subsections. Recently, a coexistence of three distinct chaotic attractors has been reported[16], when two asymmetric attractors coexist with a symmetric one. Some other coexistence phenomena, including point attractors, periodic attractors, and chaotic attractors, can be found in Ref. 17.

4.2. Chaos-chaos Intermittency and 1/f Noise

It has been known [18,19] that interaction between chaotic attractors can give rise to intermittency — random switching process between attractors after long periods of "laminar phases", when the trajectory stays near one of the attractors. A characteristic statistical property of the chaos-chaos type intermittency is the slope of its power spectrum in the low-frequency region. Such a property has also been observed[20] in Chua's circuit for parameter values near the birth of the Double-scroll attractor. The power spectrum was numerically found to follow the law $S_x(w) \approx w^{-\delta}$, $\delta = 1.1 \pm 0.1$, i.e. the graph on the double logarithmic scale clings to the ideal $1/f$ line corresponding to $\delta = 1$. The $1/f$ spectrum has been observed previously in many processes of different origin, e.g., the fluctuations of the current in electron devices, the fluctuations of the Earth's rotation frequency, the fluctuation of the muscle rhythms in the human heart, etc., and has been found to obey the above universal law. The intermittency phenomenon can be used as a $1/f$ noise generator and can lead to a better understanding of the ubiquitous yet still poorly understood $1/f$ phenomenon.

4.3. Stochastic Resonance from Chua's Circuit

The phenomenon of stochastic resonance (SR) is observed in bistable nonlinear systems driven simultaneously by external noise and a sinusoidal force. In this case, the signal-to-noise ratio (SNR) increases until it reaches a maximum at some optimum noise intensity D which depends on the bistable system and on the frequency of the external sinusoidal force. In the absence of a periodic modulation signal, the noise alone results in a random transition between the two states. This random process can be characterized by the mean switching frequency w_s, depending on the noise intensity D and the height of the potential barrier separating the two stable states. In the presence of an external modulation imposed by the sinusoidal signal $A\sin(wt)$, the potential barrier changes periodically with time. The modulation signal amplitude A is assumed to be sufficiently small so that the input signal alone does not induce transitions in the absence of noise. A coherence between the modulation frequency w and the mean switching frequency w_s emerges when the system is simultaneously driven by a periodic signal and a noise source. As a result, a part of the noise energy is transformed into the energy of the periodic modulation signal so that the SNR increases. This phenomenon is qualitatively similar to the classical resonance phenomenon. However, unlike the classical circuit theory where one tunes the input frequency w to achieve resonance in an RLC circuit, here w is fixed at some convenient value and one tunes the noise intensity D to achieve SR.

In Chua's circuit, the SR phenomenon can be observed[21] in conjunction with the chaos-chaos type intermittency[20] arising in a small vicinity of the bifurcation curve in the $\alpha - \beta$ parameter space when two spiral attractors merge to form the Double-scroll attractor. In this case, the SNR of the amplified output signal is observed to be significantly greater than the SNR of the input signal — a novel phenomenon which can not be achieved with a linear amplifier.

4.4. Signal Amplification via Chaos

Apart from the stochastic resonance phenomenon described above, another mechanism for achieving voltage gain (up to 50 dB has been demonstrated experimentally) from Chua's circuit has been discovered recently[22]. The mechanism of this voltage gain is different from that of stochastic resonance because the effect is observed even when Chua's circuit is operating in a spiral attractor regime far from the bifurcation boundary where stochastic resonance takes place.

4.5. Antimonotonicity Phenomenon

Antimonotonicity — concurrent creation and annihilation of periodic orbits, or inevitable reversals of period-doubling cascades was shown to be a fundamental phenomenon for a large class of nonlinear systems[23,24]. The phenomenon was observed numerically in an RC ladder circuit[25], then investigated in the Henon map[23], and the cubic map [24]. Experimental[26] and numerical[27] evidence was given that this phenomenon is typical for a wide range of parameters in Chua's circuit and Chua's oscillator (see Sec. 8.), respectively.

4.6. Universality and Self-similarity: Two-parameter Bifurcation Studies

In the standard bifurcation scenarios described in the literature, usually only one control parameter is changed. In physics, engineering, and other fields, however, one often needs to control two or more parameters to obtain a broader view of the global geometry. Here we mention two cases of self-similar and universal structures, one for the autonomous and the other for the forced Chua's circuit.

4.6.1. Self-similar and Universal Structures in Two-parameter Study of Transition to Chaos

Using the Poincaré map technique, the exact description of the system (1) can be reduced to a two-dimensional map which, in turn, can be approximated by a one-dimensional map[28,29], generally called Chua's 1-D map in the literature. This map happens to be bimodal in certain parameter regions, which means that it has both a maximum and a minimum on an interval which is mapped onto itself. This condition is responsible for the complicated structure of the boundary of chaos in a two-parameter bifurcation diagram.

In a typical one-parameter bifurcation sequence, if we tune only one parameter in Chua's circuit, we usually see a typical period-doubling cascade, which exhibits remarkable properties of quantitative universality[30,31] and self-similarity, namely, an interval encompassing regions of different dynamical regimes reproduces itself under a change in scale by the universal factor $\delta = 4.6692...$.

If we turn to a two-parameter study, we can no longer restrict ourselves to the Feigenbaum scenario which is a codimension-1 bifurcation phenomenon. In Ref. 32 the construction of a binary tree of superstable orbits[33] is reproduced for the 1-D Chua's map to show that beside the Feigenbaum critical lines, the boundary of chaos contains an infinite number of codimension-2 critical points, defined by a set of infinite binary codes. The topography of the parameter plane near the corresponding critical points reveals a property of two-parameter self-similarity: a two-dimensional structure of regions of different behavior is reproduced under a scale change along appropriate axes in the parameter space. These self-similar two-dimensional patterns are universal (up to a linear parameter change) for all bimodal maps, and depend only on the code of the associated critical point. Moreover, two universal scaling numbers have been found for the two-parameter 1-D maps, which are a generalization of the Feigenbaum number.

4.6.2. Devil's Staircase from the Driven Chua's Circuit

One of the remarkable properties of nonlinear oscillators is their ability to lock onto certain subharmonic frequencies when driven by an external source of energy. Associated with the phase-locking property is usually the appearance of "staircases" of phase-locked states when the parameters are varied over certain range. The picturesque name *devil's staircase* was coined by Mandelbrot[34] to capture the intricate, often fractal, structure of such staircases. By adding the forcing term $I(t) = A\cos(wt)$ to the first equation in (1) and setting $B_1 = -1$, $B_2 = 0.0234168$, $\alpha = 10$, $\beta = 0.3014987$, $s_1 = 0.078573$, $s_0 = -1.25719$, $s_2 = 55.78573$ we obtain the driven Chua's circuit. With these parameter values and $|A| > 0.07$, $w \neq 0$, the system exhibits a high-relaxation, cyclical behavior. Given a frequency f_s of forcing, and the system's response frequency f_d, the corresponding *winding number* will be $W = f_s/f_d$. By varying the amplitude A and angular frequency w, the system exhibits sequences of phase-locked states, which are subject to the period-adding law[35]

$$q \rightarrow q + p \rightarrow q + 2p \rightarrow \ldots q + np \rightarrow \ldots \infty \ldots \rightarrow p.$$

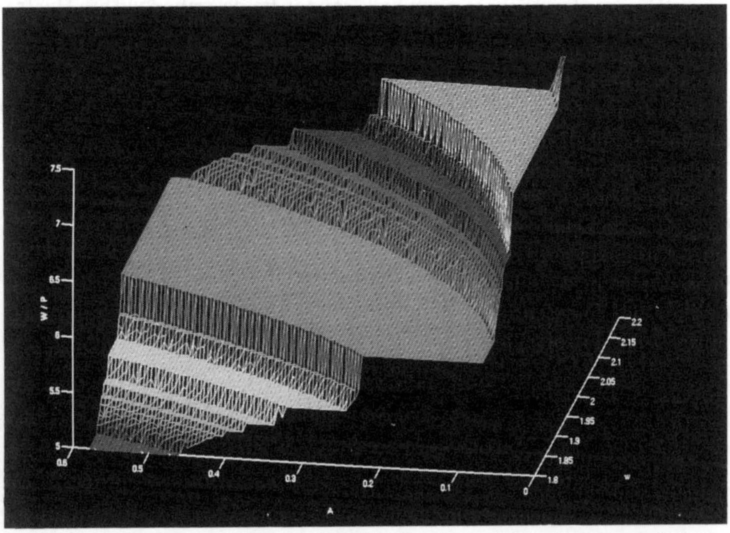

Fig. 7. Three-dimensional view of the devil's staircase.

The interpretation is that by changing the forcing frequency monotonically over a certain range, between any two successive phase-locked states with winding numbers p, q at the same staircase level , one can find an infinity of phase-locked states of ascending order, leading to infinity, before eventually dropping to phase-locked state q.

By using numerical simulations, the same law was found for period numbers (P) which can be defined as the number of local minima, per least period, in the waveform of one of the state variables. The winding and period numbers form staircases which in turn form a hierarchy of different levels. By plotting the ratio W/P as a function of A and w we obtain the familiar devil's staircase of Fig. 7. The self-similar structure of the staircase tree and the devil's staircase become apparent when magnified pictures are drawn of the portions of the devil's staircase.

5. Wave Propagation Phenomena and Spatio-temporal Chaos in Arrays of Chua's Circuits

In the preceding section we discussed dynamical phenomena arising in a single Chua's circuit and observed in various deterministic systems. However, physical experiments provide us with complex phenomena which cannot be described within the framework of a single low-dimensional dynamical system. Discrete arrays, although still greatly simplified (because of homogeneity, type of interconnection, etc.), seem to be adequate to model the basic phenomena and patterns found in different nonequilibrium media, e.g., reaction-diffusion homogeneous systems.

5.1. Autowaves and Propagation Failure in Linear Arrays

The system of equations governing the dynamics of a chain of Chua's circuits can be written, in a dimensionless form, as follows:

$$\begin{aligned} \dot{x}_k &= \alpha\,(y_k - f(x_k)) + D[x_{k-1} - 2x_k + x_{k+1}] \\ \dot{y}_k &= x_k - y_k + z_k \qquad (k = 0, 1, \ldots, l) \\ \dot{z}_k &= -\beta y_k \end{aligned} \tag{2}$$

where

$$f(x) = (1/2)\,[(s_1 + s_2)x + (s_0 - s_1)(|x - B_1| - |B_1|) + (s_2 - s_0)(|x - B_2| - |B_2|)] \tag{3}$$

is a three-segment piecewise-linear function with the breakpoints $B_1 = -1$ and $B_2 = 0.0234168$, the slopes $s_1 = 2/7$ (left), $s_0 = -1/7$ (middle), $s_2 = 1/7$, and $\alpha = 9$, $\beta = 30$. In the above equations, D represents the diffusion coefficient for the variable x. Another requirement is the zero-flux (Neumann) boundary condition.

By setting $\dot{x}_k = \dot{y}_k = \dot{z}_k = 0$, and $D = 0$, we obtain the equilibrium points $P^{\pm} = \pm[(m_1 - m_0)/m_2, 0, (m_0 - m_1)/m_2]$ of uncoupled Chua's circuit. Assume that all cells are at the state P^- initially. If some of the cells are set at the point P^+, a traveling wave can be observed for $D > 0.363$. It is shown Ref. 36 that a propagation failure occurs if the diffusion coefficient is lower than the critical value $D^* \approx 0.363$ and the phenomenon is explained by an application of the singular perturbation theory. Another phenomenon, connected with the decrease of the diffusion coefficient, is the corresponding decrease in the velocity of the wavefront as $D^* \searrow D$. Also, let us note that the propagation failure is an inherent property of the discrete systems, and is impossible to achieve in continuous, one-variable, homogeneous reaction-diffusion systems. This was proved by Keener[37,38].

The above scenario can be used to model transmission blockage in nerve impulse propagation or in a one-dimensional array of continuosly stirred tank reactors[39–41]. Another application of chains of Chua's circuits is described in Ref. 42 where results on particle motion in a one-dimensional fluctuating medium are reproduced. For a number of other phenomena, however, higher-dimensional arrays are needed to obtain appropriate models.

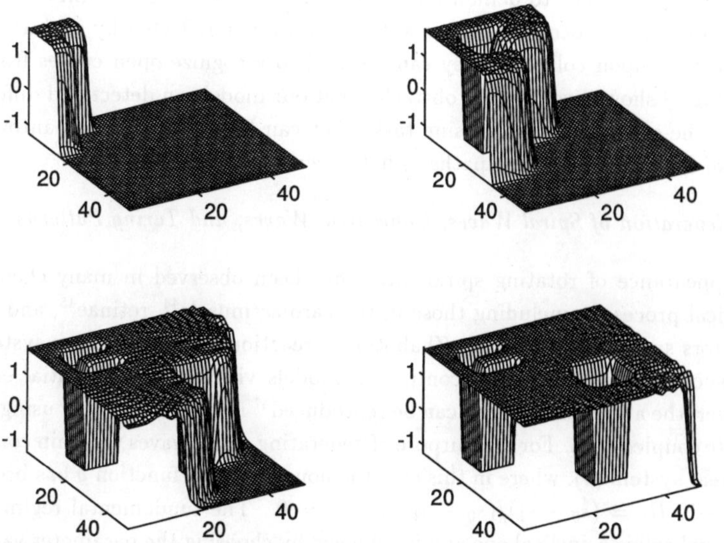

Fig. 8. Traveling wavefront in a 50 × 50 array of Chua's circuits. The snapshots were taken at time $t_1 = 32$, $t_2 = 64$, $t_3 = 96$, and $t_4 = 128$, respectively.

5.2. Autowaves in 2-D CNN Arrays

A two-dimensional Cellular Neural Network (CNN) array of resistively coupled Chua's oscillators (Sec. 8.) is modeled by state-space equations similar to those for chains:

$$\begin{aligned} \dot{x}_{i,j} &= \alpha\,(y_{i,j} - x_{i,j} - g(x_{i,j})) + D[x_{i+1,j} + x_{i-1,j} + x_{i,j+1} + x_{i,j-1} - 4x_{i,j}] \\ \dot{y}_{i,j} &= x_{i,j} - y_{i,j} + z_{i,j} \qquad (i,j = 1,2\ldots,l) \\ \dot{z}_{i,j} &= -\beta y_{i,j} - \gamma z_{i,j} \end{aligned} \tag{4}$$

The nonsymmetric three-segment PWL function $g(x) = f(x) + \epsilon$ where $f(x)$ is given by Eq. 3, and ϵ is a small constant (dc offset). Also, we impose the zero-flux (Neumann) boundary condition. With the parameter values from Sec. 5.1., breakpoints $B_1 = -1$, $B_2 = 1$, $\gamma = 0$, and $\epsilon = -1/14$, the system is bistable with two fixed points (sinks). The propagation failure phenomenon occurs at, or below, the critical value $D^* = 0.51$.

Here we also want to demonstrate the possibility of using two-dimensional CNN arrays for image processing. Since autowaves are not reflected by obstacles and do not interfere upon collision, they can be used to recognize open curves from closed ones. Fig. 8 shows two possible obstacles that our model can detect and differentiate. Among the other image processing tasks that can be solved by using autowaves are those of finding the shortest paths in flat or wrinkled labyrinths.[43]

5.3. Generation of Spiral Waves, Concentric Waves, and Turing Patterns

The appearance of rotating spiral waves has been observed in many chemical and biological processes, including those in the cardiac muscle[44], retinae[45], and chemical oscillators such as the Belousov-Zhabotinsky reaction[46]. Most of these systems have been successfully modeled by continuum models via partial differential equations. However, the above phenomena can be reproduced[47] more efficiently by using CNNs of discrete coupled cells. For the purpose of generating spiral waves we again consider the nonlinear system (4), where in this case the nonsymmetric function g has breakpoints $B_1 = -1$, $B_2 = (s_0 - s_1)/(s_0 - s_2)$, and $\epsilon = 0$. The fundamental regime in each individual cell is a cyclical one and is achieved by choosing the parameter values, e.g., as follows: $\alpha = 10$, $\beta = 0.334091$, $\gamma = 0$, $s_1 = 0.020706$, $s_0 = -0.921$, $s_2 = 15$. The strong asymmetry of function g provides for a high-relaxation character of the limit cycle, which is necessary for the dynamics to be stable. Fig. 9 shows a fully evolved spiral, started from initial conditions in Fig. 10, with the diffusion coefficient $D = 5$.

By placing a U-shaped obstacle in the path of the spiral, which is accomplished by clamping appropriate cells at fixed values, the spiral structure is gradually destroyed and a sustained train of concentric waves takes over (Fig. 10), with the obstacle as the "source". A most fascinating phenomenon found in nature is the tendency of large groups of cells or units to organize themselves into highly structured patterns from apparently disordered or random initial states. It is quite surprising that those phenomena can be successfully reproduced by using a relatively simple *deterministic* model (4) started from *random* initial state.

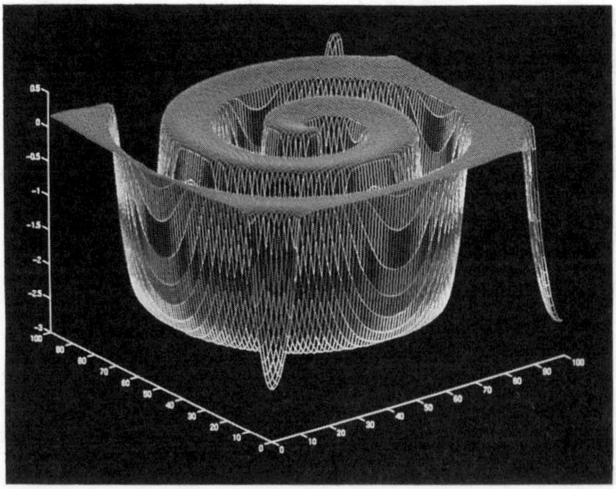

Fig. 9. Spiral wave generated in a 100×100 array of identical Chua's circuits.

In our particular example the PWL function $g(x) = f(x) + \epsilon$ was used with $\epsilon = 2$. Also, Chua's oscillator (see Sec. 8.) instead of the original Chua's circuit forms the basic cell, and coupling for both the x and y variables were used, with $\alpha = 10$, $\beta = \gamma = 0.001$, $s_0 = -1$, $s_1 = s_2 = 0.1$, $B_1 = -B_2 = 1$, $D_x = 0.5$, $D_y = 20$. The initial conditions can be chosen randomly for the variable x, while those for variable y can be random between $9/8 - 0.1$ and $9/8 + 0.1$, and $z = -9/8$. If started from such initial conditions, the system gradually evolves into a pattern of hexagonal structures (Fig. 11) which usually contain defects similar to those observed in chemical structures. Such a stable condition corresponds to a fixed point in the high-dimensional state space. The structure in Fig. 11, usually called the Turing structure, can be subject to external forcing, e.g., by means of a spiral wave, and form different patterns. A more complete description will appear in Ref. 48.

Computer simulations of three-dimensional arrays have already been reported in the literature[49,50]. The above parameter values, used for the generation of spirals have also been used recently[51] to generate so-called *scroll waves* in three-dimensional arrays of Chua's circuits.

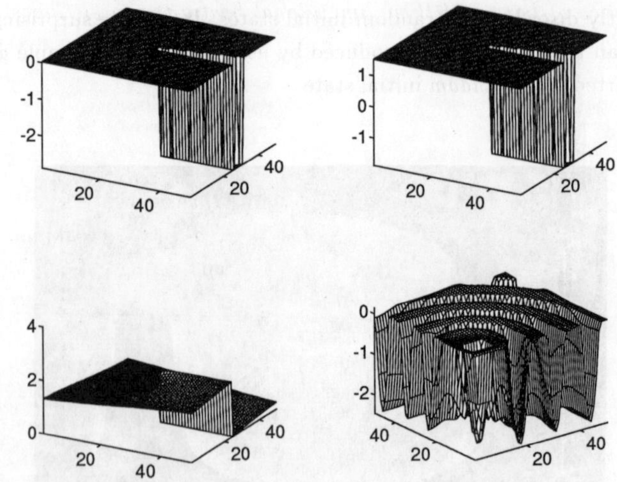

Fig. 10. Initial conditions for the generation of spiral wave (Fig. 9), and a concentric wave pattern (lower right).

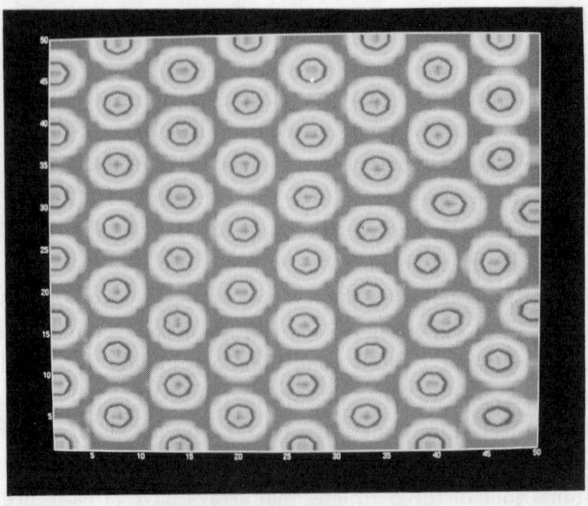

Fig. 11. Hexagonal structures formed spontaneously in an array of 50×50 Chua's oscillators. Note the penta-hepta defects in the right part of figure.

5.4. Coexistence of Low- and High-dimensional Spatio-temporal Chaos in Chains of Chua's Circuits

One-dimensional CNNs consisting of resistively coupled Chua's circuits can exhibit a variety of spatio-temporal phenomena which are typical of spatially extended reaction-diffusion media.

An investigation of spatio-temporal dynamics was conducted[52] for the system governed by Eq. 2 with parameter values $\alpha = 9$, $\beta = 19$, breakpoints $B_1 = -1$, $B_2 = 1$, slopes $s_1 = -5/7$, $s_0 = -8/7$, $s_2 = -5/7$, and periodic boundary conditions $x_1(t) = x_{N+1}$, $y_1(t) = y_{N+1}$, $z_1(t) = z_{N+1}$, with $N \gg 1$.

The parameters for an uncoupled Chua's circuit were chosen so that it has two stable limit cycles, symmetrical with respect to origin. Despite simple local dynamics of the basic cell, the global behavior of the system can be very complicated. The dynamics were studied for two types of initial conditions. In the first type:

$$x_j(0) = 1 + 0.1\sin\frac{2\pi(j-1)}{N}, \qquad y_j(0) = z_j(0) = 0.1, \tag{5}$$

the trajectories of all cells are attracted to the same limit cycle. In the second type:

$$x_j(0) = \sin\frac{2\pi(j-1)}{N}, \qquad y_j(0) = z_j(0) = 0.1, \tag{6}$$

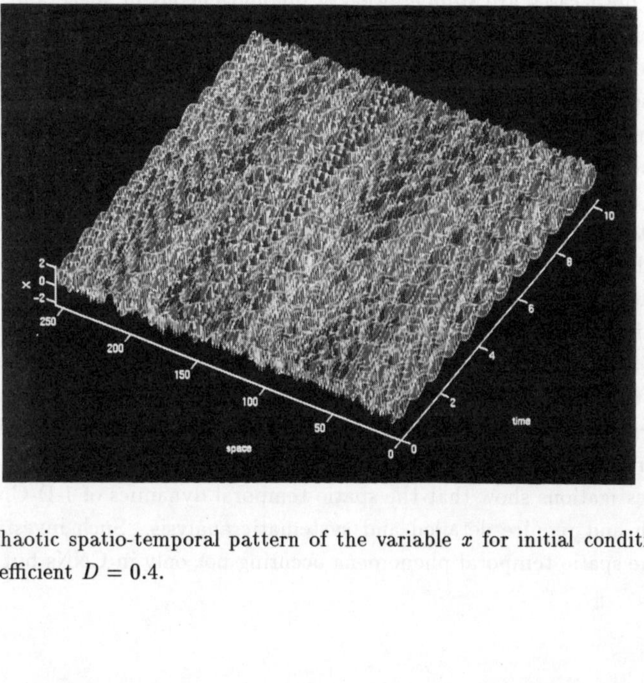

Fig. 12. Chaotic spatio-temporal pattern of the variable x for initial conditions (5) and diffusion coefficient $D = 0.4$.

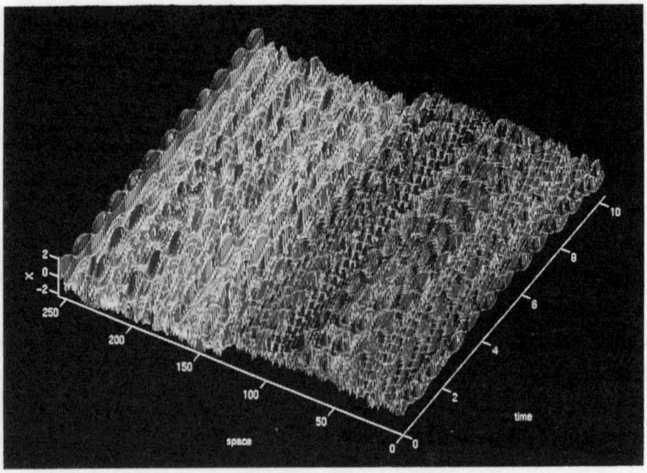

Fig. 13. Result corresponding to Fig. 12 for initial condition (6).

the trajectories of cells belong to the basins of attraction of two different limit cycles. The computations showed that for both types of initial conditions, the dynamics of the system have much in common. For weak couplings ($D \ll 1$), the spatio-temporal patterns in both cases are simple: weakly inhomogeneous in space and periodic in time. When the diffusion coefficient D exceeds the critical value D_{cr}, a pattern occurs — so called π-spatial oscillation — where almost all adjacent cells are 180° out of phase. Increasing the diffusion coefficient D further gives rise to envelope waves against a background of π-oscillations and finally, when $D \geq 0.33$, chaotic spatio-temporal patterns appear. We can see that these patterns no longer exhibit any regular structures in space and are chaotic (with broadband power spectrum) in time. Chaotic spatio-temporal patterns were further studied based on a new approach describing the properties of patterns through characteristics of associated attractors of a multidimensional dynamical system with *matrix* phase space. Using this approach it has been found that the attractors corresponding to the chaotic patterns for initial conditions (5) (Fig. 12) have large correlation dimension (> 10), whereas for initial conditions (6) (Fig. 13) the correlation dimension is low (< 5). Thus, it is shown that in the matrix phase space of CNNs, different high- and low-dimensional attractors coexist, corresponding to different initial conditions.

The investigations show that the spatio-temporal dynamics of 1-D CNNs are extremely rich and require detailed and systematic analysis. Such investigation can elucidate the spatio-temporal phenomena occuring not only in CNNs but in a broad range of spatially extended reaction-diffusion media as well.

6. Applications of Chaos from Chua's Circuit

Not a long time ago, chaotic phenomena were recognized as a curiosity phenomenon, interesting and worth studying only from an academic point of view, without a prospect of immediate practical applications. Most practically oriented studies were directed toward avoiding chaos as a phenomenon which is undesirable in real-life applications. Following some breakthrough discoveries, especially of chaotic synchronization[53] and stochastic resonance[21] phenomena, this attitude changed dramatically. In this section we outline some potential applications in which chaotic behavior plays an essential role.

6.1. Secure Communication via Chaos

The key concept in the application to secure communication is that of synchronization. Given two (or more) nonlinear systems

$$\dot{\mathbf{x}}_k = \mathbf{f}_k(\mathbf{x}_k) \quad \mathbf{x}_k \in R^n \ , \ k = 1, 2, ..., N$$

we are interested in conditions leading to the synchronization of the solutions, i.e. the convergence $(\mathbf{x}_l - \mathbf{x}_j) \to 0$, as $t \to \infty$, for $l \neq j$. For the special case of linear coupling of two systems

$$\begin{aligned} \dot{\mathbf{x}} &= \mathbf{f}_1(\mathbf{x}) \\ \dot{\mathbf{y}} &= \mathbf{f}_2(\mathbf{y}) + \Delta(\mathbf{x} - \mathbf{y}) \ , \ \ \Delta = \mathrm{diag}(\delta_1, ..., \delta_n)^T \ , \ \ \mathbf{x}, \ \mathbf{y} \in R^n \end{aligned}$$

some general conditions are known[54] to ensure synchronization. In order to employ the above synchronization effect, Pecora and Carroll[55] suggested a drive-response concept as follows:

Consider an n-dimensional autonomous system with the state equation

$$\dot{\mathbf{x}} = \mathbf{f}(\mathbf{x})$$

and divide the system into two parts, the driving subsystem D, and the response subsystem R:

$$\dot{\mathbf{x}}_D = \mathbf{g}(\mathbf{x}_D, \mathbf{x}_R)$$

$$\dot{\mathbf{x}}_R = \mathbf{h}(\mathbf{x}_D, \mathbf{x}_R).$$

By adding an identical copy of the response subsystem, we obtain

$$\begin{aligned} \dot{\mathbf{x}}_D &= \mathbf{g}(\mathbf{x}_D, \mathbf{x}_R) \\ \dot{\mathbf{x}}_R &= \mathbf{h}(\mathbf{x}_D, \mathbf{x}_R) \\ \dot{\mathbf{x}}_R' &= \mathbf{h}(\mathbf{x}_D, \mathbf{x}_R'). \end{aligned}$$

Then for $\mathbf{x}_R$ and $\mathbf{x}_R'$ to synchronize it is necessary that the conditional Lyapunov exponents (depending on $\mathbf{x}_D$) be all negative[53]. Several approaches have been suggested so far to employ the synchronization effect for data transmission by using Chua's circuit as the chaos generator.

6.1.1. Secure Communication via Chaotic Masking

The idea of chaotic masking has been proposed in Refs. 56,59, where the information-carrying signal is added to the masking chaotic signal. Such an approach was used in an experimental setup for secure communication, where Chua's circuits were used as synchronizing blocks. Fig. 14 shows a schematic diagram, in which the transmitter part is a Chua's circuit and the receiver part is two partial Chua's circuits. The first subcircuit is a decoding key and synchronizes only when exactly matched with the transmitter circuit, thus reproducing the $y1$ signal. The second subcircuit is used for obtaining the variable $x2$ needed for recovering the information signal through subtraction as shown in the diagram of Fig. 14.

6.1.2. Chaos Shift Keying

In the chaotic synchronization approach used above, the parameters in the transmitter and in the receiver must be matched exactly. Slight parameter mismatch will cause synchronization error to appear. This property is exploited in the *chaos shift keying* scheme[57,58], where a binary signal is encoded in terms of two different attractors existing for two different system parameter values. The transmitter and receiver is similar to the chaotic masking scheme, but rather than adding the information signal onto the chaotic signal, a waveform corresponding to one attractor is transmitted when a "1" occurs in the information stream and a waveform corresponding to the other attractor is transmitted when a "0" occurs in the information stream. Although the two attractors appear similar, the receiver will synchronize only to the attractor

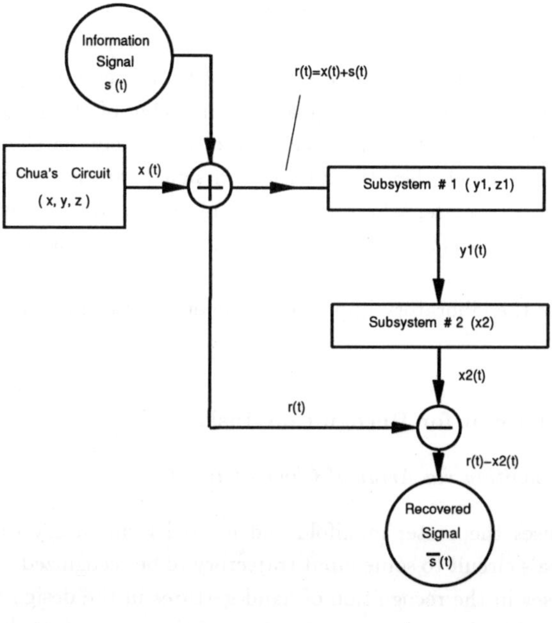

Fig. 14. Diagram for secure communication system using chaotic masking.

with the same parameter set. Whether the receiver synchronizes or not determines whether a "1" or a "0" is transmitted. For a more reliable system, a second receiver with the second parameter set can be used. The feasibility of this approach for secure communication has been demonstrated experimentally in Ref. 57.

6.1.3. Spread-spectrum Secure Communication via Chaotic Masking

Ref. 60 provides another approach for information transmission by making use of the broad spectrum of the chaotic carrier signal. The transmitter and the receiver each contain an identical Chua's circuit. In the transmitter a current signal $I(t)$ is injected into the circuit to modify the voltage across capacitor C_1. An invertible coding function c is chosen so that $I(t) = c(v_s(t))$ where $v_s(t)$ is the input signal to be transmitted. The detected output signal I_d is then decoded from $v_r(t) = c^{-1}(I_d(t))$. In the spread spectrum scheme, the coding function is multiplication with the chaotic signal $x(t)$ which spreads the spectrum of the input signal. For proper operation, it is necessary that $v_r(t) \approx v_s(t)$ and the coding function should be chosen in such a way that the transmitted signal remains chaotic. The voltage across the capacitor C_1 is transmitted to the receiver circuit and is used as a forcing voltage on the second Chua's circuit capacitor $\tilde{C}_1$. If all circuit components of the transmitter and receiver are matched exactly, we have[60] $I_d(t) \to I(t)$ and $v_r(t) \to v_s(t)$ for $t \to \infty$, which means that the current flowing into the second Chua's circuit will eventually equal the current injected into the first Chua's circuit. One of the features of this scheme is that when the transmitter and receiver are completely matched and there is no distortion in the channel, then the information signal can be recovered at the receiving end without any noise, whereas in Fig. 14, there will be some residual noise in the receiver ($\overline{s}(t) \not\to s(t)$). For a dual scheme which also has this noise-free property, see Ref. 61.

The full account of the laboratory implementation of the transmission system can be found in Ref. 60.

6.2. Using Chua's Circuit for Recognition Tasks

6.2.1. *Trajectory Recognition via Array of Chua's Circuit*

Recently, Altman[62] uses the center manifold and normal form theory to relate the local behavior of Chua's circuit to some input trajectory to be recognized. This mathematical problem arises in the recognition of hand gestures in the design of artificial intelligence, where the hand position as a function of time is used to drive Chua's circuit to an attracting surface. Since Chua's circuit is known to undergo a series of bifurcations from fixed points, to limit cycles, to a cascade of period-doubling oscillations leading to chaotic oscillations in the vicinity of the center manifold surface, the rapid entrainment of the chaotic system to an external signal having a trajectory near the center manifold surface provides the basic mechanism for trajectory recognition. The recognition of many trajectories can be achieved by using a 2-dimensional array of Chua's circuits. In this case, the variation of responses to the common input trajectory creates a *spatial pattern* which can be used to recognize the input trajectory. The above approach to trajectory recognition is both novel and fascinating.

6.2.2. *Handwritten Character Recognition Using Chua's Oscillator*

A neural network architecture and learning algorithm for associative memory storage of analog patterns, continuous sequences, and chaotic attractors via a network of Chua's oscillators has recently been designed by Baird and Hirsch[63]. Their design is used in the application to the problem of real-time handwritten digit recognition. They have demonstrated that several of the attractors from Chua's oscillator have outperformed the previously studied Lorenz attractor system in terms of both accuracy and speed of convergence.

6.3. Music from Chua's Circuit

While investigating Chua's system, several distinct features were observed[64] that can be described in musical terms: the tendency to produce pitch due to its isochronic behavior, the simultaneous presence of noise along with embedded periodic orbits, and sounds resembling the formant structures of acoustic instruments. The interplay of these factors can create steady-state waveforms that resemble steady-state portions of acoustic instrument sounds. Chua's circuit has parameter regions where noisy frequency- and amplitude-modulated sounds are generated, each of which is related to a certain transition to chaos, e.g., period-doubling, intermittency, torus breakdown, etc. Interesting, almost harmonic pitch changes can be produced through a period-adding sequence of bassoon-like sounds. Transient dynamics were found to be important in the context of percussion-like sounds.

The great variety of attractors that can be generated from Chua's circuit makes this circuit an excellent candidate for a universal signal-generating standard. These properties have been succesfully employed for sound synthesis and design[64,65] by tuning different circuit parameters[66]. Ref. 65 describes real-time simulations of Chua's circuit on a digital workstation allowing for easy experimentation with the properties and behaviors of the circuit and of the sounds. Rich and novel musical sounds have been obtained. Also, the audification of the local properties of the parameter space allowed an easy determination of very complex structures which would not be simple to determine by other ways.

In another setup[65], the time-delayed Chua's circuit was found to be able to model the basic behavior of an interesting class of musical instruments, namely those of clarinet, consisting of a massless reed coupled to a linear system. Other instruments are expected to be obtained from Chua's circuit, e.g., brass, voice, flute, and strings. A possibility of creating completely new electronic instruments is considered in Ref. 64 where a method (convex sums of vector fields) of generating new attractors that inherit the properties of parent attractors is proposed.

7. Unified Framework for Synchronization and Control

Recently, there has been much interest in the dynamics of coupled chaotic circuits and systems. Of particular interest is synchronization and control phenomena of chaotic systems. As seen in Sec. 6.1., applications based on chaotic sychronization such as secure communication systems have emerged in recent years. In Ref. 67, a unified framework was given to unify many of these ideas on chaotic synchronization and control by relating them to asymptotic stability of related systems. We give the main ideas here with regards to Chua's oscillator (see Sec. 8.). Detailed statements and proofs can be found in Ref. 67.

The main theorem can be stated as follows:

Theorem 1 *Consider the system*

$$\dot{\mathbf{x}} = \mathbf{f}(\mathbf{x}, \mathbf{x}, \mathbf{y}, t) \tag{7}$$
$$\dot{\mathbf{y}} = \mathbf{f}(\mathbf{y}, \mathbf{x}, \mathbf{y}, t) \tag{8}$$

where $\mathbf{x}, \mathbf{y} \in R^n$, *and* $\mathbf{f}$ *is defined on* $R^n \times R^n \times R^n \times R$. *Suppose that for every* $\eta_1(t)$ *and* $\eta_2(t)$ *continuous functions into* R^n, *the system*

$$\dot{\mathbf{z}} = \mathbf{g}(\mathbf{z}, t) = \mathbf{f}(\mathbf{z}, \eta_1(t), \eta_2(t), t) \tag{9}$$

is uniform-asymptotically stable. Then $\|\mathbf{y}(t) - \mathbf{x}(t)\| \to 0$ *as* $t \to \infty$.

The main idea to achieve synchronization of two chaotic systems is to "extract" the parts of the system such that the rest of the system is uniform-asymptotically stable, and use the extracted parts as coupling. More precisely, we start off with the system $\dot{\mathbf{x}} = \mathbf{g}_1(\mathbf{x}, t)$ which is chaotic. Next we decompose the arguments in the vector field $\mathbf{g}_1$ into 4 components such that it can be written as $\mathbf{g}_1(\mathbf{x}, t) = \mathbf{f}(\mathbf{x}, \mathbf{x}, \mathbf{x}, t)$. The decomposition is chosen in such a way that the system $\dot{\mathbf{x}} = \mathbf{g}_2(\mathbf{x}, t) = \mathbf{f}(\mathbf{x}, \eta_1(t), \eta_2(t), t)$ is asymptotically stable for all $\eta_1(t)$ and $\eta_2(t)$. The second and third arguments to $\mathbf{f}$ is the "extracted" part. When we couple two identical systems as in Eqs. 7-8, the two systems will synchronize so that they are decoupled at the synchronized state.

For the class of systems which has vector fields with uniformly bounded Jacobians, strong enough linear coupling between two identical systems will cause them to be globally synchronized.

In Chua's oscillator for the standard parameters values, the only active element is the nonlinear resistor. Therefore taking that part out of the system, the resulting system will consist of linear passive elements and is uniform-asymptotically stable. Applying this to two coupled Chua's oscillators we obtain the following result where the coupling is separable:

Proposition 1 *Consider the system:*

$$\begin{aligned}
\dot{x} &= \alpha\left(y - x - f(x) + g_1(\tilde{x}) - g_1(x)\right) \\
\dot{y} &= x - y + z \\
\dot{z} &= -\beta y - \gamma z \\
\dot{\tilde{x}} &= \alpha\left(\tilde{y} - \tilde{x} - f(\tilde{x}) + g_2(x) - g_2(\tilde{x})\right)
\end{aligned}$$

$$\begin{aligned}\dot{\tilde{y}} &= \tilde{x} - \tilde{y} + \tilde{z} \\ \dot{\tilde{z}} &= -\beta\tilde{y} - \gamma\tilde{z}\end{aligned}$$

Suppose $\alpha, \beta, \gamma > 0$. *If the function* $f + g_1 + g_2$ *is strictly increasing, then*

$$\begin{pmatrix} \tilde{x} \\ \tilde{y} \\ \tilde{z} \end{pmatrix} \rightarrow \begin{pmatrix} x \\ y \\ z \end{pmatrix}$$

as $t \rightarrow \infty$.

In linear diffusive coupling, g_1 and g_2 are chosen to be multiplication by a positive constant, i.e., $g_i(x) = c_i x$. If $f(x)$ is the piecewise-linear function $f(x) = bx + \frac{1}{2}(a - b)[|x+1| - |x-1|]$, then $c_1 + c_2 > \max(-a, -b)$ will cause the coupled system to synchronize.

Certain schemes of controlling trajectories of chaotic systems to lie on unstable periodic cycles can also be cast into this framework, where we take $g_1 = 0$, and take the first system to be moving along the unstable periodic cycle. Because there is no coupling from the second system to the first system, the first system can be replaced by a memory device which recalls a particular (unstable) trajectory of the system[69].

In Ref. 67 it was also shown that synchronization between two Chua's oscillators in master-slave configuration is robust. Strictly speaking, when the conditions of Proposition 1 are satisfied with $g_1 = 0$, arbitrarily small changes in the component values C_1, C_2, R, L, R_0 (which result in arbitrarily small changes in α, β, γ) in one of the systems will result in arbitrarily small synchronization error. This result suggests that practical physical implementations where parameter mismatches are unavoidable will still synchronize approximately.

8. Global Unfolding of Chua's Circuit

Despite the rich dynamical phenomena exhibited by Chua's circuit (1), there remain bifurcation phenomena so far not observed from the circuit but previously reported from other three-dimensional systems. One of the reasons is that the equations (1) do not generate all possible three-dimensional, piecewise-linear, continuous, 3-region, odd-symmetric vector fields. However, by adding a linear resistor in series with the inductor (Fig. 15) we obtain Chua's oscillator[70] which corresponds to a global unfolding of Chua's circuit, encompassing *almost all* three-dimensional, piecewise-linear, continuous, 3-region, odd-symmetric systems by being topologically conjugate, for suitable parameter values, to any such vector field (except for a set of measure zero).

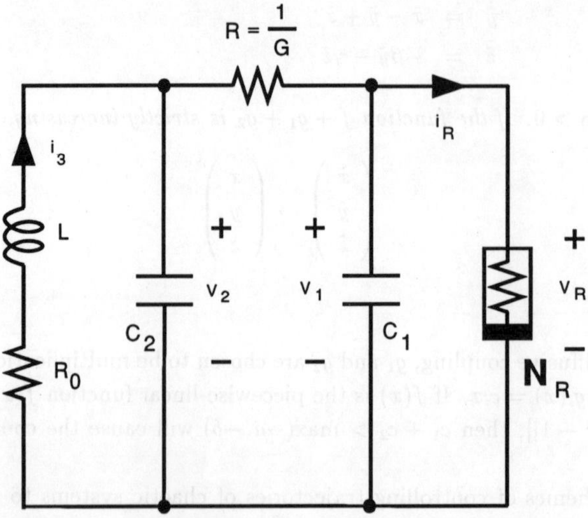

Fig. 15. Circuit diagram of Chua's oscillator.

The governing equations for the unfolded circuit, in a dimensionless form, are

$$\begin{aligned}\dot{x} &= \alpha\,(y - x - f(x)) \\ \dot{y} &= x - y + z \\ \dot{z} &= -\beta y - \gamma z \end{aligned} \tag{10}$$

Because of the generality of Chua's oscillator, many dynamical phenomena can be observed in addition to those in Chua's circuit. Beside period-doubling and torus-breakdown, intermittency route to chaos has also been observed[71]. Among other phenomena observed are period-adding[27,71], symmetry-breaking[27], and remerging[27] (antimonotonicity) of periodic attractors. A large number of distinct strange attractors have been obtained via simulations[71,72]. Moreover, Chua's oscillator has also been successfully used[71] to mimic the behavior of other three-dimensional dynamical systems, both piecewise-linear[73,74] and smooth[75,76].

Beside Chua's oscillator, individual vector fields which are continuous, odd-symmetric, 3-region, and piecewise-linear, can be generated by the members of Chua's circuit family[77]. Many members of Chua's circuit family have been synthesized and built[78–81]. Except for the circuit reported in Ref. 81, the dynamics of the above cited circuits, including the original Chua's circuit, are not sufficiently general in the sense that certain phenomena observed from one such member of Chua's circuit family can not be observed from another member, regardless of the choice of circuit parameters. From the circuit-theoretical point of view, it is desirable to synthesize the simplest circuit topology which is capable of reproducing the qualitative phenomena exhibited by every member of Chua's circuit family. Chua's oscillator was shown[70] to meet these requirements, so that its significance can be summarized as follows:

Chua's oscillator is structurally the simplest and dynamically the most complex member of the Chua's circuit family.

The significance of Chua's oscillator transcends beyond nonlinear circuit theory. Indeed, there are many publications involving systems which are not circuits but which are also described by Eq. 10., e.g., Refs. 75,76.

9. Concluding Remarks

Chua's circuit has proved to be an excellent paradigm for the generation of a multitude of different dynamical phenomena, and can thus obviate the need to consider many different models to simulate those phenomena. Chua's oscillator has unified the nonlinear dynamics of the entire 21-parameter family of piecewise-linear vector fields into a single system defined by Eq. 10, hence it is not necessary for beginners in nonlinear dynamics to study all those papers with diverse notations and jargons.

Even more significantly, arrays of Chua's circuits appear to be a suitable candidate for important applications ranging from image processing to the simulations of biological processes. The building of the monolithic IC chip of Chua's circuit is an important step toward building large arrays via VLSI technology, and will make it possible to reproduce, in real time, almost all reaction-diffusion situations, described in the literature with a relatively simple low-cost system.

Acknowledgments

This work was supported in part by the Office of Naval Research under grant N00014-89-J-1402, by the National Science Foundation under grant MIP 86-14000, and by the Josephine de Kármán Fellowship.
We would like to thank Alex Zheleznyak for his help in the preparation of the paper.

References

1. E.N. Lorenz, "Deterministic non-periodic flows," *J. Atmos. Sc.* **20** (1963),130–141.

2. T. Matsumoto, "A chaotic attractor from Chua's circuit," *IEEE Transactions on Circuits and System* **31**(12) (1984) 1055–1058.

3. G.-Q. Zhong, and F. Ayrom, "Experimental confirmation of chaos from Chua's circuit," *International Journal of Circuit Theory and Applications* **13**(1) (1985) 93–98.

4. L.O. Chua, "The genesis of Chua's circuit," *Archiv für Elektronik und Übertragungstechnik* **46** (1992) 250–257.

5. L.O. Chua, "Chua's circuit: ten years later", to appear in *Int. J. Cir. Theory and Appl.*

6. Kennedy, M. P., "Robust OP Amp realization of Chua's circuit," *Frequenz* **46**(3-4) (1992) 66-80.

7. J.A.K. Suykens and J. Vandewalle, "Generation of n-double scrolls(n=1,2,3,4,...)," *IEEE Transactions on Circuits and Systems–I: Fundamental Theory and Applications* **40**(11) (1993) 861–867.

8. E.J. Altman, "Bifurcation analysis of Chua's circuit with applications for low-level visual sensing," *Journal of Circuits, Systems, and Computers* **3**(1) (1993) 63–92.

9. A.I. Khibnik, D. Roose, and L.O. Chua, "On periodic orbits and homoclinic bifurcations in Chua's circuit with smooth nonlinearity," *International Journal of Bifurcation and Chaos* **3**(2) (1993) 363–384.

10. G.-Q. Zhong, "Implementation of Chua's circuit with a cubic nonlinearity," *Memorandum No. UCB/ERL M94/42*,(1994), UC Berkeley, USA.

11. J.M. Cruz and L.O. Chua, "A CMOS IC nonlinear resistor for Chua's circuit," *IEEE Transactions on Circuits and System* **39**(12) (1992) 985–995.

12. J.M. Cruz and L.O. Chua, "An IC chip of Chua's circuit," *IEEE Transactions on Circuits and System-II: Analog and Digital Signal Processing* **40**(10) (1993) 614–625.

13. A. Rodríguez-Vásquez and M. Delgado-Restituto, "CMOS design of chaotic oscillators using state variables: a monolithic Chua's circuit," *IEEE Transactions on Circuits and System-II: Analog and Digital Signal processing* **10**(40) (1992) 596–613.

14. L. Duchesne, "Using characteristic multiplier loci to predict bifurcation phenomena and chaos — a tutorial," *IEEE Transactions on Circuits and Systems — I: Fundamental Theory and Applications* **40**(12) (1993) 683–688.

15. V.C. Anishchenko, T.E. Vadivasova, D.E. Postnov, O.V. Sosnovtseva, L.O. Chua, and C.W. Wu, "Dynamics of the non-autonomous Chua's circuit," *unpublished.*

16. R. Lozi, and S. Ushiki, "Co-existing attractors in Chua's circuit: accurate analysis of bifurcation and attractors," *International Journal of Bifurcation and Chaos* **1**(4) (1991) 923–926.

17. C.W. Wu and L. Pivka, "From Chua's circuit to Chua's oscillator: A picture book of attractors," in *Nonlinear Dynamics of Electronic Systems*,W. Schwarz and A.C. Davies eds., (Dresden,Germany, 23–24 July, 1993),(World Scientific,Singapore).

18. F.T. Arecci, R. Badii, and A. Politi, "Low-frequency phenomena in dynamical systems with many attractors," *Phys. Rev.* **A29** (1984) 1006–1009.

19. V.S. Anishchenko,"Interaction of attractors. Intermittency of the chaos-chaos type," *Lett.J.Tech.Phys.* **10** (1984) 629–633.

20. V.S. Anishchenko, A.B. Neiman, and L.O. Chua, "Chaos-chaos intermittency and 1/f noise in Chua's circuit," *International Journal of Bifurcation and Chaos*, **4**(1) (1994) 99–107.

21. V.S. Anishchenko, M.A. Safonova, and L.O. Chua, "Stochastic resonance in the nonautonomous Chua's circuit," *Journal of Circuits, Systems and Computers*, **3**(2) (1993) 553–578.

22. K.S. Halle, L.O. Chua, V.S. Anishchenko, and M.A. Safonova, "Signal amplification via chaos: experimental evidence," *International Journal of Bifurcation and Chaos* **2**(4) (1992) 1011–1020.

23. I. Kan and J.A. Yorke, "Antimonotonicity — concurrent creation and annihilation of periodic orbits", *Bull.Am.Math.Soc.* **23**(2) (1990) 469–476.

24. P.S. Dawson, C. Grebogi, J.A. Kan, and H. Kocak, "Antimonotonicity — inevitable reversals of period-doubling cascades," *Phys. Lett.* **A162**(3) (1992) 249–254.

25. M.J. Ogorzałek, "Remerging bifurcation sequences in an autonomous electronic circuit," *Proc. 1989 European Conference on Circuit Theory and Design (Brighton)*,142-146.

26. Lj. Kočarev, K.S. Halle, K. Eckert, and L.O. Chua, "Experimental observation of antimonotonicity in Chua's circuit," *International Journal of Bifurcation and Chaos* **3**(4) (1993) 1051–1055.

27. L. Pivka and V. Špány, "Boundary surfaces and basin bifurcations in Chua's circuit," *Journal of Circuits, Systems and Computers* **3**(2) (1993) 441–470.

28. L.O Chua and I. Tichonicky, "1-D map for the double scroll family," *IEEE Transactions on Circuits and Systems* **38**(3) (1991) 233–243.

29. M. Genot, "Applications of 1-D map from Chua's circuit: A Pictorial Guide," *Journal of Circuits, Systems and Computers* **3**(2) (1993) 375–409.

30. M.J. Feigenbaum, "Quantitative universality for a class of nonlinear transformations," *J. Stat. Phys.* **19** (1978) 25–52.

31. M.J. Feigenbaum, "The universal metric properties of nonlinear transformations," *J. Stat. Phys.* **21** (1979) 669–706.

32. A.P. Kuznetsov, S.P. Kuznetsov, I.R. Sataev, and L. O. Chua, "Two-parameter study of transition to chaos in Chua's circuit: renormalization group, universality and scaling," *International Journal of Bifurcation and Chaos* **3**(4) (1993) 943–962.

33. M. Shell, S. Fraser, and R. Kapral, "Subharmonic bifurcations in the sine map: An infinite hierarchy of cusp bistabilities," *Phys. Rev.* **A28** (1983) 373–379.

34. B.B.Mandelbrot, *Fractals: Form, Chance, and Dimension* (San Francisco,CA: Freeman,1977).

35. K. Kaneko,"Transition from torus to chaos accompanied by frequency lockings with symmetry breaking," *Prog. Theor. Phys.* **69**(5) (1983) 1427–1442.

36. V. Pérez-Muñuzuri, V. Pérez-Villar, and L.O. Chua, "Traveling wave front and its failure in a one-dimensional array of Chua's circuits," *Journal of Circuits, Systems, and Computers* **3**(1) (1993) 211-215.

37. J.P. Keener, "Propagation and its failure in coupled systems of discrete excitable cells," *SIAM J. Appl. Math.* **47** (1987) 556–572.

38. J.P. Keener, "The effects of discrete gap junction coupling on propagation in myocardium," *J. Theor. Biol.* **148** (1991) 49–82.

39. J.P. Laplante and T. Erneux, "Propagation failure and multiple steady states in an array of diffusion coupled flow reactors," *Physica* **A188** (1992) 89–98.

40. J.P. Laplante and T. Erneux, "Propagation failure in arrays of coupled bistable chemical reactors," *J. Chem. Phys.* **96** (1992) 4931–4934.

41. M. Dolnik and M. Marek, "Extinction of oscillations in forced and coupled reaction cells," *J. Phys. Chem.* **92** (1988) 2452–2455.

42. V. Pérez-Muñuzuri, M. Gómez-Gesteira, V. Pérez-Villar, and L.O. Chua, "Traveling wave propagation in a one-dimensional fluctuating medium," *International Journal of Bifurcation and Chaos* **3**(1) (1993) 211–215.

43. V. Pérez-Muñuzuri, V. Pérez-Villar, and L.O. Chua, "Autowaves for image processing on a two-dimensional CNN array of excitable nonlinear circuits: flat and wrinkled labyrinths," *IEEE Transactions on Circuits and Systems*, **40**(3) (1993) 174–181.

44. M.A. Allesie, F.I.M. Bonke, and T.Y.G. Schopman, "Circus movement in rabbit atrial muscle as a mechanism in tachycardia," *Circulation Res.* **33** (1973) 54–62.

45. J. Bures, V.I. Koroleva, and N.A. Gorelova, "Leao's spreading depression, an example of diffusion-mediated propagation of excitation in the central nervous system," in *Autowaves and Structures Far from Equilibrium*, ed. V.I. Krinsky (Springer-Verlag, 1984),180–183.

46. S.C. Müller, T. Plesser, and B. Hess, "Two-dimensional spectrophotometry of spiral wave propagation in the Belousov-Zhabotinskii reaction I. Experiments and digital repersentation. II. Geometric and kinematic parameters," *Physica* **D24** (1987) 71–96.

47. A. Pérez-Muñuzuri, V. Pérez-Muñuzuri, and V. Pérez-Villar, "Spiral waves on a two-dimensional array of nonlinear circuits," *IEEE Transactions on Circuits and Systems–I: Fundamental Theory and Applications* **40**(11) (1993) 872-877.

48. V. Pérez-Muñuzuri, M. Gómez-Gesteira, A. Pérez-Muñuzuri, L.O. Chua, and V. Pérez-Villar, " Sidewall forcing of hexagonal Turing patterns: Rhombic patterns," submitted to *Physica* **D**.

49. N.F. Otani and R. Hwa, "Topological considerations in the generation of scroll waves in excitable and cyclical media," submitted to *Physica* **D**.

50. R.R. Klevecz, J.Bolen, and O. Durán, "Self-organization in biological tissues: Analysis of asynchronous and synchronous periodicity, turbulence and synchronous chaos emergent in coupled chaotic arrays," *International Journal of Bifurcation and Chaos* **2**(4) (1992) 941–953.

51. A.L. Zheleznyak, *personal communication.*

52. A.L. Zheleznyak and L.O. Chua, "Coexistence of low- and high-dimensional spatio-temporal chaos in a chain of dissipatively coupled Chua's circuits," to appear in *International Journal of Bifurcation and Chaos.*

53. L.M. Pecora and T.L. Carroll, "Synchronization in chaotic systems," *Phys. Rev. Lett.* **64** (1990) 821–824.

54. Lj. Kocarev, A. Shang, and L.O. Chua, "Transitions in dynamical regimes by driving: A unified method of control and synchronization of chaos," *International Journal of Bifurcation and Chaos* **3**(2) (1993) 479–483.

55. L.M. Pecora and T.L. Carroll, "Driving systems with chaotic signals," *Phys. Rev.* **A44** (1991) 2374–2383.

56. A.V. Oppenheim, G.W. Wornell, S.H. Isabelle, and K.M. Cuomo, "Signal processing in the context of chaotic signals," *Proc. IEEE ICASSP* **4** (San Francisco, CA),(1992), pp. IV–117–IV–120.

57. U. Parlitz, L.O. Chua, Lj. Kocarev, K.S. Halle, and A. Shang, "Transmission of digital signals by chaotic synchronization ," *International Journal of Bifurcation and Chaos* **2**(4) (1992) 973–977.

58. H. Dedieu, M.P. Kennedy, and M. Hasler, "Chaos Shift Keying: Modulation and demodulation of a chaotic carrier using self-synchronizing Chua's circuits," *IEEE Transactions on Circuits and Systems–II: Analog and Digital Signal Processing* **40**(10) (1993) 634–642.

59. Lj. Kocarev, K.S. Halle, K. Eckert, L.O. Chua, and U. Parlitz, "Experimental demonstration of secure communications via chaotic synchronization," *International Journal of Bifurcation and Chaos* **2**(3) (1992) 709–713.

60. K.S. Halle, C.W. Wu, M. Itoh, and L.O. Chua, "Spread spectrum communication through modulation of chaos," *International Journal of Bifurcation and Chaos*, **3**(2) (1993) 469–477.

61. C.W. Wu and L.O. Chua, "A simple way to synchronize chaotic systems with applications to secure communication systems," *International Journal of Bifurcation and Chaos*, **3**(6) (1993) 1619–1627.

62. E.J. Altman, "Normal form analysis of Chua's circuit with applications for trajectory recognition," *"IEEE Transactions on Circuits and Systems–II: Analog and Digital Signal Processing* **40**(10) (1993) 675–682.

63. B. Baird, and M. Hirsch, and F. Eeckman, "A neural network associative memory for handwritten character recognition using multiple Chua attractors," *"IEEE Transactions on Circuits and Systems–II: Analog and Digital Signal Processing* **40**(10) (1993) 667–674.

64. G. Mayer-Kress, I. Choi, N. Weber, R. Bargar, and A. Hubler, "Musical signals from Chua's circuit," *IEEE Transactions on Circuits and Systems–II: Analog and Digital Signal Processing* **40**(10) (1993) 688–695.

65. X. Rodet, "Models of musical intruments from Chua's circuit with time delay," *"IEEE Transactions on Circuits and Systems–II: Analog and Digital Signal Processing* **40**(10) (1993) 696–701.

66. G.-Q. Zhong, K.S. Halle, R. Bargar, "Circuits for adjusting the parameters of Chua's circuit by voltage tuning," *unpublished.*

67. C.W. Wu and L.O. Chua, "A unified framework for synchronization and control of dynamical systems," *International Journal of Bifurcation and Chaos* **4**(4) (1994).

68. G. Chen, "Controlling Chua's global unfolding family," *IEEE Transactions on Circuits and Systems–I: Fundamental Theory and Applications* **40**(11) (1993) 829–832.

69. K. Pyragas, "Predictable chaos in slightly perturbed unpredictable chaotic systems," *Phys. Lett.* **A181** (1993) 203–210.

70. L.O. Chua,"Global unfolding of Chua's circuit," *IEICE Transactions on Fundamentals of Electronics, Communications and Computer Sciences*, **E76-A**(11) (1993) 704–734.

71. L.O. Chua, C.W. Wu, A.S. Huang, and G.-Q. Zhong, "A universal circuit for studying and generating Chaos, Part I,II," *IEEE Transactions on Circuits and Systems–I: Fundamental Theory and Applications*, **40**(10) (1993) 732–761.

72. R.N Madan, ed., *Chua's circuit: A paradigm for chaos.* (World Scientific, Singapore, 1993).

73. R.W. Brockett, "On conditions leading to chaos in feedback systems," in *Proc. IEEE Conf. Decision Contr.* (1982) 932–936.

74. C.T. Sparrow, "Chaos in three-dimensional single loop feedback system with a piecewise-linear feedback function," *J. Math. Anal. Applications* **83** (1981) 275–291.

75. A. Arnéodo, P. Coullet, and C. Tresser, "Possible new strange attractors with spiral structure," *Commun. Math. Phys.* **79** (1981) 573–579.

76. A.S. Dmitriev and V.Y. Kislov, "Stochastic oscillations in a self-excited oscillator with a first-order inertial delay," *Radiotekhnika i elektronika* (29) (1984) p. 2389 (in Russian).

77. S. Wu, "Chua's circuit family," *Proc. of the IEEE* **75**(8) (1987) 1022–1032.

78. P. Bartissol and L.O. Chua, "The double hook," *IEEE Transactions on Circuits and Systems* **35**(12) (1988) 1512–1522.

79. A.S. Huang, "A study of the chaotic phenomena in Chua's circuit," in *Proceedings of the ISCAS*, (Helsinki, Finland), (1988) 273–276.

80. Y. Nishio, N. Inaba, and S. Mori, "Chaotic phenomena in an autonomous circuit with nonlinear inductor," in *Proceedings of the ISCAS*,(1990).

81. L.O. Chua and G.-N. Lin, " Canonical realization of Chua's circuit family," *IEEE Transactions on Circuits and Systems* **37**(7) (1990) 885–902.

CONTROLLING CHAOS IN ELECTRONIC CIRCUITS

G. A. JOHNSON and E. R. HUNT
Department of Physics and Astronomy
Condensed Matter and Surface Sciences Program
Ohio University, Athens, OH 45701

Abstract

The occasional proportional feedback (OPF) control technique, originally designed to control chaos in the diode resonator system, has been found to be applicable to a wide variety of systems. It has stabilized fixed points and unstable periodic orbits, including high-period orbits, in higher-dimensional and autonomous systems. We review the method and its application to three electronic circuits: systems that undergo the period-doubling route to chaos, both driven and autonomous, and the quasiperiodic route. The circuits are the diode resonator, a self-driven oscillator referred to as Chua's circuit, and a pair of coupled diode resonators.

1. Introduction

1.1. Controlling Chaos

The control of chaos has become a major subfield in the study of nonlinear dynamics. Most of the current work in the field is based on the pioneering work of Ott, Grebogi, and Yorke[1] (OGY) in 1990, and the experimental work of Ditto, Rauseo, and Spano[2]. At the time of this writing, more than 250 papers have been published on the subject, as tabulated by Chen[3], in journals of physics, math, chemistry, engineering, and others. In addition, recent review articles on the subject have appeared in *Scientific American*[4], *Nature*[5], and *IEEE Spectrum*[6]. This chapter is intended to provide a review of the occasional proportional feedback (OPF) control method[7] and its application to different types of electronic circuits which exhibit chaos. Controlling electronic circuits is an important part of the field because the universal properties of chaos can allow one to consider simple circuits as representations of more experimentally complicated systems, thus making electronics a testing ground for chaos control. Evidence for this is the application of the OPF method by Roy *et al.*[8] to control a multimode laser system, and by Peng *et al.*[9] to control the Belousov-Zhabotinsky chemical reaction.

Because of its random-like behavior, the presence of chaos in physical systems is not generally advantageous, hence applications of chaos are relatively few. However, instead of avoiding chaos altogether, a recent surge of research demonstrates that chaotic systems can be easily and efficiently forced to behave periodically. In fact, the ease with which a chaotic system can be stabilized in and switched between a number of states, can provide a high degree of flexibility and a reason to intentionally build chaos into some systems. To achieve the goal of stabilization, a number of methods have emerged, the most widely employed stemming from the technique described by OGY, in which a feedback signal is generated to perturb a system parameter in a manner which stabilizes an unstable periodic orbit embedded in a chaotic attractor. The main challenge of this work is to preserve the integrity of the system by making only tiny parameter adjustments so that the only change in the system is its periodicity. To do this, the OGY scheme employs a feedback signal which may be expressed as $\mu_n = \bar{\sigma} \cdot (\bar{\xi}_n - \bar{\xi}_F)$, where $\bar{\xi}_n$ typically represents a two-dimensional state vector of the system at the nth piercing of the Poincaré plane, and $\bar{\xi}_F$ the fixed point location of a periodic orbit. The vector $\bar{\sigma}$ contains geometrical information in the neighborhood of the fixed point which can be extracted from time-series data analysis. This feedback signal is only applied when the system comes arbitrarily close to the fixed point, thus ensuring only small perturbations.

Several variations of the OGY method are in existence. The OPF error signal has the form $\mu_n = \alpha(X_n - X_S)$ where X is a single variable and X_S the set point. Under ideal conditions the set point, a moveable reference point, coincides with the fixed point X_F. Similar methods include that of Bielawski *et al.*[10] where X_{n-1} replaces X_S, and the recursive method of Rollins *et al.*[11] which includes the OPF term in addition to a second proportional to μ_{n-1}. Continuous-time feedback methods have been applied to electronic circuits by some including Pyragas[12], Pyragas and Tamasevicius[13] as well as Chen and Dong[14], otherwise all of these techniques use discrete-time variables and are well understood in terms of one dimensional maps. For the systems discussed in this chapter, we discretize the continuous signals by taking voltage and current peaks to be the chaotic variables.

In addition to stabilizing periodic orbits in chaotic attractors, methods have been extended to stabilizing steady-state fixed points that exist in many autonomous systems. Two methods have been given recent attention: derivative feedback[15] and modified OPF[16]. The derivative method is a continuous strategy where the error signal is simply $\mu(t) = \alpha\dot{X}$, which approaches zero as the goal, $X(t) = X_{Fixed}$, is approached. A modification of the OPF system can be implemented to achieve the same goal. A possible connection between the two methods will be discussed. These methods have been successful in controlling steady-state fixed points in laser[15,16], chemical[17], and electronic[18] systems.

This chapter reviews the control of the diode resonator and the extension of the OPF method to control a simple autonomous circuit as well as a system comprised of coupled diode resonators. In all circuits periodic orbits are stabilized, including high-period orbits, and Hopf bifurcations are suppressed with minimal correction signals in the latter two examples.

1.2. The OPF Method

The OPF control strategy is best described in terms of the return map $X_{n+1} vs X_n$ where X is a measurable, chaotic variable. Ideally, a parameter is available such that changes in the parameter result in the return map shifting along the diagonal. A simple relationship can be found between a small parameter change Δp and the horizontal displacement, δX, of the map. For small parameter changes, the relationship is approximately linear such that

$$\delta X = \kappa \Delta p \tag{1}$$

where κ is a constant of the system. Provided that the system responds quickly to parameter changes, the control scheme can be implemented to manipulate X_{n+1} using the measured value of X_n. Assuming at least a nearly one-dimensional map, we can express X_{n+1} in terms of a mapping function F:

$$X_{n+1} = F(X_n, p). \tag{2}$$

From this simple expression the strategy becomes clear: measure X_n, change p to get the desired X_{n+1}, and repeat as necessary.

To stabilize the period-1 fixed point (the intersection of the map with the diagonal), ΔX is measured with respect to the fixed point X_F, and the perturbation:

$$\Delta p = \alpha \Delta X = \alpha(X_n - X_F) \tag{3}$$

is applied for one iteration. Since Eq. 1 is valid only for small Δp, no action is taken if the magnitude of ΔX is greater than some threshold value referred to as the control window.

To illustrate, an example is given in Fig. 1. The return map for the diode resonator is sketched at two different parameter settings. The maps are generated by plotting peaks of a chaotic current which will be discussed later in this chapter. The quantity ΔI replaces ΔX and is measured along the horizontal axis with respect to the period-1 fixed point I_F. Suppose we measure the current at I_n on the lower map and wish to stabilize the fixed point on the same map. We immediately apply a change to the parameter according to Eq. 3. The system now resides on the upper attractor, and I_{n+1} coincides with the desired fixed point. The parameter is returned to its original setting just as the current reaches I_F. In this example, the fixed point is reached in one step, which is accomplished when the multiplier, α, is exactly κ^{-1} of Eq. 1. Note that no system equations appear in the OPF scheme. All necessary information is contained in κ, which is obtained experimentally.

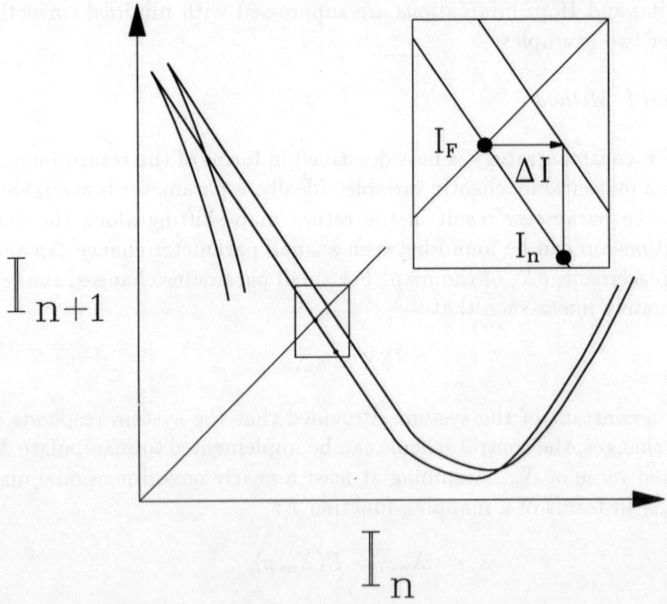

Figure 1: To stabilize the period-1 fixed point, I_F, we measure I_n to get ΔI so that the control parameter can be adjusted to move the system to the upper map where the same I_n maps into I_F of the lower map.

When controlling a system on an orbit other than period-1 in the OPF scheme, a guess is made at the location of the fixed point (referred to as the set-point) and a small perturbation is made any time the system trajectory visits the control window of the set-point. The perturbations generally modify the control parameter just a few percent of the original setting. When the set-point coincides with the fixed point of the stabilized orbit, the corrections approach zero. Experimentally, the correction signal can be reduced to a level that is indistinguishable from the typical noise on an oscilloscope trace. For high-period orbits, each perturbation is applied for one *drive* cycle, and generally more than one perturbation is applied as the longer trajectories may pierce the control window multiple times. Again, in this and similar control schemes, a detailed knowledge of the location of the fixed points or system equations is nonessential. Upon implementing control, if no orbit becomes stabilized, the feedback gain can be adjusted or inverted; the window width can be modified; or the set-point can be shifted. In effect, we let the system locate the unstable fixed points of the high-period orbits through a trial-and-error search.

1.3. Circuit Implementation

The circuit which generates the error signal in the OPF scheme is shown in Fig. 2. The chaotic signal is input to the circuit as a voltage signal, V_{in}. The initial input signal is referenced to the set-point voltage controlled by the variable resistance at the first op amp. This *difference signal*, $V_{in} - V_{set}$, is sent to the window comparator (two LM319 comparators) to determine when the chaotic signal visits the control window of the set-point. The synchronizing signal input is delayed and shaped so that the output of the lower left one-shot (74123) is a narrow square pulse centered about the peaks of the input signal, so that the signal essentially acts as a peak indicator. If the difference signal is in the window and the peak indicator is high, the second one-shot is triggered to send the pictured pulses to the sample/hold and the gate. The difference signal (at its peak) is held by the sample/hold on the negative transition of the pulse shown. The held voltage is then passed through the gate when the gate pulse shown is high, otherwise the gate output is ground. The final stage is the gain control supplied by the op amp inverter with variable gain. The signal from the inverter is sent to the modulation input of the control parameter.

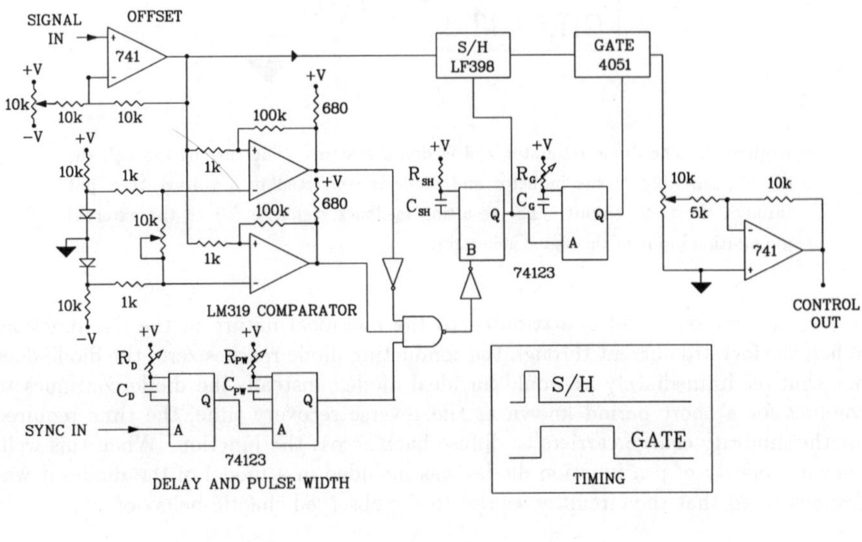

Figure 2: The OPF control circuit. See text for a description of its operation.

2. The Diode Resonator

2.1. The Circuit

The first system to be controlled using the OPF circuit was the diode resonator[19]. The system consists of a p-n junction diode in series with an inductor (10-100mH), as in Fig. 3, and is sinusoidally driven between 50 and 100kHz. The current through the diode undergoes the period-doubling route to chaos as the drive voltage is increased. A current to voltage converter is employed to view the current, and the peak forward current is taken to be a discrete chaotic variable. The mechanism giving rise to chaos

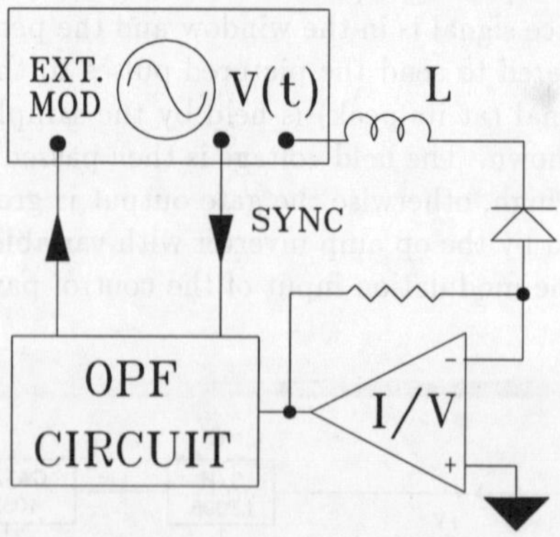

Figure 3: The diode resonator and feedback systems. The current through the circuit consisting of the inductor and diode is converted to a voltage then put through the OPF circuit. The resulting feedback signal is fed to the external modulation input of the signal generator.

is well understood[20] and is attributed to the non-ideal nature of the p-n junction. When the forward current through the conducting diode reaches zero, the diode does not shut off immediately as would an ideal diode. Instead, the diode continues to conduct for a short period known as the reverse recovery time, the time required for the minority charge carriers to diffuse back across the junction. When this well-known property of p-n junction diodes was included in a model of the diode, it was demonstrated that the circuit gives rise to the observed chaotic behavior.

2.2. Controlling the Diode Resonator

To control periodic orbits in the diode resonator system, as in Fig. 4, we make use of the system response to changes in drive voltage, and take $I_{n+1} = F(I_n, V)$. The two maps of Fig. 1 represent the system at two slightly different drive amplitudes,

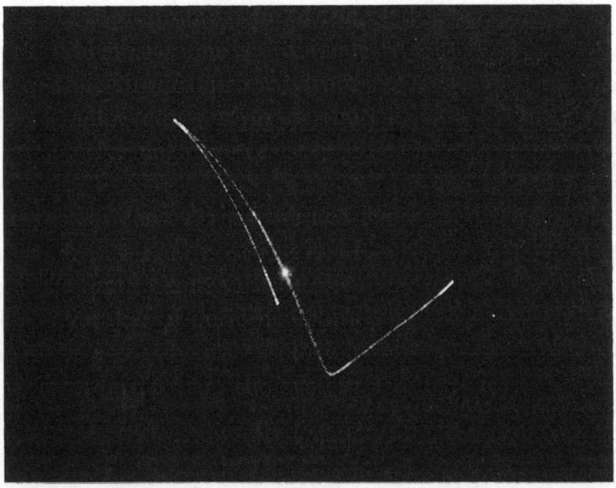

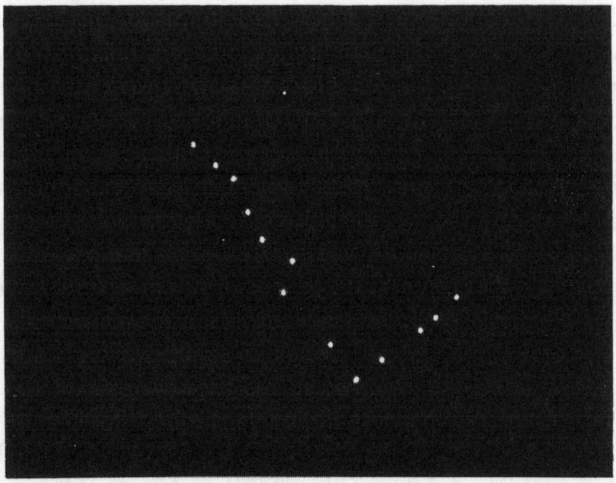

Figure 4: The stabilized period-1, bright spot, of the diode resonator superimposed on the uncontrolled attractor (upper) and a typical high-period orbit (lower) stabilized with two or more corrections.

demonstrating the desired movement along the diagonal as discussed in *1.2*. The feedback signal is applied to the external amplitude modulation input of the signal generator as in Fig. 3. In Fig. 1 it is shown how to reach the fixed-point in a single iteration. The feedback system is somewhat forgiving, however, in that if the feedback gain α is near to but not precisely at the right value, the system will not reach the period-1 fixed point in one step, yet the point becomes attracting so that each iteration brings the system closer than the last.

The somewhat unexpected result of the diode resonator experiment was the stabilization of high-period orbits of lengths up to 23 drive cycles. The control of these orbits was achieved by essentially a trial-and-error search. The control window position and width, as well as feedback gain, were all varied until orbits became stabilized. The controlled period-1 and a high-period orbit are shown in Fig. 4.

3 Controlling an Autonomous System

3.1. Chua's Circuit

Chaotic behavior in what is popularly referred to as Chua's circuit has been the subject of extensive research efforts in nearly all areas of chaos[21]. For more details of the circuit, see the chapter by Leon Chua in this book. The circuit contains only one nonlinear device, a three-segment, piecewise-linear, negative resistance. This negative resistance, built with active components, supplies power to the circuit and drives the oscillations through a linear, dissipative resistance of nearly equal magnitude. The chaotic oscillations arise due to the unstable power balance between the positive and negative resistances and are sustained by an LC tank circuit. Chua's circuit presents a rich array of dynamical behavior, and the voltages across the inductor and negative resistance are easily accessible variables. From a d.c. equilibrium, the system undergoes a Hopf bifurcation and a period-doubling sequence to one of two Rössler-type attractors, and lastly to the Double-Scroll strange attractor as either resistance is varied.

3.2. Modulation

Applying the OPF method to an autonomous system such as Chua's circuit[22] may present problems finding an accessible control parameter. For a driven circuit, drive voltage amplitude is an easily manipulated parameter, as in the case of the diode resonator. However in Chua's circuit, the parameters are the values of the components of the circuit. Therefore, in order to perturb the system, we take the negative resistance to be our control parameter as it is most accessible and does not significantly alter the natural frequency of the system. Most importantly, by changing the negative resistance, the return map shifts in the manner necessary to employ the OPF method.

The negative resistance between V_R and ground is supplied by the op-amp circuit of Fig. 5, with a magnitude equal to the resistance from the negative terminal to

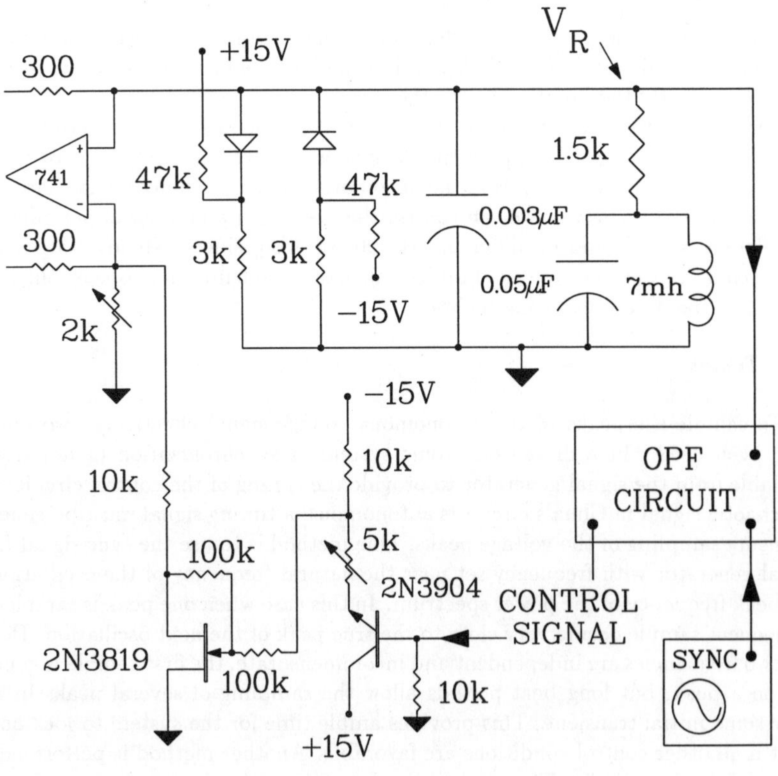

Figure 5: **The piecewise-linear negative resistance of Chua's circuit is supplied by the op-amp, diodes, 300 Ω, 3 kΩ and 47 kΩ resistors, and the 2 kΩ variable resistor. We modify this by placing a voltage-variable resistance (10 kΩ + drain-source resistance) in parallel with the 2k pot. The emitter follower on the front end biases the FET circuit. In this configuration the net negative resistance becomes a function of the correction signal voltage. The sync signal for the control circuit can be provided by an independent oscillator as shown.**

ground. To modulate the resistance we set up R_{Neg} as a parallel combination of the 2 kΩ pot resistor and a voltage-variable resistance. This variable resistance is provided by a field-effect transistor (FET), which has a variable resistance from source to drain depending on the gate voltage. Modulation of the system is achieved by applying the OPF error signal to the gate of the FET, changing the effective negative resistance and consequently the input power to the circuit.

For the chaotic variable we use the peak voltage, nearest to zero, across the negative resistance, V_{Rpeak}. Dropping the peak subscript, the error signal is of the form $\mu = \alpha(V_R - V_{Set})$ provided V_R is in the neighborhood of V_{Set}. Since Chua's circuit displays a variety of behavior we can control the system on a number of periodic orbits in different regimes and on different attractors. Also, since a steady-state solution to the equations exists, we can stabilize this fixed point in order to keep the system oscillation-free in all operating regimes.

3.3. *Timing*

To sample the peaks of the autonomous voltage signal effectively, two methods have been used. In a driven electronic system, a synchronization (sync) signal is available from the signal generator to provide the timing of the control circuit. Since the chaotic signal in Chua's circuit is autonomous, a timing signal must be generated to ensure sampling of the voltage peaks. One method is to use the sync signal from a signal generator with frequency set near the natural frequency of the oscillations, or the peak frequency of the power spectrum. In this case when one peak is sampled, the subsequent sample occurs very close to the true peak of the next oscillation. Because the two frequencies are independent and incommensurate, the first sample is generally not on a peak, but long beat periods allow the sampling of several peaks in a row after some initial transient. This provides ample time for the system to lock onto an orbit if all other control conditions are favorable. Another method is performed with a peak detecting circuit. The method involves taking the time derivative of V_R, and using a comparator to generate a pulse whenever the derivative crosses zero in one direction. This pulse signal is then used as the timing signal to dictate the sampling so that the control circuit samples only the peaks of V_R.

3.4. *Single-Basin Orbits*

When control is to be attempted in a single basin, the system is set up so that the oscillations take place in the chaotic regime yet confined to one of the two Rössler-type attractors. Control is switched on, and the control variables are adjusted until the system locks in on a stabilized orbit. Periods 1, 2, and 4, remnants of the period doubling cascade, are easily stabilized, and the control signal is very small. In Fig. 6 we show the return map in this regime generated by plotting the subsequent peaks of the voltage V_R. The upper photo is the unperturbed system and the lower

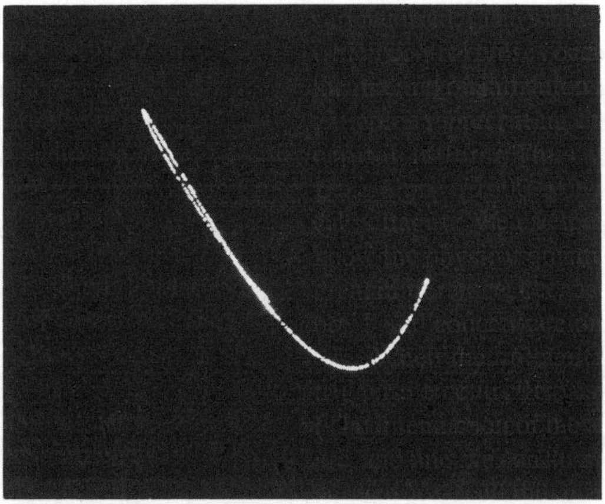

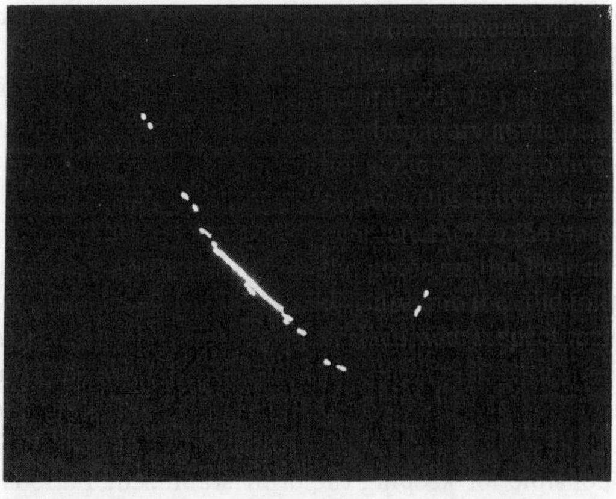

Figure 6: The unperturbed (upper) and perturbed (lower) return maps for Chua's circuit. The perturbations reduce the magnitude of the slope about the period-1 fixed point such that it nearly becomes attracting. The return maps are generated by plotting $V_{R_{n+1}}$ *vs* V_{R_n}.

demonstrates the result of applying a feedback signal not quite strong enough to stabilize the period-1 fixed point. The control window can be seen as the bright portion of the map centered about the fixed point. The slope of the map inside the window is seen to be flattened as compared to the unmodified map, so that when the trajectory falls near the fixed point, the system moves away much slower than usual, pops out of the control window for a few cycles, then repeats. If the slope is reduced such that the magnitude is less than 1, the fixed point becomes stable.

All orbits up to and including period-8 have been stabilized with a single correction per complete orbit. Note that the correction pulse itself lasts slightly less than one period of the natural frequency, not the entire orbit length as is the case with normal OGY feedback. The divergent nature of the chaotic system limits the length of the single-correction orbits. Nearly identical trajectories will have a separation exceeding the window width after a number of oscillations depending on the Lyapunov exponents of the system.

When a signal generator is implemented to provide a timing signal, the frequency of oscillation in the circuit is slightly shifted as it becomes synchronous with the signal generator upon stabilization of an orbit. Single-correction orbits can be initially controlled only near the frequency of the peak in the power spectrum, around 5 kHz in our circuit. Once locked in, however, the control frequency can be varied over a range without losing control. This range depends on the length of the orbit. The bandwidth of this control is nearly 200 Hz for period-1, whereas period-4, for example, has a control bandwidth of only about 15 Hz.

3.5. High-Period Orbits

When controlling high-period orbits, multiple corrections are usually made. Figure 7 shows a period-11 orbit with two corrections. Many of these high-period orbits are not likely to exist as unstable orbits, rather they are combinations of a number of orbits that may become mixed as a result of the perturbations. The correction pulses slightly deform the attractor to create new orbit paths and a tiny ripple on top of the square pulses is responsible for keeping the orbit stabilized. Periods 9 through 19 in Chua's circuit have been stabilized this way.

3.6. Double-Scroll Orbits

The Double-Scroll attractor is formed when the two Rössler-type attractors intersect so that the system trajectories bounce between these two regions. To stabilize orbits in this operating regime, generally corrections need to be applied on both sides. To achieve this, two independent controllers are used, so that a control window is placed in each of the two regions. By summing the outputs of the two controllers, we are able to apply corrections during visits to both attractors during one stabilized orbit. Although less common, some orbits become stabilized with corrections on one

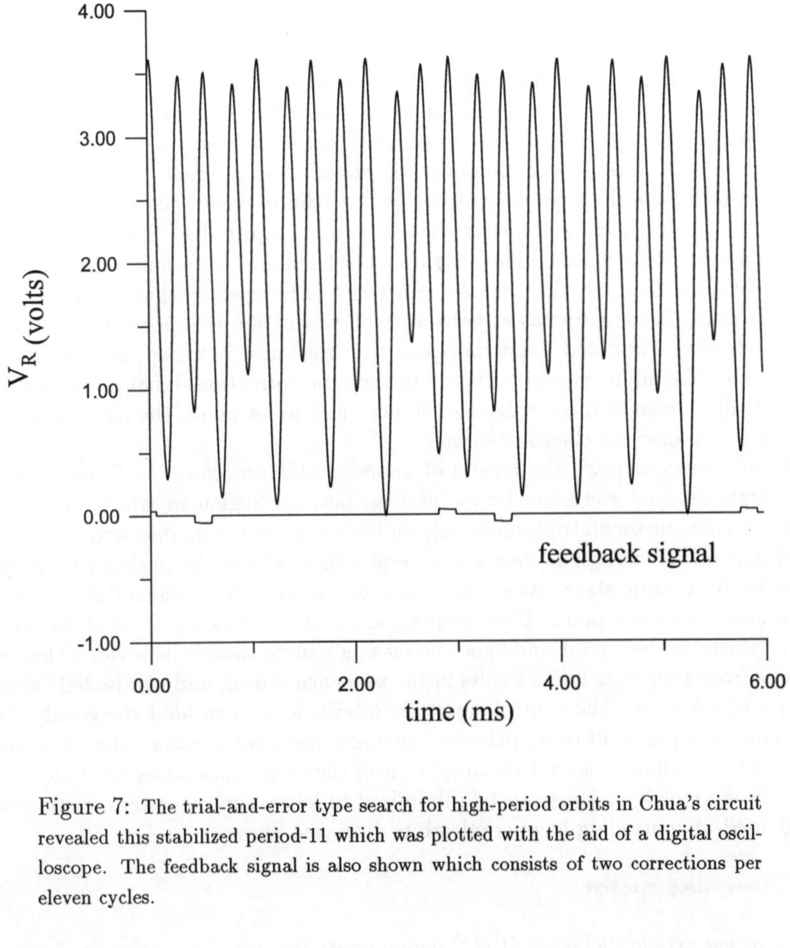

Figure 7: The trial-and-error type search for high-period orbits in Chua's circuit revealed this stabilized period-11 which was plotted with the aid of a digital oscilloscope. The feedback signal is also shown which consists of two corrections per eleven cycles.

side only. In this case the control conditions are more critical as the orbits can be quite long.

3.7. Steady-State Control

3.7.1. OPF Method

To control the non-oscillating state of a laser system, Gills *et al.*[16] made use of the OPF technique combined with the continuation method described by Carroll *et al.*[23]. We apply the same to Chua's circuit in the following manner. The negative resistance is set so that the steady state is stable, i.e. there is only a d.c. voltage across the negative resistance. A sync frequency is provided by a signal generator which is set somewhat above the frequency of the period-1 limit cycle. The control circuit is switched on while V_R is steady, then the set-point is adjusted so that the correction signal is approximately zero, indicating that the fixed and set points coincide. Once this condition is reached, the negative resistance is reduced so that the fixed point would normally become unstable, yet the Hopf bifurcation is suppressed so that V_R remains steady. The fixed point moves as the resistance is varied, so that the set point needs to be adjusted appropriately to keep the correction signal minimized. By repeated adjustment of the negative resistance and the set-point, the fixed point may be tracked well into the chaotic regime.

Figure 8 demonstrates the results of applying this procedure to Chua's circuit. The unstable steady state can be maintained into the region in which the system would otherwise be oscillating chaotically on both sides of the double-scroll attractor. The uncontrolled attractor is shown at several stages between its period-1 limit cycle and its highly chaotic state. At each of these operating values, the stabilized steady state is also shown as a point. Close inspection shows that the d.c. level of the steady state increases as the circuit undergoes increasingly more chaotic behavior. This level increases from 1.40 volts to 2.24 volts in the sequence shown, and can be held steady in excess of 3.5 volts. The control system is robust, as we can hold the steady state for an indefinite period of time without changing control parameters. The remarkable feature of this technique is that the amplitude of the corrections necessary to keep the system in the steady state is can be minimized to levels very near zero, comparable to that used in controlling low-period orbits discussed above.

3.7.2. Derivative method

In a recent article Bielawski *et al.*[15] demonstrate the use of a continuous feedback signal to stabilize the steady-state fixed point. The continuous signal is proportional to the time derivative of the chaotic signal. The method was employed in stabilizing the steady state of a optical-fiber laser where the light intensity is the chaotic variable. To explore this method, we apply the same strategy to Chua's circuit. We use a simple analog differentiator circuit to generate the continuous feedback signal proportional

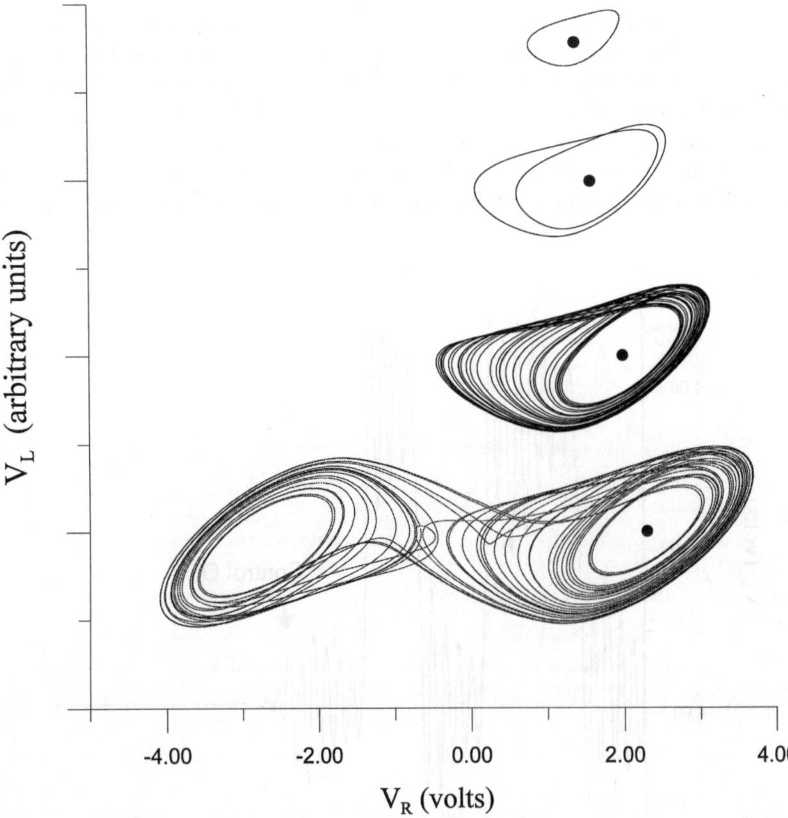

Figure 8: The steady state in Chua's circuit moves as the dynamics change as shown here by stabilizing the fixed point in several regimes using the modified OPF method. V_L is the voltage measured across the LC tank circuit.

to $\dot{V}_R$. The control parameter R_N (the negative resistance) may be expressed as

$$R_N(t) = R_{No} + \beta \dot{V}_R, \tag{4}$$

where R_{No} is the unperturbed resistance which determines the regime of the circuit dynamics. The Jacobian evaluated at the steady-state fixed point and hence eigenvalues of the system become altered such that the fixed point becomes stable. Once the fixed point is reached, the error signal approaches zero and the Jacobian resumes its original form. Ref. 17 addresses these issues in more detail.

In the given example, R_{No} is set so that the system oscillates chaotically in the double-scroll attractor regime. In Fig. 9 prior to $t = 0$, the signal measured across the

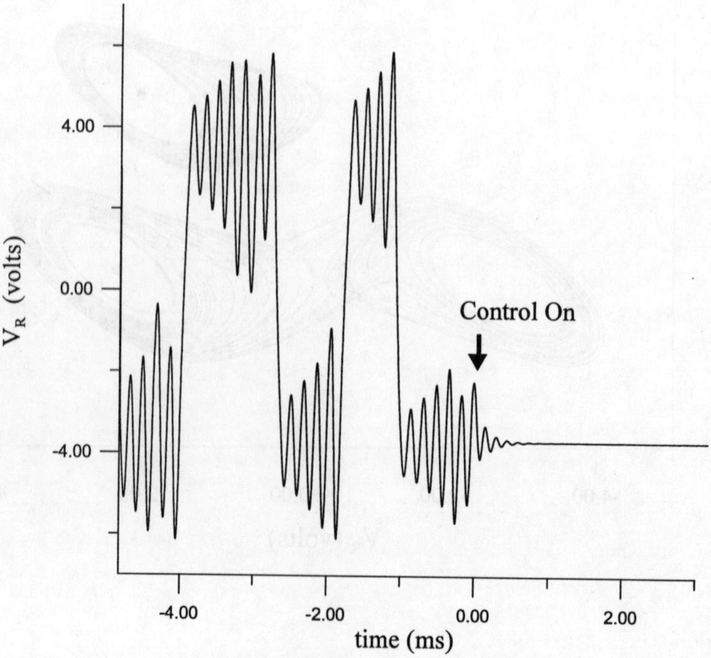

Figure 9: The derivative feedback signal is applied at t=0, quickly stabilizing one of the two fixed points. The other fixed point is stabilized by simply switching the sign of the feedback.

negative resistance oscillates chaotically about two unstable steady state fixed points. At $t = 0$ a switch connecting the derivative feedback with the variable negative resistance is closed and the system quickly settles onto the desired unstable fixed point. Reversing the polarity of the feedback has the result of stabilizing the other unstable fixed point which lies in the positive V_R region. The time required to bring the oscillations down to the noise level is determined by the magnitude of the constant β. If β is reduced below some critical value, which depends on the system parameters, the fixed point remains unstable, and the system settles onto a limit cycle. It should be emphasized that the system is in the chaotic regime when the derivative feedback is applied. This is unlike the OPF method, in which case the fixed point is tracked into the chaotic regime from its stable state.

3.7.3. Connections

One peculiar result of the OPF method is that the frequency at which the wave form is sampled and the feedback pulses are applied must be at least roughly 10% higher (or lower) than the frequency of the stable limit cycle. If we consider a sinusoidal waveform to approximate a limit cycle of the system we find that the correction signal generated by the OPF circuit has a component that is proportional to the derivative of the input. Specifically, this component is proportional to $Sin^2(\omega T_S)Sin^2(\frac{1}{2}\delta\omega T_S)$ where ω is the frequency of the input, T_S is the period of the wave generator sync signal, and δ is the width of the correction pulse as a fraction of T_S. We find that the ability to control the steady state is strongly correlated to the size of this component.

We consider here a few special cases. Experimentally we are unable to control the steady state when the frequency of the correction pulses is set very near to the frequency of the oscillations, i.e. $\omega T_S = 2\pi$. The analysis shows that the derivative component approaches zero in this case. A second case to consider is one which results in easy control of the steady state, for example $\omega T_S = \frac{4}{3}\pi$. The derivative component can be maximized with adjustment of the pulse width δ, which we confirm with aid of a lock-in detector. We find experimentally that $\delta = \frac{3}{4}$ gives robust control and note that this value also gives the maximum calculated value for the derivative component. These results suggest that the control of the steady state using the modified OPF method may be attributed to the derivative component of the feedback signal. This derivative component appears to contribute a damping term to the system as in the case of the derivative control.

4. Controlling a Quasiperiodic System

4.1. The Coupled Diode Resonator System

The coupled diode resonator system[24] is a parallel combination of two diode resonators in series with a resistor, which provides a coupling between the currents through the two diodes, as in Fig. 10. Taking the two currents as the chaotic variables, we observe that the system period doubles once, then undergoes a Hopf bifurcation into a quasiperiodic state as the drive voltage is increased. Continuing, the quasiperiodicity gives way to mode-locked states and chaos. For certain conditions, a second Hopf bifurcation may be observed prior to the chaotic regime. To view the dynamics

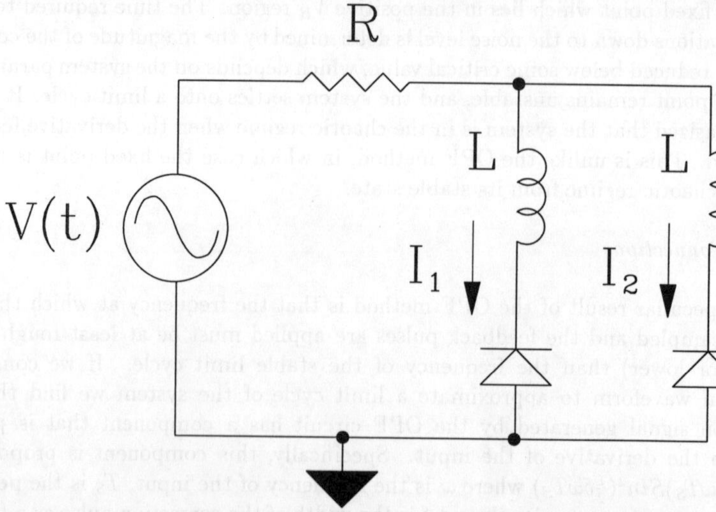

Figure 10: The coupled diode resonator circuit.

of the system, we produce a display of the Poincaré section, i.e. the peak currents through one branch versus the peak currents through the other. The section for this period-doubled quasiperiodic system appears as two circles immediately following the Hopf bifurcation. A smooth rotation of the points making up the circles occurs in the quasiperiodic regime giving way to stationary points in the phase-locked regime. As the critical line is approached, the circles develop kinks signalling the break-up of the two-torus. To control the coupled resonator circuit, we take either of the two currents as the chaotic variable and apply the OPF method outlined in part 1. The perturbation in this system is applied in the form of amplitude modulation of the signal generator. Since the control window can be placed in only one circle of the Poincaré section at a time, it is convenient to take a section every other cycle to reduce the system to a single circle. In the same sense, we have found that applying the perturbations for two drive cycles (one Poincaré cycle) results in the most robust control. A number of different orbits can be stabilized using this method.

4.2. The Pre-Hopf Bifurcation State

As described in *1.1*, a number of systems which undergo a Hopf bifurcation have been stabilized in their steady states by two different methods, the OPF method and

and derivative feedback. In these cases, the systems are autonomous and the state preceding the Hopf bifurcation is a non-oscillating state of steady light intensity or d.c. voltage. In contrast, we emphasize here that the spontaneous oscillation is suppressed while retaining the preceeding period-doubled state. This state can be stabilized in either the quasiperiodic or chaotic regimes. The double exposure photograph in Fig. 11 shows the stabilized orbit embedded in the chaotic regime. With proper positioning of the set-point, this state is stable with a feedback signal which approaches zero down to the normal noise level.

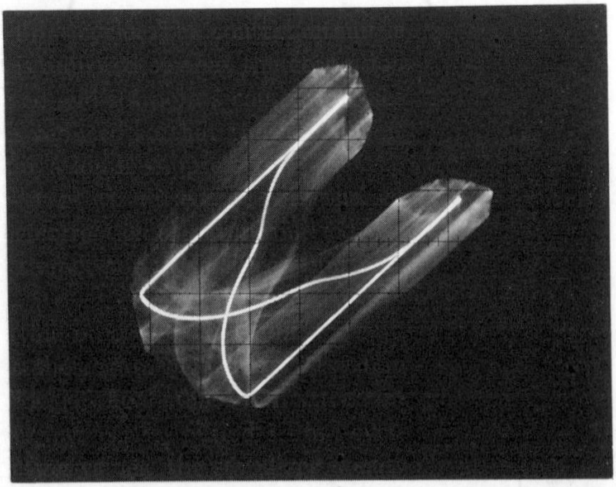

Figure 11: The pre-Hopf period-2 state is shown stabilized in the chaotic regime in this double exposure photograph of I_1 vs I_2.

The transition from quasiperiodicity to chaos is marked by the loss of invertibility of the circle map which is the map of θ_{n+1} vs θ_n. The angle θ_n is measured with respect to some polar coordinate system inside the circular Poincaré section. The map gives smooth curves for both quasiperiodic and chaotic motion, yet only becomes noninvertible in the chaotic regime. As noted by Brandstater and Swinney[25], it is difficult to determine with much accuracy, the point at which the map becomes noninvertible, because the appearance of the map depends on the choice of the origin of the coordinate system. By making use of the OPF stabilization technique, we are provided with the position of the unstable pre-Hopf fixed point, the proper origin for measuring the polar angle and radius. Figure 12 shows the Poincaré section and circle map near the critical line. The invertible map generated with the fixed point origin confirms the quasiperiodicity of the system. In this way, the control of these types of orbits may provide another tool for gaining information and insight near the transition to chaos.

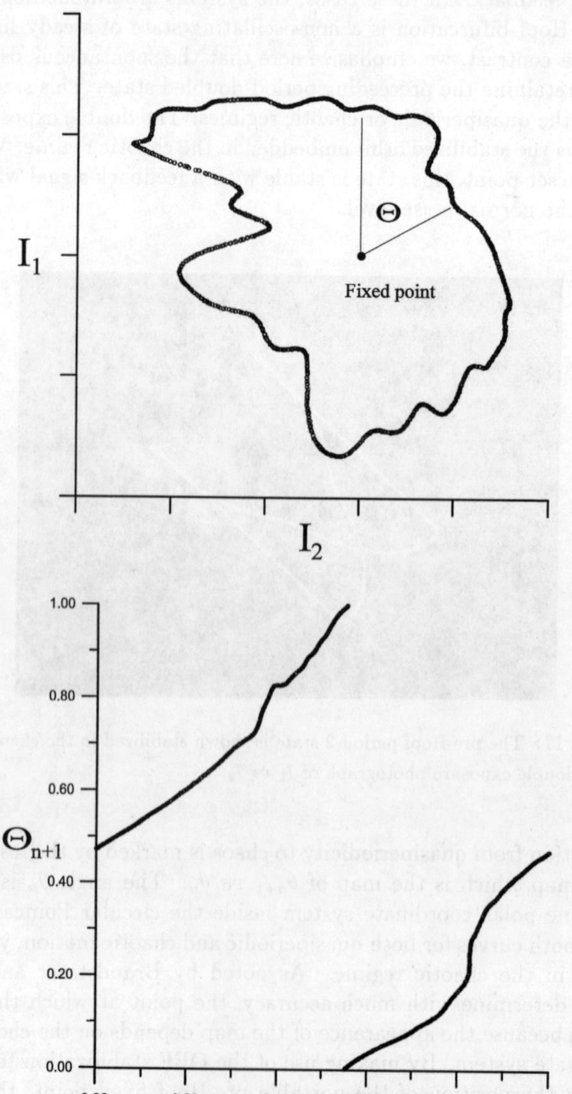

Figure 12: The Poincaré section (upper) of the double resonator circuit. Using the unstable fixed point, we can measure the angular displacement around the distorted circle to generate a circle map (lower).

4.3. Destabilized Orbits

Depending on the coupling resistance, drive frequency and drive amplitude, a number of different mode-locked states may appear below the chaotic regime. For example, with a given set of parameters, a period-14 mode-locked state appears prior to the chaotic regime. As the system is moved into the chaotic regime, the mode-locked state becomes unstable. These pre-existing orbits, similar to the pre-Hopf bifurcation orbit, can be stabilized with very small corrections resulting in less than 1% modulation of the drive voltage. The period-14 orbit shown in Fig. 13 is retrieved from the chaotic attractor with two tiny corrections. This is analogous to stabilizing remnants of the period-doubling cascade (such as periods 1, 2 and 4) in the diode resonator and Chua's circuit in the chaotic regime. Here too the orbits exist as unstable orbits and require only very small nudges to keep them stabilized.

Figure 13: The destabilized period-7 (14 drive cycles) recovered from the chaotic regime of the coupled resonator circuit.

4.4. High-period Orbits

The OPF method was shown to stabilize high-period orbits in the single diode resonator with periods as long as 23 drive cycles. We demonstrate in Fig. 14 that

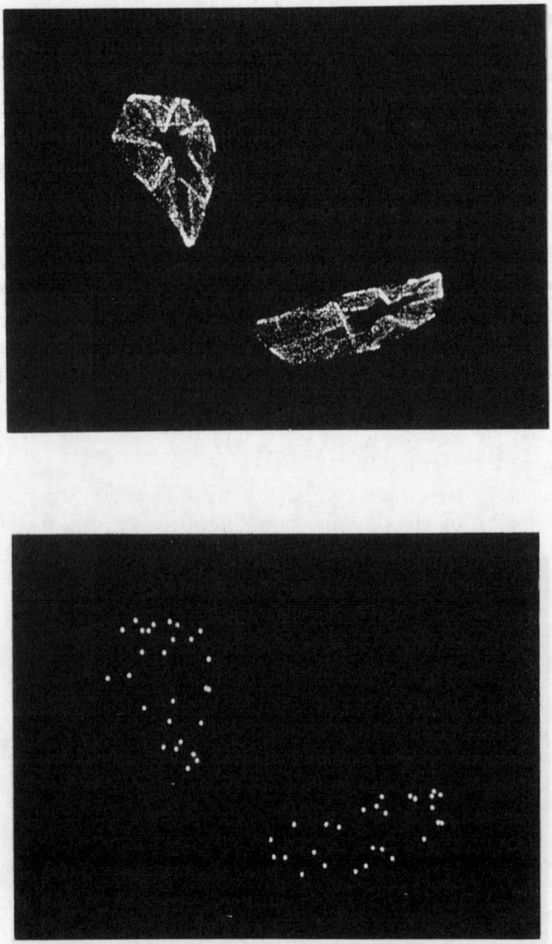

Figure 14: The chaotic Poincaré section and one of the extremely long orbits stabilized in the coupled resonator system. This one is a Period-54.

the orbits stabilized in the coupled resonator circuit can reach lengths of 54 drive cycles, and others which are longer (>60) have been stabilized as well. Figure 14a is the Poincaré section in the chaotic regime prior to control, while Fig. 14b shows the stabilized period-54 orbit.

These orbits are controlled with multiple corrections, as the nature of chaotic systems limits the length of single correction orbits. The correction pulses deform the attractor somewhat to create new orbit paths and a tiny ripple on top of the square pulses is responsible for keeping the orbit stabilized. It should be emphasized that the deformation of the attractor in this circuit is minimal in the stabilization of many of the high-period orbits. The percent of drive modulation for these orbits is generally less than 3%.

5. Conclusion

A review of the OPF method of controlling chaos in electronic systems was presented. It was demonstrated that the method is robust in controlling nonautonomous and autonomous systems and systems that have undergone the quasiperiodic route to chaos. Suppression of Hopf bifurcations has been demonstrated as well as the stabilization of orbits which are remnants of the stable orbits outside the chaotic regime. These states were stabilized in either the quasiperiodic or chaotic regimes with very small feedback perturbations. Also in the chaotic regime, new high-period orbits were shown to be effectively stabilized with multiple corrections and without significant attractor distortion. With these and further explorations in controlling the dynamics of electronic circuits, the limits of the presented control techniques can be determined, allowing the proper application of these findings to more physically complex and real-world systems.

6. Acknowledgements

We wish to acknowledge invaluble assistance in the form of insightful discussion, technical advise, and editing from the following people: M. Löcher, R. Rollins, P. Parmananda, A. Rhode, D. Cigna, and T. Tigner.

7. References

1. E. Ott, C. Grebogi, and J. A. Yorke, *Phys. Rev. Lett.* **64** (1990) 1196.
2. W. L. Ditto, S. N. Rauseo, and M. L. Spano, *Phys. Rev. Lett.* **65** (1990) 3211.
3. G. Chen, "Control and synchronization of chaotic systems (bibliography)," EE Dept, Univ of Houston, TX.
4. W. L. Ditto and L. M. Pecora, *Scientific American* Aug. 1993, 78.

5. T. Shinbrot, C. Grebogi, E. Ott, and J. A. Yorke, *Nature* **363** (1993) 411.
6. E. R. Hunt and G. Johnson, *IEEE Spectrum* Nov. 1993, 32.
7. E. R. Hunt, *Phys. Rev. Lett.* **67** (1991) 1953.
8. R. Roy, T. W. Murphy, T. D. Maier, Z. Gills, and E. R. Hunt, *Phys. Rev. Lett.* **68** (1992) 1259.
9. V. Petrov, V. Gaspar, J. Masere, and K. Showalter, *Nature* **361** (1993) 240.
10. S. Bielawski, D. Derozier, and P. Glorieux, *Phys. Rev.* A**47** (1993) 2492.
11. R. W. Rollins, P. Parmananda, and P. Sherard, *Phys. Rev.* E**47** (1993) 780.
12. K. Pyragas, *Phys. Lett.* A**170** (1992) 421.
13. K. Pyragas and A. Tamasevicius, *Phys. Lett.* A**180** (1993) 99.
14. G. Chen and X. Dong, *J. Circ. Syst. Comput.* **3** (1993) 139.
15. S. Bielawski, M. Bouazaoui, D. Derozier, and P. Glorieux, *Phys. Rev.* A**47** (1993) 3276.
16. Z. Gills, C. Iwata, R. Roy, I. B. Schwartz, and I. Triandaf, *Phys. Rev. Lett.* **69** (1992) 3169.
17. P. Parmananda, A. Rhode, G. A. Johnson, R. W. Rollins, H. D. Dewald, and A. J. Markworth, *Phys. Rev.* E (July 1994).
18. G. A. Johnson and E. R. Hunt, *IEEE Trans. Circ. Syst.* **40** (1993) 833.
19. See the chapter by Paul Lindsay in this book.
20. R. W. Rollins and E. R. Hunt, *Phys. Rev. Lett.* **49** (1982) 1295.
21. *Chua's Circuit: A Paradigm for Chaos*, Editor R. N. Madan (World Scientific, Singapore, 1993).
22. G. A. Johnson and E. R. Hunt, *Int. J. Bifurcation and Chaos* **3** (1993) 789.
23. T. Carroll, I. Triandaf, I. B. Schwartz, and L. Pecora, *Phys. Rev.* A**46** (1992) 6189.
24. Z. Su, R. W. Rollins, and E. R. Hunt, *Phys. Rev.* A**40** (1989) 2689, 2698.
25. A. Brandstater and H. L. Swinney, *Phys. Rev.* A**35** (1987) 2207.

Using Chaos for Digital Communication

SCOTT HAYES[a]
U. S. Army Research Laboratory, Adelphi, MD 20783

CELSO GREBOGI[b]
University of Maryland, College Park, MD 20742

ABSTRACT

We have verified experimentally that chaos in electrical oscillators can be controlled to produce digital communication signals by use of extremely small current pulses to guide the oscillator state space trajectory. Symbolic (digital) information is encoded in large-scale features of the waveform by use of small perturbations to control the *symbolic dynamics*. A digital signal can thus be produced directly at the transmission stage, with no need for subsequent amplification. We have used this technique, which we call ***micropulse injection symbolic control***, to directly control a chaotic electrical circuit for digital information transmission, periodic orbit hopping, subharmonic pulse formation, and control of the signal power spectral density. We stress that this is a fundamentally new physical mechanism for information transmission and waveform production. We envision the application of this technique for complex controlled behavior well within the strongly nonlinear region of device operation in such systems as high-power microwave sources, lasers, rf transmitters, and other waveform sources. Because chaotic behavior occurs in simple systems, and because of the extremely low power requirement of the control mechanism, the high-power device can be simple and efficient, and all the control and guidance electronics can remain at the microelectronic level. Our control strategy also provides an algorithmic approach to the high-speed control of chaotic dynamics in systems where information transmission is not the goal.

1. Introduction

Chaos is a complex dynamic behavior that occurs in many nonlinear physical systems, including such important devices as microwave sources, lasers, and electronic circuits. The presence of chaos causes the time-evolution of the system to be exquisitely sensitive to small perturbations; thus long-range prediction of a chaotic dynamical system is impossible. It was recently realized[1] that it is precisely this sensitivity to perturbations,

---

[a]Also at Department of Physics and Astronomy, University of Maryland, College Park, MD 20742.
[b]Laboratory for Plasma Research, Institute for Physical Science and Technology, and Department of Mathematics.

long thought to be primarily a barrier to prediction, that permits chaos to be controlled with extremely small amounts of energy. Small perturbations applied at strategic times are used to gently nudge the system dynamics into a steady state, or to periodic motion. Although this concept is now generally called "control of chaos," the physical mechanism involved is more akin to "guidance of chaos." The power used to control the system can be comparable in power to the *background noise* in the system. Ideally, the control process is a very low-power noiselike process itself, and serves to cancel the noise that is constantly perturbing the system from a desired behavior.

We have recently demonstrated that small-perturbation control can be used to do much more than stabilize the dynamics of a chaotic system, and that the natural complexity of chaotic oscillations can be used to advantage.[2] It has long been known that there is a deep formal connection between the theory of dynamical systems and information theory.[3] (This connection is the basis of a branch of mathematics known as ergodic theory, which has diverged from the practical theory of information transmission, but is formally very similar.) Our work shows[2] that this connection can be used in a way that is more than formal—that in fact a chaotic electrical oscillator is a natural source of *digital* communication signals. Thus, chaos itself naturally generates a digital communication signal. The key to the practical use of this connection is the application of small guiding current pulses, on the order of the background system noise, to direct the system through a complex sequence of pulses that carry information. This physical mechanism is interesting because an *analog* oscillator produces a signal that is fundamentally *digital* in nature (that is, the information is carried in discrete packets), but the oscillation obeys deterministic continuous-time differential equations.

2. The Lorenz System as a Communication Channel

The well-known Lorenz[4] equations provide a particularly simple framework for describing the basic idea of controlling symbolic dynamics with small perturbations. If one uses direct perturbations to the trajectory of the system, controlling the symbolic dynamics is very straightforward. The characteristic of the Lorenz system that makes it a paradigm for controlled chaos communication is that, when uncontrolled, it *approximates* (there is some history dependence) a source of binary symbols that emits 0's and 1's with equal probability and independent of its history. Thus the uncontrolled Lorenz system approximates a memoryless *binary symmetric information source*, the simplest possible information source. If one controls this system using small perturbations to transmit binary symbols, it can be viewed as an almost ideal *binary symmetric communication channel.* It is a *channel* in this sense because it is through the chaotic system that one must produce the waveform. (Alternatively, we could view this as controlling the output of the Lorenz information *source*. We do not consider viewing the chaotic system as a channel to be inherently better or worse than viewing it as a source; both viewpoints are useful conceptually.)

We use here the Lorenz system of differential equations, given by $\dot{x} = -\sigma x + \sigma y$, $\dot{y} = Rx - y - xz$, and $\dot{z} = -bz + xy$ with the standard parameter values $\sigma = 10$, $R = 28$, and $b = \frac{8}{3}$. The state coordinate $(x(t), y(t), z(t))$ of the Lorenz system moves on a chaotic attractor in a three-dimensional state space. A two-dimensional projection of a trajectory of the Lorenz system is shown in Fig. 1. We have labeled one lobe of the attractor with the symbol 0, and the other lobe with a 1. If the computer algorithm is run uncontrolled, the state coordinate jumps from lobe to lobe, traveling about each lobe one or more times before switching to the other. This sequence of oscillations thus defines a binary sequence. For the equations and parameters as given, this binary sequence approximates the statistics of a coin toss, with some deviation. Thus, the Lorenz system with the standard parameter values is a chaotic oscillation which *approximates* a source of random equiprobable bits—a binary symmetric information source. This fact is remarkable in that the Lorenz equations originally derive from convection in fluid dynamics, a problem that is normally thought to be far removed from information theory and random process theory.

The lines shown intersecting the lobes in Fig. 1 are two half-planes viewed edge-on, and represent the two branches of a Poincaré surface of section. The half-planes are defined by the equations $y = \pm\sqrt{b(R-1)}$ and $|x| \geq \sqrt{b(R-1)}$. (The points $(x,y,z) = \left(\pm\sqrt{b(R-1)}, \pm\sqrt{b(R-1)}, R-1\right)$ are the unstable fixed points, or foci of the outward spiraling trajectories on either lobe.) The direction of motion where trajectories cross the surface of section is indicated for each lobe in Fig. 1. Because the Lorenz equations represent a flow that is highly dissipative and therefore strongly contracting along the stable direction, the intersection of the attractor with the surface forms an essentially one-dimensional arc. We define surface of section coordinates as $\zeta = z - (R-1)$ and $\xi = |x| - \sqrt{b(R-1)}$, with the y coordinate of the flow fixed along the surface of section at the previously given values.

We now describe how to generate a description of the symbolic dynamics of the Lorenz system. This involves determining how state-space coordinates on the surface of section are related to the symbol sequences generated after the state coordinates cross the surface of section. Because the system is deterministic, the symbol sequence produced after this crossing is determined by the point of crossing. Suppose the state coordinate passes through the 0 branch of the surface of section and generates the binary symbol sequence $b_1b_2b_3$.... We use the *binary fraction* to represent this symbol sequence by the real number $0.b_1b_2b_3...$, where the nth place behind the decimal point has value 2^{-n}. Thus, each future symbol sequence is mapped into the real number $r = \sum_{n=1}^{\infty} b_n 2^{-n}$. We refer to the real number r, specifying the future symbol sequence, as the *symbolic state* of the system. Assigning a real number r on the interval $[0,1]$ to each possible future symbol sequence, and associating this real number with a corresponding generalized Poincaré coordinate x on the surface of section, defines a function $r(x)$ relating the symbolic state to the Poincaré coordinate. Henceforth, we refer to the value of the state coordinate on the surface of section as the *statepoint*. The *inverse* of this function $x(r)$ for the Lorenz system is shown in Fig. 2. In this figure the statepoint has been normalized so that $x = \xi / L$,

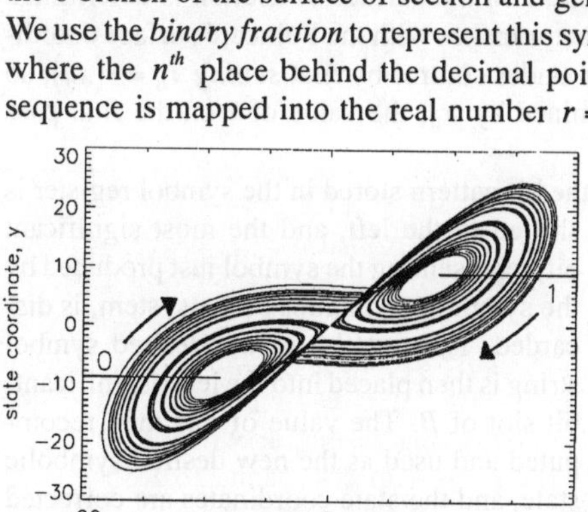

Figure 1. Lorenz trajectory projected on x-y plane showing two branches of surface of section and corresponding binary symbols.

where $L = 10.322$ is the approximate distance in across the attractor on the surface of section. (This x is of course not the same as the state coordinate x in the Lorenz equations.) We call the function $r(x)$ the *symbolic waveform coding function*, or just the *coding function*. (The process of relating waveform basis functions to discrete symbols output by an information source is sometimes called waveform coding in classical communication theory; thus we adopt a similar terminology.) To estimate this function in practice, we divide the range of x into M bins, and average the value of r produced by multiple passes through each bin to improve the estimate. This type of averaging works for the Lorenz system because the function $r(x)$ is *approximately* continuous and differentiable. We determine the value of r on each pass through a bin by storing the future symbol sequence in a *symbol register*, and converting the symbol sequence to its binary fraction.

Once the inverse coding function $x(r)$ is known for the Lorenz system, the control of the system using *direct trajectory perturbations* is straightforward. (Controlling the system with *parameter* perturbations is also straightforward, but slightly more involved. Direct trajectory perturbations are simplest for computer algorithms, but parameter perturbations offer an advantage in real-time applications because they need not be applied instantaneously, and one does not need to perturb the state coordinates directly.) We assume that one converts the first N bits of the binary string specifying the desired symbolic dynamics to its corresponding binary fraction r_0. Thus, on a typical computer, one might use the first 16 bits to be "transmitted," store this in the symbol register as the integer binary number B_0, and convert B_0 to its binary fraction $r_0 \cong B_0 / 2^{16}$. One can then simply run the computer algorithm without control until the state coordinate passes through the surface of section near $x_0 = x(r_0)$, and then apply a perturbation to put the state coordinate at the value of x that produces the desired binary string. The perturbation should be applied to move the statepoint to the proper value of x on the intersection of the attractor with the surface of section. This involves perturbing both Poincaré coordinates by the proper amount. The vector direction of the required perturbation on the surface of section is determined from the angle that the intersection of the attractor makes with the Poincaré surface coordinates. One can also quickly bring the system under control by first *setting* $r_0 = r(x_0)$, so that the first desired symbolic state is determined by x_0, the value of x on the first pass through the surface of section.

After the first pass through the surface, the bit pattern stored in the symbol register is shifted to the left, and the most significant bit, representing the symbol just produced by the symbolic dynamics of the system, is discarded. The next bit of the desired symbol string is then placed into the least-significant-bit slot of B. The value of r is then recomputed and used as the new desired symbolic state, and the state coordinates are corrected again so that the symbolic state is corrected when the state coordinate next crosses the surface of section. (Note that the bitwise complement of the bit pattern in the symbol

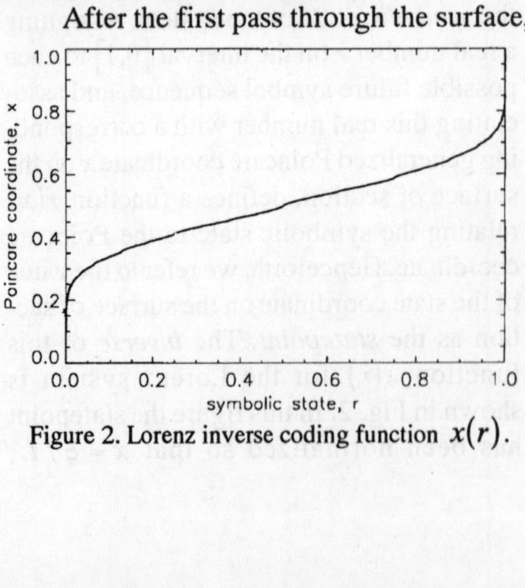

Figure 2. Lorenz inverse coding function $x(r)$.

register must be used when the state coordinate crosses the lobe 1 surface.) This process, repeated indefinitely, controls the symbolic dynamics through the desired binary sequence.

Now, in Fig. 1, the oscillations of the state coordinate about the 0 and 1 attractor lobes can be seen to correspond to negative and positive maxima, or spikes, in the x projection of the state coordinate. If this x projection $x(t)$ is used as the transmit signal, then we can extract the message by simply observing the sequence of spikes in the waveform. Thus, the Lorenz system offers probably the simplest example of how symbolic dynamics can be used to transmit a message. The sequence of oscillations on the two lobes of the attractor approximates a coin toss, and can be controlled with small perturbations so that the sequence of oscillations matches a previously specified sequence of bits. For a physical implementation of this technique, the simple procedure described above might need to be modified, because direct trajectory perturbations must be applied almost instantaneously.

Figure 1 is actually a controlled Lorenz trajectory, and in this case the source of bits that was used to specify the desired symbolic dynamics was a random bit generator constrained to never produce less than two of the same symbols in a row. With this constrained symbolic dynamics, the state-space dynamics is also constrained in such a way that a fractal structure is produced on the attractor lobes. We have produced a similar fractal structure in our experiment. Because the topological and metric entropies of the chaotic system give the channel capacity and information rate when the system is used for digital transmission, the dimension of the fractal structure is also related to the information rate.[5]

3. Experiment

Figure 3 is a schematic diagram of the electrical circuit used in our experiment. Controlled current pulses are injected by the source i_p , shown as dotted lines in the figure. A complete list of nominal component values can be found in the paper by Matsumoto.[6] The nonlinearity comes from a nonlinear negative resistance represented by the voltage v_R. The nominal frequency of oscillation for this circuit is approximately 5 kHz; the frequency varies slightly depending upon the region of operation. The oscillator was constructed with a variable air-plate capacitor in parallel with the capacitor C_1 so that we can tune the circuit for various types of behavior. (Andrea Mark, an electronics engineer at the Army Research Lab, constructed the oscillator for this experiment.) To measure signals and control the oscillator, we used a Hewlett Packard model 3567A signal analyzer. This system provides a powerful active measurement system for the study of control of chaos in general. The analyzer contains a signal processor to execute the control program, digitizers to sample the incoming signal, and digital-to-analog converters to form control pulses. The signal processor is programmable in the C programming language, and the programs executed in the signal processor are developed on a workstation and downloaded to the signal processor. We refer to this system henceforth as the *controller*. We thus have the flexibility of implementing the control procedures in easily changeable software, and we can simulate the performance of different possible integrated microelectronic control and guidance chips by merely changing the C program.

We first tuned the circuit to produce a single Rössler chaotic attractor. Figure 4 shows a measured trajectory on the Rössler-type attractor produced by the circuit, and the Poincaré surface used. (The white swath through the trajectory shows the location on the attractor at which control pulses occur and is for later reference.) The coordinates used in the figure are the internal quantization levels in the controller normalized by division by 1000. We denote variables representing the quantized values by small letters, and normalized values with capital letters. Thus we denote the quantized and normalized value of v_{C_1} by $Qv_{C_1} = qv_{C_1} / 1000$, where q is the scaling factor relating internal quantizer states to physical voltages.[7] We use these coordinates because for the control of the system, one need not know the physical quantities themselves, just the internal values in the controller. This is of great importance, because it allows us to conceptually regard the chaotic system as a "black box," and concentrate on the empirically derived model internal to the controller.

When the voltage v_{C_2} passes through zero in the negative direction, we externally trigger the controller (using the trigger circuit in an oscilloscope) to take a sample of v_{C_1}. (The intersection of the attractor with the surface $v_{C_2} = 0$ is very close to being a single thin arc, and thus we can approximately describe the dynamics by one coordinate on the Poincaré surface.) We call the quantized value of v_{C_1} the *statepoint*, that is, $x = qv_{C_1}$. By convention then, $X = Qv_{C_1} = x / 1000$.

A *first return map* for the Poincaré surface used in Fig. 3 is shown in Fig. 5. The return map was generated by the embedding of 10,000 coordinate pairs (X_n, X_{n+1}). The use of X as a coordinate on the Poincaré surface is an excellent description of the dynamics on the Poincaré surface: The value of X_n alone gives the next statepoint X_{n+1} almost exactly. The natural way to partition the Poincaré surface into symbolic regions is to choose the partition boundary at the peak of the first return map; this is shown in Fig. 5. (If the boundary is not at the peak, then two different statepoints can generate the same symbol sequence.) The symbol 0 is thus generated when the statepoint is to the left of the boundary, and a 1 is generated when the statepoint is to the right of the boundary. The partition boundary is also the point on the Poincaré surface such that trajectories passing through the point travel directly into the fold in the Rössler band. The symbolic partition is shown on the Poincaré surface with respect to the statepoint that passes into the fold in Fig. 4.

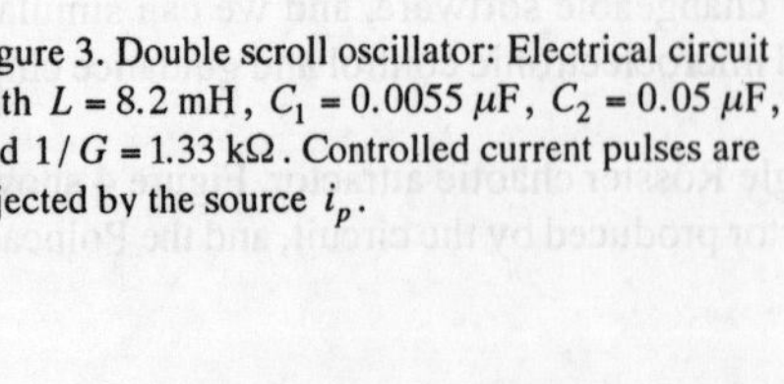

Figure 3. Double scroll oscillator: Electrical circuit with $L = 8.2$ mH, $C_1 = 0.0055\ \mu$F, $C_2 = 0.05\ \mu$F, and $1/G = 1.33$ kΩ. Controlled current pulses are injected by the source i_p.

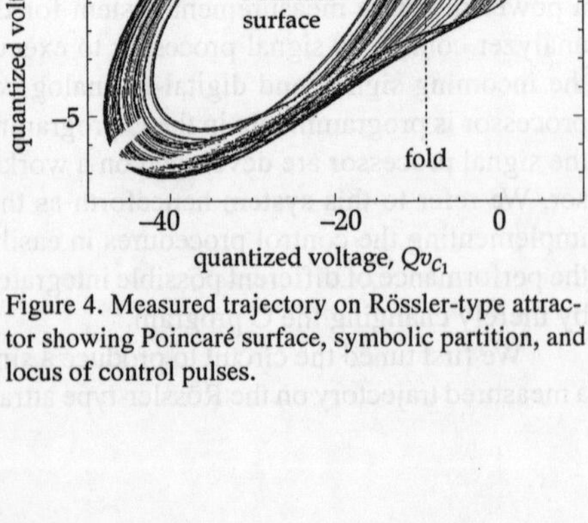

Figure 4. Measured trajectory on Rössler-type attractor showing Poincaré surface, symbolic partition, and locus of control pulses.

The first step towards controlling the symbolic dynamics is to experimentally obtain a description of the symbolic dynamics of the free-running oscillator. When the full system state coordinate crosses the Poincaré surface, the controller takes a sample and stores the value of the statepoint. The controller then records the sequence of symbols generated by the system after the crossing. We chose a 10-bit symbol sequence length to obtain adequate resolution on the Poincaré surface without needing to collect a large number of sequences, but this number is variable. The symbols are shifted sequentially into a *symbol register*. This procedure estimates the statepoint that produces a given symbol sequence for all symbol sequences that are allowed by the system. Statepoints corresponding to the same symbol sequence are averaged to give a better estimate of the source statepoint for each symbol sequence. We can thus estimate the *coding function* $r(x)$ that gives the symbol sequence r that evolves from a given statepoint x. (We actually estimate the inverse coding function $x(r)$, but inversion yields $r(x)$.) The coding function is a mapping of the *state space* into the *symbol space*. For our experiment, it is convenient to take the (discrete) symbol space to be the set of integers between 0 and 1023, representing the binary value of a 10-bit symbol sequence. We take the statepoint to be the quantized value of the voltage v_{C_1} on the Poincaré surface. We can then perform integer operations in the controller, as there is no need to convert from the internal representation to actual voltages, and integer operations are much faster.

The experimentally accumulated *direct representation* of the coding function for the oscillator is shown in Fig. 6. In this representation, we show the binary value of 10-bit symbol sequences versus the statepoint that *sources* each symbol sequence. The topology of the first return map for this system is such that the symbol sequences are ordered according to a *Gray code* ordering.[8] Some sequences are never produced by the oscillator. This is typical of the symbolic dynamics of chaotic systems and of communication channels in general, and is known as a constraint on the *grammar*, or, in the context of communication, it is known as a *channel constraint*. The effect of the constraint is most easily seen if the symbol sequences are reordered so that the ordering is monotonic with increasing statepoint. This is done as follows: The first (leftmost and most significant) bit of the reordered sequence R is taken to be the same as the first bit of the Gray-ordered sequence r. If the next bit in r (moving to the right) is a 0, the next bit in R is the same as the previous bit; if it is a 1, then the next bit in R changes value from the previous bit. In Fig. 7 we show a reordered *inverse* coding function $x(R)$. (We refer to the inverse coding function from here on, because for control we must quickly find the statepoint that evolves a desired symbol sequence; thus we must store $x(r)$ or $x(R)$.) The constraint reveals itself as gaps in the possible values of R. The constraint in values of R (in the reordered symbol space) that occur is a generalized type of *runlength constraint*. A simple runlength constraint places a maximum and

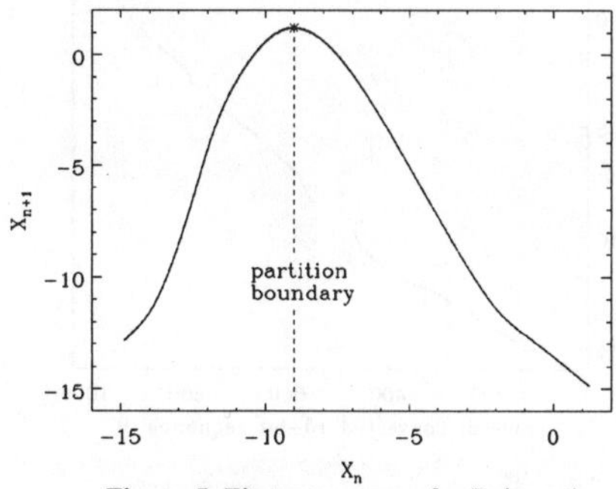

Figure 5. First return map for Poincaré surface in Fig. 4.

minimum on sequential runs of 1's and 0's. The type of constraint here can be satisfied by the use of an overly restrictive simple runlength constraint. A more restrictive constraint will suffice to satisfy the grammar for the experiment (and is easier to explain): A sufficient constraint on direct-ordered symbol sequences is that no more than one 0 is ever produced in a row.

To control the system, we must describe the effect of perturbations. A digital-to-analog converter (DAC) in our controller is connected to the positive terminal of capacitor C_2 (see Fig. 3) in series with a 25-kΩ buffer resistor R_B, thus providing a simple but effective method for injecting small current pulses. The output of the DAC is normally grounded, so current is always flowing through the resistor to ground, but the momentary application of a nonzero voltage (a pulse) for a duration of approximately 4 μs causes a change in the current flow during this period. This is equivalent to the injection for a period of 4 μs of a *differential current* equal to v_p / R_B, where v_p is the output voltage of the DAC when the pulse is on. To determine the effect of these pulses on the system, we chose a reference pulse amplitude out of the DAC of $v_r = 1.5$ V, which means that the reference differential current averaged over one cycle of the oscillator (about 200 s) is $I_r = 1.2\ \mu$A. Typically, the differential current during control is even smaller than the reference current. The rms control current was 0.2 μA for the example to be presented; compare this to circuit currents on the order of a few milliamperes.

The reference pulses occur about 80 μs after the statepoint sample is taken. The controller needs this time delay to bring in the statepoint, perform some computations, and output a pulse. The actual locus on the attractor at which the pulses occur is shown in Fig. 4 by the white swath at the bottom of the figure. The distance around the attractor at which a pulse occurs varies depending on the average velocity of the flow along a given orbit. The controller first accumulates an array of statepoints corresponding to the symbol sequences generated by the uncontrolled system. We denote this array by $s[r]$, where r is the binary value of the 10-bit symbol sequence. This procedure is then repeated, but the controller applies a reference pulse immediately after the first statepoint is stored. In this way, we accumulate an array $q[r]$, which in essence is the altered coding function in the presence of the reference perturbations. Because the pulses are small, and are applied early, the full

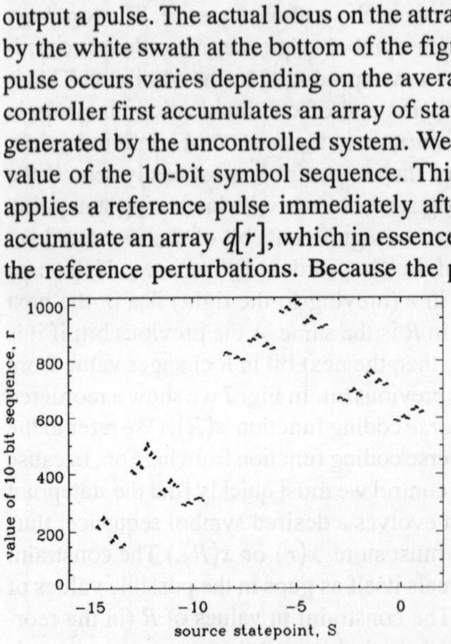

Figure 6. Binary coding function in direct ordering.

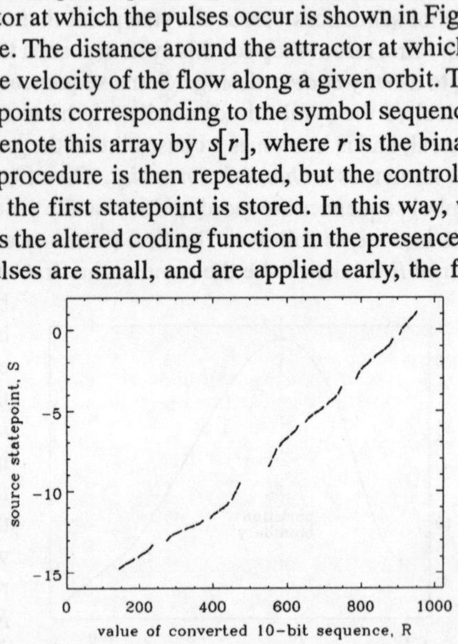

Figure 7. Inverse binary coding function in Gray-to-binary reordering.

system state coordinate falls approximately back to the attractor by the time the next statepoint is measured. The controller then computes an array $d[r] = p_r / (q[r] - s[r])$, where p_r is the integer value of the reference pulse amplitude stored in the controller. Because the pulses are small, the effect of a pulse is approximately linear with pulse amplitude. We can thus compute the *required* pulse amplitude to effect a *required* change in the value of the statepoint by the equation $p = d[r](x - s[r])$, where x is the actual value of the statepoint upon crossing the Poincaré surface, and r is the *desired* future symbol sequence, which is specified in advance. The controller can thus compute the required pulse amplitude using only one integer subtract and one integer multiply—a highly efficient algorithm that could be used in a digital microelectronic controller at much higher frequencies than in this experiment.

When the sample statepoint becomes available, the symbol register is loaded with the symbol sequence that will naturally evolve without control. This value is given by $r[x]$, where x is the first sample statepoint. (We have stored the coding function as well as its inverse so that we can do this.) As each successive statepoint x becomes available, the controller pulls $s[r]$ out of memory by direct indexing with the bit pattern given by r, where r is the desired symbol sequence in the symbol register. A pulse of amplitude $p = d[r](x - s[r])$ is then applied, the contents of the symbol register are shifted left, and the next desired code bit is placed in the symbol register in the rightmost (least significant) slot. This procedure is repeated continuously. Because the symbol register is loaded at first with the statepoint that naturally occurs without control, the first 10 iterates of the control procedure rapidly target the system from the uncontrolled chaos and into the beginning of the sequence of desired symbols.

As previously discussed, for the purpose of demonstration, a sufficient restriction on symbol sequences that will satisfy the constraints of the grammar is to require that no runs of more than one 0 occur in a row. We use the following code to satisfy the grammar so that a message consisting of an arbitrary stream of binary symbols can be encoded: Message bits map to code bits according to the rule $0 \rightarrow 01$, $1 \rightarrow 11$, and thus no more than one 0 will ever occur in a row. (This code is itself more restrictive than necessary—we could use the variable-length code $0 \rightarrow 01$, $1 \rightarrow 1$—but we are interested here in pointing out the effects of the additional restriction on the signal.) Using the standard 7-bit ASCII representation of the characters and mapping the binary digits according to the above code, we have encoded the following quotation:

> Yea, verily, I say unto you: A man must have Chaos yet within him to birth a dancing star. I say unto you: You have yet Chaos in you.
>
> —Friedrich Nietzsche, *Thus Spake Zarathustra*.[9]

A statepoint sequence corresponding to the repeated transmission of this message is shown in Fig. 8. This sequence is the sampled digital communication signal. Halfway through the sequence, we switched the information source from the encoded message data stream to a pseudo-random source in which all finite binary sequences that never have more than one 0 in row are possible. Both symbol streams cause the sequence of values to lie approximately on a *Cantor set*.[5] The additional restriction imposed by the code used for the message is now apparent, because the signal becomes confined to narrower bands than it is when all binary sequences with no runs of more than one 0 occur. This is because the encoded message contains even fewer possible sequences than the random stream. The sequences corresponding to different bands in the signal are labeled in Fig. 8. One interesting aspect of this type of signal is that if one has sufficient resolution in a detector, more than one information bit can be extracted from a single sample. Thus, although this signal is *binary* in the sense that only one new code bit becomes available at each sample point, more than one code bit can be extracted from one sample. A signal such as this could thus be detected and decoded by sampling only once every three cycles of the signal, for example, which would give the detector three code bits per sample. In this respect, the signal is not binary: in fact, it conforms to no previously known type of communication signal.

If one specifies periodic orbits (which produce line spectra and have a single frequency component in the discrete time domain) by symbol sequences, the 10-bit symbolic state for a period-1 orbit is given by 1111111111. A period-2 orbit is given by the symbolic state 1010101010 and by 0101010101. One can control the period-1 orbit by sequentially shifting 1's into the symbol register, and the controller will maintain the system on a period-1 orbit, audible on a loudspeaker as a tone with characteristic overtones. If alternating 0's and 1's are then shifted in, the system will switch to a period-2 orbit, without going through chaos in between.[10] We have used this method to rapidly switch periodic orbits.[11] (The closest existing engineering terminology available to describe this is "frequency hopping," and we can do this over only a few cycles of the system by gently guiding the natural dynamics of the system from orbit to orbit.) A statepoint sequence during a rapid switching experiment is shown in Fig. 9. In this figure, we have programmed the controller to dwell on each periodic orbit for 100 cycles of the system, which means that switches occur about 50 times per second. In the frame shown in Fig. 9, the sequence of orbit periodicities is 1-2-4-1-3-4-2-4. Note that only a few points occur between switches. We have controlled much shorter dwell times and switched to longer period orbits. We have also controlled the oscillator from period-1 through a sequence of subharmonic bursts lasting for less than 10 cycles of the system, and then back to period-1: this is wideband subharmonic pulse formation.

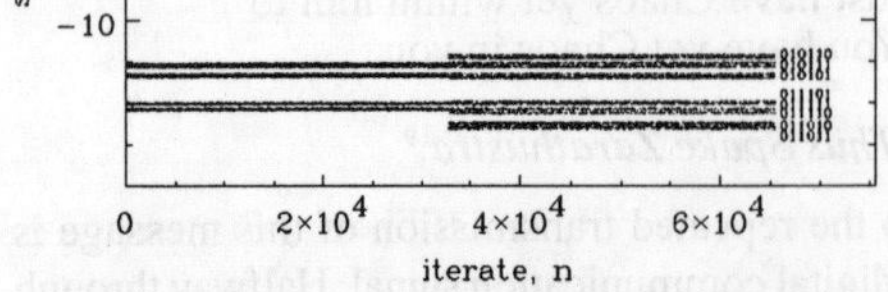

Figure 8. A digital communication signal on a fractal: Statepoint sequence shows Cantor set structure produced by constrained code during control with two different symbol streams. Note that more than one encoded binary symbol can be extracted from a single sample of waveform.

4. Conclusion

Traditionally, to produce a complex waveform at the high-power level in an electrical system, one must amplify a small signal with inefficient, bulky, and costly linear or quasi-linear circuitry. By using small perturbations to control the complex dynamics *inherent* in chaotic behavior, we can produce a complex waveform *directly* at the

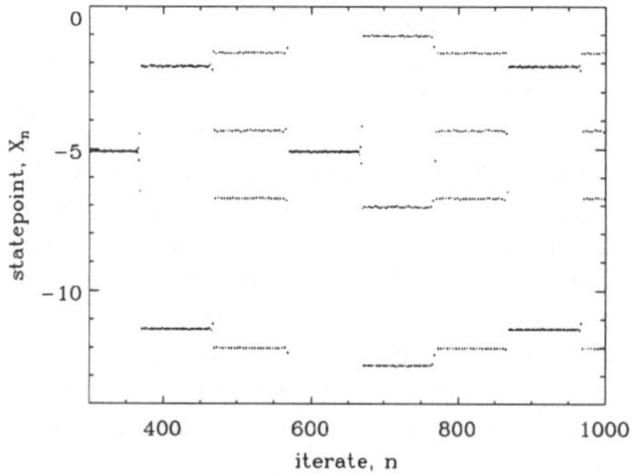

Figure 9. Fast switching between periodic orbits (similar to frequency hopping) using symbolic control.

high-power level, with all the control circuitry remaining at the microelectronic level, and with no intervening stages of amplification. This basic method is applicable to many important devices, including lasers, microwave sources, and many other electrical and mechanical oscillators.[12] Without the requirement for linearity, or even static stability, we can operate devices in strongly nonlinear regimes. Furthermore, for covert signaling, certain types of chaos signals are ideal, because they reveal their structure only in certain system-dependent signal space embeddings. We are also developing the general adaptive control theory needed to control chaotic dynamics in a wide variety of physical systems. From an engineering perspective, these developments mark a significant departure from traditional design philosophy.

References

1. E. Ott, C. Grebogi, and J. A. Yorke, *Controlling Chaos*, Phys. Rev. Lett. **64** (1990), 1196. F. J. Romeiras, C. Grebogi, E. Ott, and W. P. Dayawansa, *Controlling Chaotic Dynamical Systems*, Physica D **58** (1992), 165. For further references, see the review article by T. Shinbrot, C. Grebogi, E. Ott, and J. A. Yorke, *Using Small Perturbations to Control Chaos*, Nature **363** (1993), 411. For an overview of the experimental work, see W. L. Ditto and Louis M. Pecora, *Mastering Chaos*, Sci. Am. **269** (1993), 78.
2. The central idea involved in controlling chaos for communication is described in S. Hayes, C. Grebogi, and E. Ott, *Communicating with Chaos*, Phys. Rev. Lett. **70** (1993), 3031. The experimental verification was first reported in S. Hayes, C. Grebogi, and E. Ott, *Controlling Symbolic Dynamics for Communication*, Proc. 2nd Experimental Chaos Conference (1993). A different use of chaos in communication is the use of chaos synchronism for the purpose of secure (i.e., difficult to decipher by an interceptor) transmission of information; see L. M. Pecora and T. L. Carroll, Phys. Rev. Lett. **64** (1990), 821; K. M. Cuomo and A. V. Oppenheim, Phys. Rev. Lett. **71** (1993), 65.
3. The concept of entropy for a dynamical system, based formally on a generalization of Claude Shannon's entropy (C. Shannon, *A Mathematical Theory of Communication*, Bell Tech. J. **27** (1948), 379, 623. C. E. Shannon and W. Weaver, *The Mathematical Theory of Communication*, University of Illinois Press, 1962) for information sources, originated in the papers by Kolmogorov and Sinai: A. N. Kolmogorov, *A New Metric Invariant of Transitive Dynamical Systems and Automorphisms in Lebesgue Spaces*, Dokl. Akad. Nauk **119** (1958), 861–864. A. N. Kolmogorov, *Entropy Per Unit Time as a Metric Invariant of Automorphisms*, Dokl. Akad. Nauk **124** (1959), 754–755. Ja. G. Sinai, *On the Concept of Entropy for a Dynamical System*, Dokl. Akad. Nauk **124** (1959), 768–771. The concept of topological entropy is discussed in R. L. Adler, A. G. Konheim, and M. H. McAndrew, *Topological Entropy*, Trans. Am. Math. Soc. **114** (1965), 309.

4. Edward N. Lorenz, *Deterministic Non-Periodic Flow*, J. Atmos. Sci. **20** (1963), 130.
5. For a brief overview of the relationship between communication entropies and dynamical entropies, see S. Hayes and C. Grebogi, Proc. SPIE **2038** (1993), 153. The Kolmogorov-Sinai and topological entropies of a dynamical system are the information rate and the channel capacity, respectively, when the system is used for information transmission. The box-counting and information dimensions of the symbol space are the channel capacity and information rate if the proper state-space partition is being used. Because the symbol space and state space are related by the coding function, the *dimension* of the fractal structure on the attractor during control is also related to the information transmission rate. The use of a code for information transmission that is more restrictive than is absolutely necessary produces a Cantor-set structure on the attractor. These connections and a full overview of the theory of using dynamical systems for communication, including signal generation, coding, and detection, will be discussed in a longer paper, to be published.
6. Several articles about this circuit appear in Proc. IEEE **75**, No. 8 (1987): e.g., T. Matsumoto, p. 1033.
7. Except for fine-scale stepping, there is an approximately linear relation $x = qv$, where $1/q = [1/2^{15}D)]10^{r+o/20}$, and for the settings used in our controller for this experiment, $o = 2$, and $r = 6$. D is a digital filter correction factor and is approximately 0.46066. Thus there are about 6000 levels per volt. The trajectory in Fig. 4 was recorded on a different scale, and converted to the scale used for the rest of the figures for consistency.
8. E. N. Gilbert, Bell Syst. Tech. J. **37** (1958), 815.
9. We have this liberal yet appealing translation of the German "Ich sage euch: man muß noch Chaos in sich haben, um einen tanzenden Stern gebären zu können. Ich sage euch: ihr habt noch Chaos in euch." from the frontispiece of the book by Manfred Schroeder, *Fractals, Chaos, Power Laws — Minutes from an Infinite Paradise* (W. H. Freeman and Company, New York, 1991). A more literal translation by Thomas Common (Random House Modern Library, New York) is "I tell you: one must still have chaos in one, to give birth to a dancing star. I tell you: ye have still chaos in you."
10. The concept of *targeting* as such was first described by T. Shinbrot, E. Ott, C. Grebogi, and J. A. Yorke, Phys. Rev. Lett. **65** (1990), 3215, but the algorithmic approach was rather cumbersome, and not suited for high-speed control. Symbolic control provides an efficient algorithm for targeting.
11. Typically, switching between periodic orbits involves turning off control of the first orbit and establishing control of the second when the trajectory wanders close to the desired orbit.
12. Our experiment produced a *baseband* signal. An interesting aspect of chaos in high-frequency systems such as microwave sources and lasers is that the chaos is *bandpass* in nature, that is, the signal power spectral density is confined to a small fraction of the total bandwidth. Thus, symbolic control in these cases would produce a communication signal somewhat similar to a classical *modulation* signal, and this signal could be transmitted through an antenna or bandwidth-limited medium such as an optical fiber.